Many uses of lavender

USED for more than 2,000 years by the Egyptians, Romans and reputedly even by Mary Magdalene to anoint Jesus, lavender soared in popularity during Victorian times and was also used as a wound dressing for soldiers in World War I, due to its antibacterial properties.

Today it can be used in many ways. For example, it is an excellent first aid remedy for small cuts, scratches, burns, insect bites, wasp and bee stings and can also be used to treat spots.

Add a few drops to bath water to reduce stress and ease away muscular aches and pains, or to a sponge or flannel when showering.

Lavender oil makes an excellent insect repellent and a few drops on a dog or cat collar will help to keep away fleas. When diluted it also makes a good household disinfectant for washing floors, surfaces and inside the fridge.

If you suffer from headaches, dab a few drops on your temples. And place drops under the pillow case to help you sleep. Add a few drops to the final rinse when washing clothes or to shoes to remove any odours. Lavender really is versatile!

Picture: HEALTH-BUZZ

Lavender

Medicinal and aromatic plants – industrial profiles

Individual volumes in this series provide both industry and academia with in-depth coverage of one major medicinal or aromatic plant of industrial importance.

Edited by Dr Roland Hardman

Volume 1
Valerian
Edited by Peter J. Houghton

Volume 2
Perilla
Edited by He-ci Yu, Kenichi Kosuna and Megumi Haga

Volume 3
Poppy
Edited by Jenö Bernáth

Volume 4
Cannabis
Edited by David T. Brown

Volume 5
Neem
Edited by H.S. Puri

Volume 6
Ergot
Edited by Vladimír Křen and Ladislav Cvak

Volume 7
Caraway
Edited by Éva Németh

Volume 8
Saffron
Edited by Moshe Negbi

Volume 9
Tea Tree
Edited by Ian Southwell and Robert Lowe

Volume 10
Basil
Edited by Raimo Hiltunen and Yvonne Holm

Volume 11
Fenugreek
Edited by Georgios Petropoulos

Volume 12
Gingko biloba
Edited by Teris A. Van Beek

Volume 13
Black Pepper
Edited by P.N. Ravindran

Volume 14
Sage
Edited by Spiridon E. Kintzios

Volume 15
Ginseng
Edited by W.E. Court

Volume 16
Mistletoe
Edited by Arndt Büssing

Volume 17
Tea
Edited by Yong-su Zhen

Volume 18
Artemisia
Edited by Colin W. Wright

Volume 19
Stevia
Edited by A. Douglas Kinghorn

Volume 20
Vetiveria
Edited by Massimo Maffei

Volume 21
Narcissus and Daffodil
Edited by Gordon R. Hanks

Volume 22
Eucalyptus
Edited by John J.W. Coppen

Volume 23
Pueraria
Edited by Wing Ming Keung

Volume 24
Thyme
Edited by E. Stahl-Biskup and F. Sáez

Volume 25
Oregano
Edited by Spiridon E. Kintzios

Volume 26
Citrus
Edited by Giovanni Dugo and Angelo Di Giacomo

Volume 27
Geranium and Pelargonium
Edited by Maria Lis-Balchin

Volume 28
Magnolia
Edited by Satyajit D. Sarker and Yuji Maruyama

Volume 29
Lavender
Edited by Maria Lis-Balchin

Lavender
The genus *Lavandula*

Edited by
Maria Lis-Balchin

London and New York

First published 2002
by Taylor & Francis
11 New Fetter Lane, London EC4P 4EE

Simultaneously published in the USA and Canada
by Taylor & Francis Inc,
29 West 35th Street, New York, NY 10001

Taylor & Francis is an imprint of the Taylor & Francis Group

© 2002 Taylor & Francis

Typeset in 10/12 Garamond by
Newgen Imaging Systems (P) Ltd, Chennai, India
Printed and bound in Great Britain by
TJ International Ltd, Padstow, Cornwall

All rights reserved. No part of this book may be reprinted or reproduced or utilised in any form or by any electronic, mechanical, or other means, now known or hereafter invented, including photocopying and recording, or in any information storage or retrieval system, without permission in writing from the publishers.

Every effort has been made to ensure that the advice and information in this book is true and accurate at the time of going to press. However, neither the publisher nor the authors can accept any legal responsibility or liability for any errors or omissions that may be made. In the case of drug administration, any medical procedure or the use of technical equipment mentioned within this book, you are strongly advised to consult the manufacturer's guidelines.

British Library Cataloguing in Publication Data
A catalogue record for this book is available from the British Library

Library of Congress Cataloging in Publication Data
A catalogue record has been requested

ISBN 0-415-28486-4

Contents

List of figures	vii
List of tables	ix
List of contributors	xi
Preface to the series	xiii

1 General introduction to the genus *Lavandula* — 1
   MARIA LIS-BALCHIN

2 The taxonomy of the genus *Lavandula* L. — 2
   TIM UPSON

3 History of usage of *Lavandula* species — 35
   JO CASTLE AND MARIA LIS-BALCHIN

4 History of nomenclature of *Lavandula* species, hybrids and cultivars — 51
   MARIA LIS-BALCHIN

5 Lavender growing in England for essential oil production — 57
   MARIA LIS-BALCHIN AND HENRY HEAD

6 The retail lavender nursery — 60
   SIMON CHARLESWORTH

7 Lavender growing in Australia — 76
   ROSEMARY HOLMES

8 Naming and misnaming of lavender cultivars — 80
   MARIA LIS-BALCHIN

9 Phytochemistry of the genus *Lavandula* — 86
   JEFFREY B. HARBORNE AND CHRISTINE A. WILLIAMS

10 Distillation of the lavender type oils: theory and practice — 100
   E.F.K. DENNY

11 Lavender essential oil: standardisation, ISO; adulteration and its detection using GC, enantiomeric columns and bioactivity — 117
   MARIA LIS-BALCHIN

12	Lavender oil and its therapeutic properties GERHARD BUCHBAUER	124
13	Pharmacology of *Lavandula* essential oils and extracts *in vitro* and *in vivo* STEPHEN HART AND MARIA LIS-BALCHIN	140
14	The psychological effects of lavender MICHAEL KIRK-SMITH	155
15	Antimicrobial properties of lavender volatile oil STANLEY G. DEANS	171
16	Lavender oil and its use in aromatherapy MARIA LIS-BALCHIN	180
17	Perfumery uses of lavender and lavandin oils RHONA WELLS AND MARIA LIS-BALCHIN	194
18	Miscellaneous uses of lavender and lavender oil: use in hair products, food flavouring, tissanes, herbal pillows and medicinal products MARIA LIS-BALCHIN	200
19	New research into *Lavandula* species, hybrids and cultivars MARIA LIS-BALCHIN	206
20	Further research into *Lavandula* species: cell cultures of *L. vera* and rosmarinic acid production MLADENKA PAUNOVA ILIEVA-STOILOVA, ATANAS IVANOV PAVLOV AND ELENA GEORGIEVA KOVATCHEVA-APOSTOLOVA	214
21	*Lavandula x heterophylla* and *L. x allardii*: a puzzling complex SUSYN ANDREWS	227
22	Comparative study of essential oil quantity and composition from ten cultivars of organically grown lavender and lavandin DENYS J. CHARLES, ERICA N.C. RENAUD AND JAMES E. SIMON	232
23	Chemical profiles of lavender oils and pharmacology MARIA LIS-BALCHIN AND STEPHEN HART	243
24	Chemical composition of essential oils from different species, hybrids and cultivars of *Lavandula* MARIA LIS-BALCHIN	251
	Index	263

Figures

2.1	The diversity of leaf shapes and forms found within the genus *Lavandula*	5
2.2	Diagrammatic representations of the structure of the many-flowered (bicicinnus) and single-flowered cymes	6
2.3	Variation in corolla morphology within the genus *Lavandula*	7
2.4	Examples of variation in the calyces within the genus *Lavandula*, illustrating differences in the calyx lobes	8
2.5	Examples illustrating the diversity of bract shapes in the genus *Lavandula*	9
2.6	*L. angustifolia* 'Hidcote' – cultivated at Cambridge University Botanic Garden	13
2.7	*L. latifolia* – France, Col de Ferrier nr. Grasse. View of flower spike showing linear bracts and bracteoles	14
2.8	*L. x intermedia* 'Grosso' – cultivated at Norfolk Lavender, UK	15
2.9	*L. dentata* var. *dentata* – Morocco	16
2.10	*L. stoechas* subsp. *stoechas*	18
2.11	*L. multifida* – close up showing flower spike, Morocco near Oued Laou	21
2.12	*L. canariensis* – plant in full bloom, Tenerife near Chio	22
2.13	*L. antineae* – cultivated at Cambridge University Botanic Garden	24
2.14	*L. subnuda* – cultivated at Cambridge University Botanic Garden. Section Subnudae	28
2.15	*L. aristibracteata* – cultivated at Cambridge University Botanic Garden. Section Subnudae	29
2.16	*L. bipinnata* – close up of flower, cultivated at Cambridge University Botanic Garden. Section Chaetostachys	31
5.1	'Old Major' helps to collect the harvest, around 1940	57
6.1	Out-buildings and sales area fronted by the display of lavenders	60
6.2	*L. stoechas* subsp. *Stoechas* 'Kew Red'	70
6.3	*L. stoechas* × *viridis* 'Willow Vale' *L. angustifolia* Hidcote	71
6.4	*L. stoechas* × *viridis* 'Helmsdale'	72
6.5	*L. minutolii*	73
9.1	*Lavandula* monoterpenes	87
9.2	*Lavandula* triterpenoids	90
9.3	*Lavandula* flavonoids	91
9.4	*Lavandula* hydroxycinnamic acid esters	97
10.1	The lavender oil glands: oil glands on a calyx of ripe lavender	100
10.2	Harvesting at about two tons lavender flowers per hour	101
10.3	Sketches illustrating the modern cylindrical still system	102
10.4	Effects of steam moisture: oil and water on an absorptive surface	105

List of figures

10.5	The two orthodox still schemes mentioned in distillation equipment	108
10.6	Still for small-scale operation and (parameter) tests	110
10.7	Multi-tube condenser: vapour-tube type	111
10.8	Condensers for standard temperate zone conditions	112
10.9	Receiver–separator for oils less dense than water	113
13.1	The spasmolytic effect of commercial *Lavandula* oils on the electrically-stimulated guinea-pig ileum *in vitro*	143
13.2	Mode of action of lavender oil on the guinea-pig ileum *in vitro*	145
13.3	The enhancement of spasmolysis following the application of Trequensin, a phosphodiesterase inhibitor, suggesting that cAMP is involved	146
13.4	Investigating the possibility of calcium channel involvement using calcium-free buffers	148
13.5	The effect of lavender oil on the spontaneously contracting uterus of the rat *in vitro*, showing the inhibition of contractions, which at higher lavender oil concentrations actually cease altogether	149
13.6	The action of lavender oil on skeletal muscle using the chick biventer muscle preparation *in vitro*	151
13.7	(a) The action of linalyl acetate on the skeletal muscle showing a spasmolytic action (b) the action of thyme oil for comparison, showing a rise in baseline but no decrease in the size of the contractions	151
20.1	HPLC of the methanolic extract of cell biomass from *L. vera* MM	216
20.2	The basic physiological characteristics of the *L. vera* MM cell suspension	218
20.3	Time course of growth (a) and RA biosynthesis (b) by *L. vera* MM cell suspension culture	219
20.4	Effect of sucrose concentration in the nutrient medium on yields of the cell biomass (g DB/g sucrose) and RA (mg RA/g sucrose)	220
20.5	Time course of growth (a) and RA biosynthesis (b) by *L. vera* MM cell with different concentration of ammonium ions	221
20.6	Time course of growth (a) and RA biosynthesis (b) by *L. vera* MM cell with different concentration of nitrate ions	221
20.7	Time course of growth (a) and RA biosynthesis (b) by *L. vera* MM cell suspension culture cultivated in LS nutrient medium with different amounts of KH_2PO_4	222
21.1	*Lavandula x allardii* 'African Pride'	229
21.2	*Lavandula x allardii* 'Clone B'	230

Tables

9.1	The major constituents of lavender, lavandin and spike lavender oils	88
9.2	A comparison of the main constituents of lavender oil from different countries	88
9.3	The major essential oils of some *L. latifolia* and some wild *Lavandula* taxa from Spain and Portugal	89
9.4	Flavones detected in *Lavandula* species, subspecies and varieties	93
9.5	Some flavones detected in *Lavandula* species	95
9.6	The distribution of the major flavone glycosides in *Lavandula* sections *Stoechas, Dentata, Pterostoechas* and *Subnuda*	95
9.7	The distribution of different flavone subclasses in the six sections of the genus *Lavandula*	96
9.8	Anthocyanins of *Lavandula* flowers: pigments present	96
9.9	Distribution of hydroxycinnamic acid esters in the genus *Lavandula*	96
11.1	*L. angustifolia* P. Miller ISO 3515 1987	117
11.2a	Lavandin abrialis ISO 3054 1987 *L. angustifolia* P. Miller $\times$ *L. latifolia* (L.f) Medikus	118
b	Lavandin grosso ISO 8902 1986 *L. angustifolia* P. Miller $\times$ *L. latifolia* (L.f) Medikus light yellow	118
11.3	Major compounds of lavender oil produced by hydrodiffusion and microwave extraction in hexane	121
11.4	Composition of lavandin grosso oil from steam distillation, oil from CO_2 extraction and an absolute	121
11.5	Changes in main chemical components of natural *L.* hybrids collected in Tanaro Valley, Italy	121
11.6	Correlation between high linalool or linalyl acetate content of commercial essential oils and bioactivity	122
12.1	Effect of lavender and other EOs and fragrance compounds on motility in mice after 1 h of inhalation	126
13.1	Evidence for contraction, C and or relaxation, R for extracts and essential oils in electrically-stimulated guinea-pig ileum *in vitro*	144
13.2	Evidence for Ca^{++} channel blocking by different extracts of *Lavandula* species	148
15.1a	Antibacterial activity of lavender volatile oil against twenty-five test bacteria	174
15.1b	Antibacterial activity of lavender oil constituents against twenty-five test bacteria	175
15.2	Correlation between linalool and linalyl acetate content per cent of lavender volatile oils and antimicrobial activity	176

16.1	Clinical studies on aromatherapy	181
16.2	Sedative and stimulant EOs	184
16.3	Evidence for transfer of components of EO into blood/brain when applied to skin, orally or by inhalation	186
17.1	Modern lavender water	195
17.2	Old English lavender	196
17.3	Eau de cologne (1834)	197
17.4	Fougère-type perfume	198
18.1	Lavandin, spike lavender and lavender uses as natural food flavours	202
20.1	Callogenesis of *L. vera* on LS media, supplemented with different concentrations of the growth regulators	215
20.2	Content of phenolic acids in cell biomass of *L. vera* MM	217
20.3	Modified nutrient medium for the production of RA from *L. vera* MM	223
22.1	Growth and development of ten cultivars of organically grown lavandin and lavender (1999) in Norway, Iowa	237
22.2	Essential oil and quality control evaluation of ten cultivars of organically grown lavandin and lavender over 2 years in Norway, Iowa	238
22.3	Comparative evaluation of the oil composition of ten cultivars of organically grown lavender, 1999, Norway, Iowa	238
22.4	Enantiomeric distribution of linalool and linalyl acetate in hydrodistilled oils of lavandin and lavender grown in Norway, Iowa	239
22.5	Analysis of authentic and commercial lavender oils	240
22.6	Radiocarbon and stable isotope ratio analyses	241
23.1	The predicted effect on guinea-pig ileum based on the percentage of components at different retention time intervals	246
23.2	Comparison of the actual effect of essential oils on guinea-pig ileum and the predicted effects	247
23.3	The predicted effect on guinea-pig ileum based on the percentage of components at different retention time intervals	248
23.4	Comparison of the predicted effect of essential oil blends on clients/patients by aromatherapists and by the chemical composition with their actual effect on guinea-pig ileum	249
24.1	Composition of French lavender (ISO 3515) (*L. angustifolia* P. Miller) compared to lavandin hybrids	251
24.2	Chemical composition of lavandin and cultivars	252
24.3	Chemical composition of *Lavandula* species and cultivars	253
24.4	Chemical composition of *L. latifolia*	253
24.5	Chemical composition of *L. stoechas* species and their chemotypes	254
24.6	Chemical composition of *L. luisieri* and its chemotypes	255
24.7	Chemical composition of *L. viridis* oil	255
24.8	*L. pinnata* L. il. *var. pinnata* grown on Madeira	256
24.9	Aroma profiles of selected *Lavandula* species	258
24.10	Aroma profiles of selected *Lavandula* species	259
24.11	*Lavandula angustifolia* aroma profile compared to Bulgarian Lavender Oil	260
24.12	Composition of different floral parts of a lavender (*L. angustifolia*) clone	261

Contributors

Maria Lis-Balchin, South Bank University, Borough Road, London, SE1 OAA, UK. E-mail: lisbalmt@sbu.ac.uk

Tim Upson, Cambridge Botanic Gardens, Cory Lodge, Bateman Street, Cambridge, CB2 IJF, UK. E-mail: tmu20@cam.ac.uk

Jo Castle, 15, Coleman Street, Brighton, East Sussex, BN2 2SQ, UK. E-mail: Jo.Castle@btopenworld.com

Henry Head, Norfolk Lavender Caley Mill, Heacham, Norfolk PE31 THE. E-mail: admin@Norfolk-lavender.co.uk

Simon Charlesworth, Downderry Nursery, Pillar Box Lane, Hadlow, Tonbridge, Kent, TN11 9SW. E-mail: simon@downderry-nursery.co.uk

Rosemary Holmes, Yuulong Lavender Estate at Mt. Egerton, near Ballarat in Victoria, Australia. E-mail: yuulong@tpgi.com.au

Jeffrey B. Harborne, Department of Botany, School of Plant Sciences, The University of Reading, Whitekights, Reading, RG6 6AS, UK

Christine A. Williams, Department of Botany, School of Plant Sciences, The University of Reading, Whitekights, Reading, RG6 6AS, UK. E-mail: Christine.williams@reading.ac.uk

E.F.K. Denny, Denny, Mckenzie Associates, PO Box 42, Lilydale, Tasmania 7268, Australia. E-mail: timdenny@southcom.com.au

Gerhard Buchbauer, Institute of Pharmaceutical Chemistry, University of Vienna, A1090 Vienna, Althanstrasse 14, Austria. E-mail: Gerhard.buchbauer@univie.ac.uk

Stephen Hart, Messenger & Signalling Research Group, School of Biomedical Sciences, King's College London, Guy's Campus, London, SE1 9RT, UK. E-mail: Stephen.hart@kcl.ac.uk

Michael Kirk-Smith, University of Ulster, Newtown Abbey, Northern Ireland, BT 37 OQB. E-mail: mks@ulst.ac.uk

Stanley G. Deans, Aromatic & Medicinal Plant Group, Food Systems Division, SAC Auchincruive, South Ayrshire KA6 5HW, Scotland, UK. E-mail: stan@clan-deans.fsnet.co.uk

Rhona Wells, Charabot et Cie, London

Mladenka Paunova Ilieva-Stoilova, Group of Applied Microbiology and Biotechnology, Department of Microbial Biosynthesis and Biotechnology, Institute of Microbiology, Bulgarian Academy of Sciences, 26 Maritza Blvd., Plovdiv 4002, Bulgaria. E-mail: stoilov@plovdiv.techno-link.com

Atanas Ivanov Pavlov, Group of Applied Microbiology and Biotechnology, Department of Microbial Biosynthesis and Biotechnology, Institute of Microbiology, Bulgarian Academy of Sciences, 26 Maritza Blvd., Plovdiv 4002, Bulgaria

Elena Georgieva Kovatcheva-Apostolova, Department of Analitical Chemistry, Higher Institute of Food and Flauver Industries, 26 Maritza Blvd., Plovdiv 4002, Bulgaria

Susyn Andrews, Royal Botanic Gardens, Kew, Richmond, Surrey, TW9 3AB. E-mail: S.Andrews@rbgkew.org.uk

Denys J. Charles, Frontier Natural Products Co-op, PO Box 299, Norway, IA, 52318, USA. E-mail: denys.charles@frontierherb.com

Erica N.C. Renaud, Frontier Natural Products Co-op, PO Box 299, Norway, IA, 52318, USA

James E. Simon, New Use Agriculture and Natural Plant Products Program, Rutgers University, Cook College, Foran Hall, New Brunswick, NJ, USA

Preface to the series

There is increasing interest in industry, academia and the health sciences in medicinal and aromatic plants. In passing from plant production to the eventual product used by the public, many sciences are involved. This series brings together information which is currently scattered through an ever increasing number of journals. Each volume gives an in-depth look at one plant genus, about which an area specialist has assembled information ranging from the production of the plant to market trends and quality control.

Many industries are involved such as forestry, agriculture, chemical, food, flavour, beverage, pharmaceutical, cosmetic and fragrance. The plant raw materials are roots, rhizomes, bulbs, leaves, stems, barks, wood, flowers, fruits and seeds. These yield gums, resins, essential (volatile) oils, fixed oils, waxes, juices, extracts and spices for medicinal and aromatic purposes. All these commodities are traded worldwide. A dealer's market report for an item may say 'Drought in the country of origin has forced up prices'.

Natural products do not mean safe products and account of this has to be taken by the above industries, which are subject to regulation. For example, a number of plants which are approved for use in medicine must not be used in cosmetic products.

The assessment of safe to use starts with the harvested plant material which has to comply with an official monograph. This may require absence of, or prescribed limits of, radioactive material, heavy metals, aflatoxin, pesticide residue, as well as the required level of active principle. This analytical control is costly and tends to exclude small batches of plant material. Large-scale contracted mechanised cultivation with designated seed or plantlets is now preferable.

Today, plant selection is not only for the yield of active principle, but for the plant's ability to overcome disease, climatic stress and the hazards caused by mankind. Such methods as *in vitro* fertilization, meristem cultures and somatic embryogenesis are used. The transfer of sections of DNA is giving rise to controversy in the case of some end-uses of the plant material.

Some suppliers of plant raw material are now able to certify that they are supplying organically-farmed medicinal plants, herbs and spices. The Economic Union directive (CVO/EU No 2029/91) details the specifications for the *obligatory* quality controls to be carried out at all stages of production and processing of organic products.

Fascinating plant folklore and ethnopharmacology leads to medicinal potential. Examples are the muscle relaxants based on the arrow poison, curare, from species of *Chondrodendron*, and the anti-malarials derived from species of *Cinchona* and *Artemisia*. The methods of detection of pharmacological activity have become increasingly reliable and specific, frequently involving enzymes in bioassays and avoiding the use of laboratory animals. By using bioassay linked fractionation of crude plant juices or extracts, compounds can be specifically targeted which, for example, inhibit blood platelet aggregation, or have anti-tumour, or anti-viral, or any other required activity. With the assistance of robotic devices, all the members of a genus may be readily screened. However, the plant material must be *fully* authenticated by a specialist.

The medicinal traditions of ancient civilisations such as those of China and India have a large armamentaria of plants in their pharmacopoeias which are used throughout South-East Asia. A similar situation exists in Africa and South America. Thus, a very high percentage of the world's population relies on medicinal and aromatic plants for their medicine. Western medicine is also responding. Already in Germany all medical practitioners have to pass an examination in phytotherapy before being allowed to practise. It is noticeable that throughout Europe and the USA, medical, pharmacy and health related schools are increasingly offering training in phytotherapy.

Multinational pharmaceutical companies have become less enamoured of the single compound magic bullet cure. The high costs of such ventures and the endless competition from 'me too' compounds from rival companies often discourage the attempt. Independent phytomedicine companies have been very strong in Germany. However, by the end of 1995, eleven (almost all) had been acquired by the multinational pharmaceutical firms, acknowledging the lay public's growing demand for phytomedicines in the Western World.

The business of dietary supplements in the Western World has expanded from the health store to the pharmacy. Alternative medicine includes plant-based products. Appropriate measures to ensure the quality, safety and efficacy of these either already exist or are being answered by greater legislative control by such bodies as the Food and Drug Administration of the USA and the recently created European Agency for the Evaluation of Medicinal Products, based in London.

In the USA, the Dietary Supplement and Health Education Act of 1994 recognised the class of phytotherapeutic agents derived from medicinal and aromatic plants. Furthermore, under public pressure, the US Congress set up an Office of Alternative Medicine and this office in 1994 assisted the filing of several Investigational New Drug (IND) applications, required for clinical trials of some Chinese herbal preparations. The significance of these applications was that each Chinese preparation involved several plants and yet was handled as a *single* IND. A demonstration of the contribution of efficacy, of *each* ingredient of *each* plant, was not required. This was a major step forward towards more sensible regulations in regard to phytomedicines.

My thanks are due to the staffs of Harwood Academic Publishers and Taylor & Francis who have made this series possible and especially to the volume editors and their chapter contributors for the authoritative information.

Roland Hardman

1 General introduction to the genus *Lavandula*

Maria Lis-Balchin

Lavandula species (Labiatiae, syn. Lamiaceae) are mainly grown for their essential oils, which are used in perfumery, cosmetics, food processing and nowadays also in 'aromatherapy' products. The dried flowers have also been used from time immemorial in pillows, sachets etc. for promoting sleep and relaxation. Numerous lavender plants are also sold as ornamental plants for the garden; these include *L. latifolia, L. pinnata, L. lanata, L. dentata* and *L. stoechas* and their numerous cultivars.

Lavender oil, distilled from *L. angustifolia* was used extensively in Victorian times as a perfume and applied in numerous cosmetic products, but now it is used mainly in combination with other essential oils and aromachemicals. This species and numerous hybrids/cultivars, for example, *Lavandin* 'grosso' were originally grown in the South of France, but are now grown virtually round the world. True lavender oil, consisting mainly of linalool and linalyl acetate, has a very variable composition due to the genetic instability of the oil-producing plants and variations due to temperature, water quantity, altitude, fertilizers, time of year, geographic distribution etc. The chemical composition also varies in the numerous hybrids, which produce larger plants with a higher essential oil yield and which are therefore grown more often.

The essential oil of lavender is often adulterated with other oils or some fractions derived from plants containing linalool and linalyl acetate, or with the synthetic components, or the original oil can be acetylated. There is a problem with recognition of such adulterations, although enantiomeric columns have been a useful tool in modern detection.

Aromatherapists consider the oil from *L. angustifolia* as the most beneficial, together with wild-grown cultivars at high altitude; as yet scientific evidence is lacking for this and all the numerous medicinal claims made, other than for a possible general relaxing effect after inhalation, produced via the Limbic system. Pharmacological studies have shown a relaxation of smooth muscles *in vitro* using animal tissues, with an initial small contraction exhibited by *L. angustifolia*; the spasmolytic action was apparently mediated through the secondary messenger cyclic AMP. Studies with animals *in vivo* have shown a decline in movement after inhalation; in man, there was a slowing down of mental and physical activities. The main components were found in the blood after inhalation and these were also active in their own right when inhaled or massaged into the skin.

Lavandula species have a variable antimicrobial effect; Spike lavender, containing camphor, is the most potent; some species have a moderate antifungal action while the antioxidant activity is very variable. Some species have an acaricidal effect and have low general insecticidal properties. *Lavandula* has a low toxicity: even the strong undiluted essential oil can be used for some burns, with, anecdotally, beneficial effects on healing, however, cases of allergic airborne contact dermatitis have been reported.

New research on *Lavandula* species has indicated a wide diversity of applications, for example, the usefulness of *L. angustifolia* essential oil in the treatment of alopecia.

2 The taxonomy of the genus *Lavandula* L.

Tim Upson

Introduction

In the introduction to the *Natural History of the Lavenders* published in 1826 the author, Baron Gingins de la Sarraz, wrote *'continuing progress in the understanding of natural history would seem to require also a constant revision of families and genera most familiar to us'*. While much progress has been made in our understanding of the genus *Lavandula* his sentiments are still true today. The genus is currently subject to ongoing research into its taxonomy and systematics being undertaken by the author and colleagues at the University of Cambridge, University of Reading and the Royal Botanic Gardens, Kew. This treatment thus represents a provisional rather than a definitive account on the taxonomy of the genus. While some aspects of the research is nearing completion there are still some problematic areas where existing treatments may not be adequate and our understanding of the taxonomy is incomplete and these are indicated and discussed where appropriate.

This treatment recognises thirty-two species of *Lavandula* which have been described in the literature plus a number of infraspecific taxa and hybrids, although the number of species is likely to be higher once all the revisionary work has been completed. The genus has a distribution stretching from the Canary Islands, Cape Verde Islands and Madeira, across the Mediterranean Basin, North Africa, South West Asia, the Arabian Peninsula and tropical NE Africa with a disjunction to India.

Some species have been widely cultivated since ancient times and are familiar garden plants and hence there are many legends and folklore associated with these plants. The essential oils, principally harvested from *L. x intermedia* and *L. angustifolia*, are of economic importance in the perfumery and fragrance industry, some are widely used in aromatherapy and are known to have antiseptic and antifungal qualities. A number of species and their hybrids are horticulturally desirable and are cultivated in both northern and southern hemispheres. The Latin name *Lavandula* comes from the ancient use of this plant to perfume water for bathing, being derived from the Latin word *lavare,* meaning to be washed.

Historical perspective: a brief history of the taxonomy of the genus

It is clear that *Lavandula* was known to the earliest botanical writers and the first written accounts of lavenders can be found in the writings of the early Greek scholars such as Theophrastus (*c.* 370–285 BC). The genus is frequently mentioned in many herbals and other botanical books although the first monograph of the genus, *De Lavandula*, was not published until 1780 (Lundmark, 1780). This work recognised five species and eight varieties.

The second monograph of the genus *Histoire Naturelle des Lavandes* (Gingins, 1826), was of great significance and is still a valuable work today. His monograph enumerated twelve species along with descriptions, geographical distributions, properties and uses. His most important contribution was the recognition of groupings of species within the genus and the erection of an infrageneric classification of three sections.

By the time of the third and most recent monograph, A *Taxonomic Study of the Genus Lavandula* (Chaytor, 1937), a substantial number of new species and varieties had been described and her revision bought much of this information together for the first time. This account recognises twenty-eight species plus many infraspecific taxa arranged in five sections. A new species, *L. somaliensis*, was described, many new combinations were made and a new section *Subnuda* was erected.

Since Chaytor's account there has been no full generic treatment although a number of useful accounts of various groups have been published. Collectively these works have contributed much to our knowledge of the genus and have been incorporated into this account. Some of the most notable works include, a revision of section *Stoechas* (Rozeira, 1949), the genus *Lavandula* in Arabia and tropical NE Africa (Miller, 1985) describing five new species from the area and on the taxa native to the Iberian Peninsula (Suarez-Cervera and Seoane-Camba, 1986, 1989).

Generic status and relationships

Lavandula is a member of the Lamiaceae (Labiatae) family and belongs to the subfamily Nepetoideae. Within the Nepetoideae a number of tribes are recognised and *Lavandula* is currently treated as a distinct and isolated group of its own, that is, the tribe Lavanduleae (Endl.) Boiss containing just the single genus *Lavandula* (Cantino et al., 1992).

What makes a lavender a lavender

The genus in terms of its general morphology is a rather mixed and divergent group. It is defined by the nectary lobes being opposite the ovary's rather than alternate (which is the case in all other Lamiaceae). The combination of a compact terminal spikes of flowers usually borne on a long peduncle (flower stalk); the declinate stamens (stamens curved downwards) borne within the corolla tube and persistent bracts are characteristic.

Many species are highly aromatic due to the presence of essential oils that are borne in glands covering much of the plant. In habit they vary from woody shrubs up to a metre in height, to perennial woody-based shrubs or annual herbs. The leaves can be entirely or deeply dissected, and are often absent in some of the Arabian species. The flowers (spike) consists of cymes, a branching determinate inflorescence with a flower at the end of each branch, either an opposite decussate arrangement (each pair of flower whorls at right angles to the pair above or below) or an alternate spiral arrangement. The cymes are subtended by bracts, which vary in their size, shape and nervation, which can be diagnostic for many species. The cymes can be single flowered usually without bracteoles or many flowered (3–9 flowers per cyme) with bracteoles. The bracteoles are small, often minute, bracts borne at the points of branching within each flower whorl. In some species, such as *L. stoechas* and *L. dentata*, the bracts at the apex are enlarged, coloured and sterile and known as a coma. The calyx is (two-lipped) bilabiate and varies in the number of nerves, lobe shape, presence of an appendage, colour and provides many important characters used to diagnose both sections and species. The corolla is usually bilabiate, tubular and with five lobes varying in size, colour, shape and markings.

Taxonomic treatment

The species and infraspecific taxa are arranged according to sections. Those whose sectional position is uncertain are dealt with at the end. Hybrids have been dealt with under each section except for intersectional hybrids which are placed after the sections. Keys to the species are given under each section. Only the most important synonyms are given as appropriate for each species and placed in brackets **as appropriate**. Author abbreviations follow Brummitt & Powell (1992).

Sectional classification

Although the species exhibit a wide variation of morphological characters, it is evident that natural groupings of related taxa can be recognised. These groups have been classified as sections, of which there are six presently recognised. It seems likely that further groupings need to be recognised and current work is presently aiming to confirm this. The sections can be distinguished by differences in habit, leaf shape, the arrangement of the flowers in the verticils, bract, calyx and corolla characters (Figures 2.1–2.5).

Nomenclatural notes and clarification on sectional classification

The naming of all plants is governed by the International Code of Botanical Nomenclature which sets out a series of rules on the naming of plants. The code provides an internationally agreed set of rules and standards to which all names must comply to be accepted. While the code may require what appears to be some frustrating name changes it replaced a situation in which there was no agreed code and individuals constantly changed names to the greater confusion of everyone. Names will be stable when using this international standard once these changes and corrections have been made. In accordance with this code a number of changes have been made over recent years to the section names and hence it seems appropriate to explain and clarify these changes as follows:

1. Section *Spica* Ging. to section *Lavandula* – Article 22 of the code requires that the name of a subdivision of a genus that includes the type of the genus must repeat the name of the genus unaltered. The type concept is a central element of botanical nomenclature. Any name must have a type specimen (this is usually a herbarium specimen) which essentially acts as a permanent reference point that determines the application of that name. This rule allows the subdivision containing the type of the genus to be instantly identified. In the case of *Lavandula* the type of the genus is a specimen of *L. spica* L. (although the correct name in general usage is *L. angustifolia*). This type species was placed by Gingins (1826) in section *Spica* and hence the name of this section was changed to *Lavandula* so as to repeat generic name as required by Article 22.
2. The names of plants must also follow certain grammatic rules of botanical Latin, again defined in the code. In the case of section *Subnuda* and *Dentata* the names as originally published are incorrect according to Article 21.2 as this requires the epithet to be a plural adjective agreeing in gender with the generic name. In practice this requires the correction of the endings to *Subnudae* and *Dentatae*, respectively.

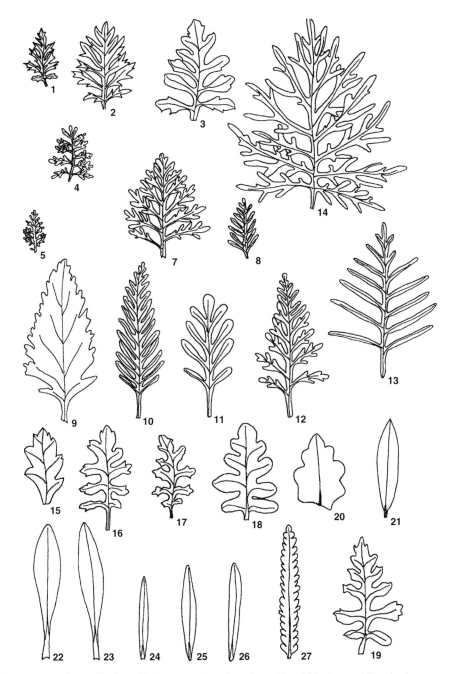

All leaves are shown life size with the exception of numbers 20 and 21 shown at 2 × life size.

1. L. mairei var. mairei
2. L. pubescens
3. L. citriodora
4. L. maroccana
5. L. antineae
7. L. multifida
8. L. coronopifolia
9. L. rotundifolia
10. L. minutolii var. minutolii
11. L. pinnata
12. L. canariensis
13. L. buchii var. gracile
14. L. bipinnata
15. L. subnuda
16. L. dhofarensis
17. L. nimmoi
18. L. somaliensis
19. L. galgalloensis
20. L. hasikensis
21. L. atriplicifolia
22. L. lanata
23. L. latifolia
24. L. angustifolia
25. L. stoechas subsp. stoechas
26. L. viridis
27. L. dentata var. dentata

Figure 2.1 The diversity of leaf shapes and forms found within the genus *Lavandula*.

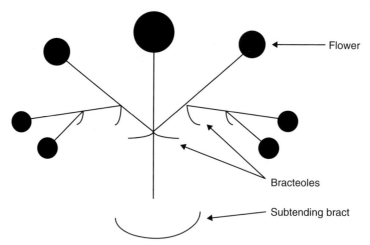

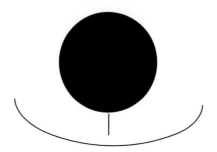

Figure 2.2 Diagrammatic representations of the structure of the many-flowered (bicicinnus) and single-flowered cymes in *Lavandula*.

Generic description

Lavandula L. (Synonyms – *Stoechas* Mill., *Fabricia* Adans., *Chaetostachys* Benth., *Sabaudia* Buscal. and Muschl., *Isinia* Rech. f.)

Shrubs, woody-based perennials or short-lived herbs; often aromatic, glabrous or with a variable indumentum. Leaves variable in shape, entire or dissected, sessile or petiolate. Inflorescence a congested terminal spike, occasionally with a coma, usually with distinct peduncle, simple or branched, verticils either a many-flowered (3–9) cyme with minute bracteoles or a reduced single-flowered cyme usually without bracteoles, subtended by persistent bracts variable in form, which maybe opposite or spirally arranged on the inflorescence axis. Calyx persistent, regular or two-lipped, upper with three-lobed, the lower two-lobed, $\pm$ equal in size or with the posterior lip larger or modified into an appendage, eight-, thirteen- or fifteen-nerved, the nerves in the lower sepals all borne to the apex. Corolla tube either just exceeding or up to 3 × longer

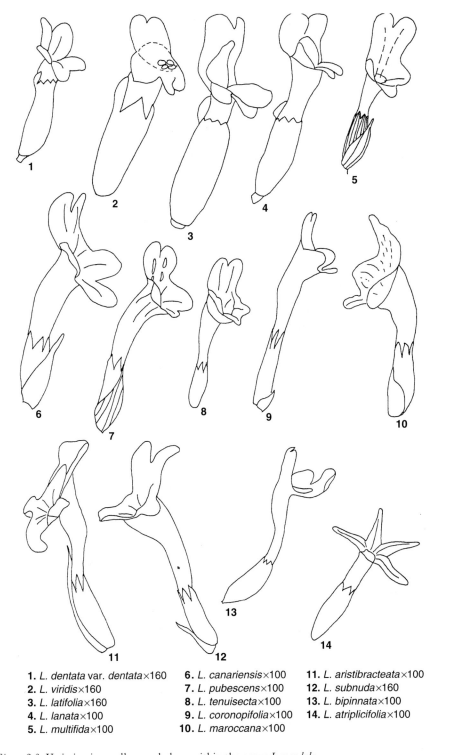

1. *L. dentata* var. *dentata* ×160
2. *L. viridis* ×160
3. *L. latifolia* ×160
4. *L. lanata* ×100
5. *L. multifida* ×100
6. *L. canariensis* ×100
7. *L. pubescens* ×100
8. *L. tenuisecta* ×100
9. *L. coronopifolia* ×100
10. *L. maroccana* ×100
11. *L. aristibracteata* ×100
12. *L. subnuda* ×160
13. *L. bipinnata* ×100
14. *L. atriplicifolia* ×100

Figure 2.3 Variation in corolla morphology within the genus *Lavandula*.

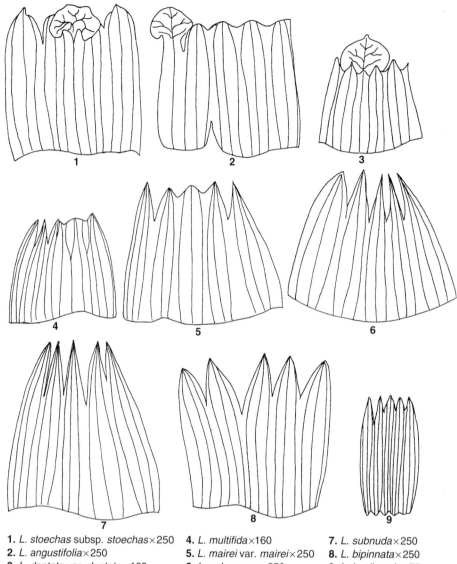

1. *L. stoechas* subsp. *stoechas*×250
2. *L. angustifolia*×250
3. *L. dentata* var. *dentata*×160
4. *L. multifida*×160
5. *L. mairei* var. *mairei*×250
6. *L. pubescens*×250
7. *L. subnuda*×250
8. *L. bipinnata*×250
9. *L. hasikensis*×75

Figure 2.4 Examples of variation in the calyces within the genus *Lavandula*, illustrating differences in the calyx lobes.

than calyx, weakly or strongly two-lipped, upper lip with two lobes, the lower three-lobed, the lobes variable in size. Four stamens declinate (curving downwards), usually didynamous (two pairs of stamens unequal in length), the anterior pair longer, included within tube. Stigma single, bilobed or capitate. Nectary lobes borne opposite the ovaries. Nutlets variable in shape, colour and size, with either a small basal scar or lateral scar $0.25-0.75 \times$ length of nutlet, mostly mucilaginous.

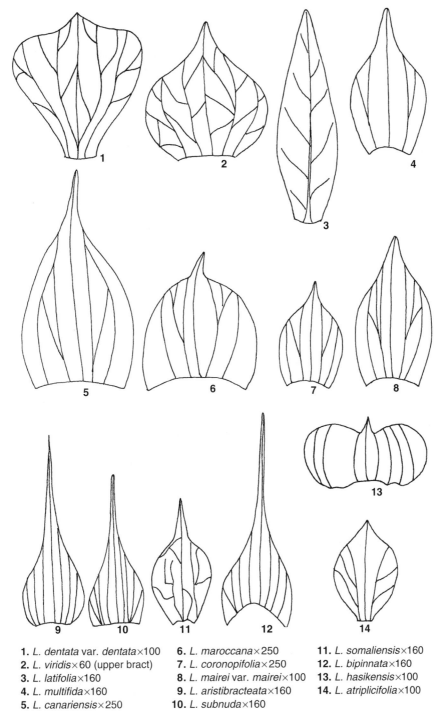

1. *L. dentata* var. *dentata* ×100
2. *L. viridis* ×60 (upper bract)
3. *L. latifolia* ×160
4. *L. multifida* ×160
5. *L. canariensis* ×250
6. *L. maroccana* ×250
7. *L. coronopifolia* ×250
8. *L. mairei* var. *mairei* ×100
9. *L. aristibracteata* ×160
10. *L. subnuda* ×160
11. *L. somaliensis* ×160
12. *L. bipinnata* ×160
13. *L. hasikensis* ×100
14. *L. atriplicifolia* ×100

Figure 2.5 Examples illustrating the diversity of bract shapes in the genus *Lavandula*.

Synopsis of sectional classification and taxa recognised in this treatment

1. Section *Lavandula* (= section *Spica* Ging.)	Section *Pterostoechas* (continued)
1. *L. angustifolia* Mill. subsp. *angustifolia* 　　subsp. *pyrenaica* (DC) Guinea 2. *L. lanata* Boiss. 3. *L. latifolia* Medik. Hybrids 　　*L. x intermedia* Emeric. ex. Loisel. 　　*L. lanata* Boiss. × *L. angustifolia* Mill.	12. *L. maroccana* Murb. 13. *L. tenuisecta* Coss. ex. Ball 14. *L. maroccana* Murb. 15. *L. mairei* Humbert var. *mairei* 　　　　var. *antiatlantica* Maire 16. *L. antineae* Maire 17. *L. coronopifolia* Poir.
2. Section *Dentatae* Suarez-Cerv. & Seoane-Camba	18. *L. pubescens* Decne. 19. *L. citriodora* A.G. Miller
4. *L. dentata* L. var. *dentata* 　　forma *rosea* Maire 　　forma *albiflora* Maire *L. dentata* L. var. *candicans* Batt.	Hybrids 　　*L. x christiana* Gattef. & Maire 　　*L. x murbeckiana* Emb. & Maire 　　*L. canariensis* Mill. × *L. buchii* Webb & Berthol. var. *buchii*
3. Section *Stoechas* Ging.	5. Section *Subnudae* Chaytor
5. *L. stoechas* L. subsp. *stoechas* 　　forma *leucantha* Ging. 　　forma *rosea* Maire 　　subsp. *pedunculata* (Mill.) Samp. ex Rozeira 　　subsp. *sampaiana* Rozeira 　　subsp. *lusitanica* (Chaytor) Rozeira 　　subsp. *luisieri* (Rozeira) Rozeira 　　subsp. *atlantica* Braun-Blanq. 　　subsp. *maderensis* (Benth.) Rozeira 　　subsp. *cariensis* (Boiss.) Rozeira 6. *L. viridis* L'Hér. Hybrids 　　*L. stoechas* L. × *L. viridis* L'Hér.	20. *L. subnuda* Benth. 21. *L. macra* Baker 22. *L. dhofarensis* A.G. Miller 　　subsp. *dhofarensis* 　　subsp. *ayunensis* A.G. Miller 23. *L. setifera* T. Anderson 24. *L. nimmoi* Benth. 25. *L. galgalloensis* A.G. Miller 26. *L. aristibracteata* A.G. Miller 27. *L. somaliensis* Chaytor
	6. Section *Chaetostachys* Benth.
4. Section *Pterostoechas* Ging.	28. *L. gibsonii* Grah. 29. *L. bipinnata* (Roth) Kuntze
7. *L. multifida* L. 8. *L. canariensis* Mill. 9. *L. pinnata* L. f. 10. *L. buchii* Webb & Berthel. var. *buchii* 　　var. *gracile* M.C. León 　　var. *tolpidifolia* (Svent.) M.C. León 11. *L. minutolii* Bolle var. *minutolii* 　　var. *tenuipinna* Svent.	Unclassified taxa
	30. *L. hasikensis* A.G. Miller 31. *L. atriplicifolia* Benth. 32. *L. erythreae* (Chiov.) Cufud.
	Intersectional Hybrids
	L. x allardii Hy *L. x heterophylla* Poir. *L. dentata* L. × *L. lanata* Boiss.

Key to sections

1. Cymes 3–9 flowered; nutlets with a basal scar only 2
   Cymes single-flowered; nutlets with a lateral scar 4
2. Leaves and calyx sessile, stigma capitate .... Section *Stoechas*
   Leaves and calyx with short stalk, stigma bilobed 3
3. Leaves with regular shallow dissections, corolla lobes subequal in size, corolla tube just exerted from calyx, calyx appendage c. 1.5 × width of the calyx .... Section *Dentatae*
   Leaves simple and entire, corolla lobes differing in size, corolla tube c. 2 × length of the calyx, calyx appendage the same width as the calyx .... Section *Lavandula*
4. Leaves ovate lanceolate in outline, dissected or lobed; calyx zygomorphic (bilaterally symmetrical); corolla violet blue or white, strongly zygomorphic 5
   Leaves linear and simple, calyx regular and lobes all equal; corolla yellow .... *L. atriplicifolia*/brown in colour, with equal lobes, star shaped .... *L. erythrae*
5. Cymes and bracts arranged in an opposite and decussate fashion, spike 4-seriate (quadrate) to biseriate in shape .... Section *Pterostoechas*
   Cymes and bracts arranged in an alternate, spiral fashion. Spike cylindrical or head like (capitate) in shape 6
6. Spikes elongated, leaves pinnatifid or bipinnatisect, bracts ovate spinescent 7
   Spike capitate, leaves lobed, bracts with orbicular wings .... *L. hasikensis*
7. Leaves predominantly pinnatifid to bipinnatisect, suffruticose shrubs, lateral scar on nutlets c. 0.25 × the length of nutlet. Lower corolla lip about the same size as lateral corolla lobes .... Section *Subnudae*
   Leaves very distinctly bipinnatisect, plant herbaceous, lateral scar on nutlets c. 0.75 × length of nutlet. Lower corolla lip larger than lateral lobes .... Section *Chaetostachys*

Section 1: *Lavandula* (=*Spica* Ging.)

Woody shrubs with simple leaves generally linear in shape. Cymes borne in an opposite and decussate arrangement. Each cyme many-flowered (3–)5–7(–9) with bracteoles present, the subtending bracts variable in shape with reticulate veining. Calyx tubular with a very short stalk (pedicellate), with thirteen or eight nerves, the upper middle lobe modified into a circular appendage. The corolla tube exerted from the calyx, the upper corolla lobes larger than the lateral lobes. Stigma bilobed. Nutlets bear a small to minute basal scar and produce no mucilage on wetting.

Contains three species from central and south west Europe. Commercially this is the most important section containing, *L. angustifolia* (English lavender) and the hybrid *L. x intermedia* (lavandin), which are the principal taxa cultivated for the production of essential oils and widely grown for their horticultural value. Numerous cultivars of *L. angustifolia* and *L. x intermedia* have been and continue to be selected both for oil production and ornamental value. There is no complete listing of cultivars but the following references provide useful information: Tucker and Hensen (1985); Andrews (1994); McNaughton (2000).

One of the greatest confusions in the naming of lavenders has occurred over the application of the name *L. spica* (see Green, 1932). The Swedish botanist Carl Linnaeus first used the name

L. spica, to included both lavender (*L. angustifolia*) as his *L. spica* var. α and spike lavender (*L. latifolia*) as his *L. spica* var. β. Unfortunately, subsequent authors who recognised these taxa as distinct species where not consistent in the use of the name and *L. Spica* has variously been applied to both *L. angustifolia* and *L. latifolia*. With no consistency in the use of this name the situation became completely confused and hence the use of the name *L. spica* was abandoned, the next available name being *L. angustifolia* Miller.

Key to species and major hybrids

1. Bracts subtending flowers ovate rhombic (diamond shape) 2
   Bracts subtending flowers linear to linear-lanceolate in shape 4
2. Bracts at least twice as broad as long, bracteoles minute
   (not clearly visible) .... *L. angustifolia*
   Bracts c. 3 × longer than broad, bracteoles large 1–4 mm 3
3. Flower stalks usually unbranched, flowers deep purple/mauve in colour,
   L. leaves with woolly grey to silver indumentum .... *L. angustifolia* × *L. lanata*
   Flower stalks branched, shades of purple lilac, blues or white, leaves
   with non-woolly grey to silver grey indumentum .... *L. x intermedia*
4. Calyx with eight nerves and eight-toothed. Leaves with uniform dense
   white woolly indumentum .... *L. lanata*
   Calyx with thirteen nerves and five-toothed. Leaves with
   silvery–grey indumentum .... *L. latifolia*

1. **L. angustifolia** Miller subsp. **angustifolia** (L. spica L. var. α, L. officinalis Chaix, L. fragrans Jord., L. vera DC)

Shrub to 50 cm with linear-lanceolate leaves grey tomentose when young, becoming greener with age. Inflorescence stalk usually unbranched 10–25 cm long with a compact spike 4–5(–8) cm, sometimes with a lower flower cluster distant from the main spike. Bracts broadly ovate-rhombic to obovate, bracteoles present but minute. Calyx thirteen-nerved, with small circular appendage. Corolla strongly bilaterally symmetrical, nearly twice the length of calyx with prominent lobes, shades of blue/mauve, white, rarely violet pink in colour. Flowers from mid-June to July. Native to SW and South Central Europe (Italy, France and Spain) in mountainous areas usually over 1500 m, but widely cultivated and sometimes naturalised elsewhere. The natural variation in this species across its range is not fully understood and there are many names in the literature. While only two subspecies are listed here this reflects that these are the only two whose delimitations and identification are clearly understood. Further work may identify other infraspecific taxa. This species produces the best quality oils and is a fine ornamental plant. Its mountain origins make this the hardiest species in cultivation. Of the many cultivars the following are some of the best known, for example, 'Hidcote' 30 cm very deep violet blue flowers (Figure 2.6), 'Loddon Blue' 60 cm dark violet blue, 'Munstead' 45 cm blue lilac flowers, 'Rosea' 45 cm pink flowers, 'Nana Alba' 20 cm dwarf white variant.

> *L. angustifolia* Miller subsp. *pyrenaica* (DC) Guinea – a variant from the East Pyrenees and NE Spain which has larger floral bracts which usually exceed calyx in length, a shorter and more condensed spike, deep blue/purple flowers and is of smaller stature (25–35 cm).

2. **L. latifolia** Medik (L. spica L. var. β; L. latifolia Villers)

Shrub 50–70 (100) cm. Leaves grey, linear-lanceolate to spathulate in outline. Inflorescence stalk distinctly branched usually forming a trident shaped flower spike, up to 25 cm high. Spike often

Figure 2.6 L. angustifolia 'Hidcote' – cultivated at Cambridge University Botanic Garden. (See Color Plate I.)

interrupted, 5–8 cm long. Bracts subtending cymes linear-lanceolate in shape, Bracteoles distinct to 4 mm long. Calyx thirteen-nerved, with rotund appendage. Corolla strongly bilaterally symmetrical, blue to mauve in colour. Flowers from mid-July. Native to SW and South Central Europe to *c.* 1000 m (−1200 m) (Figure 2.7).

Only occasionally cultivated both for its oil, which is of low quality and as an ornamental.

3. **L. lanata** *Boiss.*

Shrub 50–80 cm, both stems and leaves covered with dense short white woolly indumentum of branching hairs. Leaves linear to oblanceolate (tapering towards the base). Inflorescence stalk often branched to 25 cm, spike often interrupted to 8 cm long. Fertile bracts linear-lanceolate bracteoles up to 6 mm in length, calyx eight-nerved with eight lobes (four large distinct lobes alternating with four smaller lobes) with elliptic to rotund appendage. Corolla rather small exceeding calyx by only 2–3 mm, the upper lobes only slightly larger than the lower lobes, dark purple. Flowers mid-to late July. Native to mountainous areas in South Spain over 2000 m.

Sectional hybrids

L. x intermedia Emeric ex Loisel. (**L. angustifolia** × **L. latifolia**)
(**L. hybrida** Reverchon; **L. x hortensis** Hy)

A vigorous shrub from 60–150 cm, highly variable making delimitation difficult at times. Leaves linear-lanceolate to spathulate, often grey tomentose. Inflorescence stalk branched, spike usually lax and occasionally interrupted. Fertile bracts ovate-rhombic in outline but variable in

Figure 2.7 L. *latifolia* – France, Col de Ferrier nr. Grasse. View of flower spike showing linear bracts and bracteoles. (See Color Plate II.)

exact shape and size, bracteoles 1–4 mm long. Calyx thirteen-nerved, with rotund to elliptic appendage. Corolla bilaterally symmetrical, variable in colour usually shades of lilac-purple to white. Flowers from late June to July.

A natural sterile hybrid occurring in Spain, France and Italy where the two parents meet. Commonly known as lavandin with numerous cultivars selected for oil production and horticultural purposes, for example, 'Alba' to 100 cm white flowered, Dutch Group 80 cm pale blue-violet flowers and grey leaves, 'Grappenhall' 90 cm lilac-purple flowers, green foliage, 'Grosso' 75 cm with a profusion of dark violet-blue flowers (the most popular lavender for oil production) (Figure 2.8), 'Hidcote Giant' 90 cm with stout spikes of violet-blue flowers, 'Lullingstone Castle' 100 cm dark blue-violet, grey leaves, ' Old English' 100 cm violet flowers, grey-green foliage, 'Seal' vigorous to over 100 cm, violet-blue flowers.

L. lanata Boiss. ×L. angustifolia Mill

Similar to L. *lanata* in general habit 60–70 cm tall with grey to silver-grey leaves. Flower spike with purple flowers, bracts subtending the whorls narrowly rhombic (diamond) in shape, about 3 × longer than broad. Flowers violet purple. Flowers mid-June to July.

Several hybrids of this parentage are in cultivation and make fine ornamental garden plants that tend to be more robust in cultivation than L. *lanata*. Cultivars include: 'Richard Gray' – to 60 cm with rounded flower spikes and 'Sawyers' – taller form to 70 cm with large conical flower spikes, both of UK origin; 'Silver Frost' – to 50 cm with conical spikes raised in the United States; 'Joan Head' – 60–70 cm with cylindrical spikes raised in New Zealand.

Figure 2.8 L. x intermedia 'Grosso' – cultivated at Norfolk Lavender, UK. The most widely cultivated lavandin for oil production. (See Color Plate III.)

Section 2: *Dentatae* Suárez-Cerv. and Seoane-Camba

Woody shrub with distinctive leaves, linear-lanceolate in shape with regular shallow-rounded lobes. Flower spike compact and dense, topped by a tuft of enlarged and coloured sterile floral bracts (a coma). The fertile bracts broadly obovate with an acute apex, with reticulate veining. Flower whorls 5–7(–9) flowered with minute bracteoles. Calyx with a short stalk (pedicellate), thirteen-nerved, the middle calyx lobe modified into a large appendage up to 1.5 × the width of calyx tube. Corolla tube only slightly exerted from calyx *c.* 2 mm. Stigma bilobed. Nutlets elliptic in shape with a minute basal scar producing mucilage on wetting.

Contains a single species, *L. dentata*, recognised in this treatment with two varieties and two forms. This species has an interesting distribution being native to South Spain and Balearic Islands, western North Africa (Morocco, Algeria, Tunisia) with a disjunction to the SW Arabian Peninsula (Saudi Arabia and Yemen Arab Republic) and Ethiopia (Abyssinian Highlands). It is naturalised in Portugal and Corsica and widely cultivated elsewhere principally as an ornamental in frost free areas as it is not reliably hardy below 0°C.

L. dentata was previously placed in section *Stoechas*, where its position has always been rather anomalous. This was commented on by Chaytor (1937) and first taxonomically recognised by Rozeira (1949), who created a separate subsection *Dentata*, for this species within section *Stoechas*. A new section *Dentata* (=*Dentatae*), was created by Suarez-Cervera and Seoane-Camba (1986), to accommodate *L. dentata*. Its affinities also appear to lie closer to section *Lavandula* rather than section *Stoechas*.

Figure 2.9 L. dentata var. *dentata* – Morocco. (See Color Plate IV.)

4. **L. dentata** L. var. **dentata**
(**L. santolinaefolia** Spach)

An aromatic woody shrub, 50–100 cm, with an indumentum variable in density giving plants a green to grey-green appearance. Flower spike 3–6 cm, the coma consisting of enlarged sterile floral bracts, rhombic in shape, violet-blue in colour, grading down to fully fertile non-modified bracts (Figure 2.9). Corolla principally shades of violet-blue to mauve. Variants with pink or rose flowers and bracts should be treated as forma *rosea* Maire, those with white flowers forma *albiflora* Maire.

A number of cultivars have been selected including 'Royal Crown' – to 75 cm green foliage and pale violet-blue flowers, 'Linda Ligon' – to 60 cm variegated foliage selected in the United States, 'Ploughmans Blue' – to 75 cm green foliage and pale violet-blue flowers selected in New Zealand.

> var. *candicans* Batt. (silver form) – is easily distinguished by almost all parts being white tomentose. Occurs as distinct populations throughout the range of var. *dentata*.

Section 3: *Stoechas* Ging.

A distinct group of small woody shrubs with linear-lanceolate leaves. Diagnosed by their corolla lobes that are all similar in size, the corolla tube only just exerted from the calyx, the broad ovate to slightly obovate bracts and the capitate stigma, unique to this section. The flower spike is dense and compact, each whorl of flowers (3–)5–7 flowered, minute bracteoles present. Apex of spike topped by a distinctive tuft of enlarged and coloured sterile floral bracts (a coma). The coma differs from *L. dentata* (the only other taxon bearing a coma) in being fully sterile and by the shape of the bracts, which are linear lanceolate. Some of the earliest lavenders to flower from late spring. The epithet *Stoechas* is derived from the Stoechades Islands situated off southern France (now known as the Iles de Hyères) from were the plant was first described by the Greek physician Dioscorides.

Section *Stoechas*, as circumscribed here contains just two species *L. stoechas* and *L. viridis*. This hides the great diversity of taxa and variation that exists within this section as a multitude of infraspecific taxa are recognised under *L. stoechas*. It is these taxa whose identity and true relationships remain both problematic and elusive. At present there is no completely satisfactory treatment for the whole section and hence further research is being undertaken to address these questions. The treatment given here is provisional and follows the most frequently used classification purely for stability. Hence, no key is given to the subspecies of *L. stoechas* as the characters used to distinguish them are presently not fully understood and tested. A useful treatment of the Spanish and Portuguese taxa is given in Suarez-Cervera and Seoane-Camba (1989).

Key to species

Plant with green sterile bracts, leaves very glandular and sticky (viscid) . . . . *L. viridis*
Plant with coloured sterile bracts, leaves not sticky . . . . *L. stoechas* and subsp.

5. L. stoechas *L.* subsp. stoechas

Shrub 40–70 cm. Leaves often grey tomentose. Inflorescence stalk sessile or no longer than length of spike. Apex topped by enlarged sterile floral bracts obovate or spathulate 1–2 (4) cm long (Figure 2.10). Fertile bracts broadly ovate to obovate, shortly acuminate. Calyx sessile, thirteen-nerved the middle lobe of calyx modified into an appendage. Corolla black-purple to mauve in colour, white variants being named var. *leucantha* Ging. and rose red or pink flowered variants forma *rosea* Maire. Widespread throughout the Mediterranean basin and often a common element of maquis and garrigue type vegetation on acid substrates.

Grown for the extraction of essential oils on a small scale mainly in Spain for use in air fresheners and deodorants. Widely cultivated as an ornamental shrub and sometimes naturalised. A number of cultivars have been named including, 'Snowman' – white flowers and coma to 45 cm and 'Kew Red' – rosy pink flowers and coma to 45 cm. It is now classed as an invasive weed in parts of Australia where its cultivation is strictly controlled. Commonly referred to as French or Italian lavender.

Subspecies

Subsp. *pedunculata* Miller (Samp. ex Rozeira)

Bears distinctive erect branches and long narrow lanceolate leaves. Flower stalk (peduncle) long (10–)20–30 cm, spike usually quite short 2–3 cm, ovate in outline. Lower fertile bracts kidney shaped, the upper fertile bracts broadly obovate. Coma large 3–4 cm and narrowly lanceolate to spathulate in shape. Calyx appendage ± entire. Spain and Portugal on calcareous soils.

A number of cultivars have been named of which the most commonly available is 'James Compton'.

Figure 2.10 L. *stoechas* subsp. *stoechas*. (See Color Plate V.)

Subsp. *sampaiana* Rozeira

Flower spikes large and robust, twice as long as broad. Lower fertile bracts almost circular, the upper fertile bracts obovate. Coma linear obovate in shape. Calyx appendage ± entire. Western Spain, North and Central Portugal.

Subsp. *lusitanica* (Chaytor) Rozeira

Flower spikes relatively small, not much longer than wide, ovate in shape. Lower fertile bracts kidney shaped, the upper fertile bracts broadly obovate. Coma small, linear obovate in shape. Calyx appendage ± entire. Southern Spain and Portugal.

Subsp. *luisieri* (Rozeira) Rozeira

Flower spike large and robust, twice as long as wide borne on a relatively short peduncle *c.* 4–8 cm long. Basal fertile bracts almost circular, the upper fertile bracts diamond shaped. The stems appear very erect in form. The coma large, ovate-lanceolate in shape. Calyx appendage ± entire. South West Spain, Central and Southern Portugal.

Subsp. *atlantica* Braun-Blanq.

A compact shrub with a robust conical spike borne on a peduncle to *c.* 10 cm. Calyx appendage large exceeding width of calyx tube. Basal fertile bracts almost circular, the upper bracts broadly obovate and truncate. Coma large up to 4 cm, spathulate in shape. Calyx appendage lobed. Morocco over 1000 m.

Subsp. *maderensis* (Benth.) Rozeira

The peduncles are highly variable in length from 3.5 to 13 cm. Lower fertile bracts kidney shaped. The coma small *c.* 2 cm long and spathulate in shape. Madeira.

Subsp. *cariensis* (Boiss.) Rozeira

Peduncles long up to 20 cm. Lower fertile bracts broadly obovate. Coma narrowly oblong in shape. Calyx appendage lobed. Turkey (European Turkey and W. Anatolia) on acidic soils.

6. **L. viridis** *L'Hér.*

Shrub to 30–50 cm, leaves broadly linear-lanceolate, green foliage with a distinctive dense glandular indumentum giving a sticky feel with a strong smell similar to lemons. Inflorescence spike up to 8 cm. Coma broadly ovate, fairly short 2–3 cm and green. Fertile bracts obovate or broadly ovate greenish-white in colour. Flowers white borne over a long period in cultivation. Native to SW Spain and South Portugal in open maquis at fairly low altitudes and Madeira.

Hybrids (L. stoechas × L. viridis)

Many new cultivars often classified under *L. stoechas* have been introduced into the horticultural industry in recent years and are almost certainly hybrids between *L. stoechas* and *L. virdis*. Which subspecies of *L. stoechas* has been involved in these crosses is not always clear. However, some cultivars, for example, 'St. Brelade' the parent is known to be *L. stoechas* subsp. *pedunculata*. The hybrids are reminiscent of *L. stoechas*, but have a number of characteristics of *L. viridis*, most notably the distinctive strong smell; the leaves are broadly linear-lanceolate in shape and are often slightly viscid with a very glandular indumentum. The hybrids also have the long flowering period (often repeat flowering) typical of *L. viridis*.

Many of these cultivars have originated from Australia and New Zealand. Here the climate is conducive to the outdoor cultivation of both parents and it seems likely they have freely hybridised under these conditions. Numerous cultivars have and continue to be named (see McNaughton, 2000 for most comprehensive listing), for example, 'Helmsdale' – selected in New Zealand to 70 cm with red purple flowers, 'Marshwood' – selected in New Zealand, vigorous to 90 cm with large flower spikes, purple flowers and pink coma, 'Avonview' – selected in New Zealand to 60 cm with dark purple flowers and pale purple coma, and 'Fathead' – UK origin to 45 cm with short dense flower spikes, dark purple fading to pink.

Section 4: *Pterostoechas* Ging.

Woody based perennials or woody shrubs with pinnatisect to bipinnatisect leaves with distinct petioles. Bracts on axis of spike arranged in an opposite and decussate manner, subtending a single-flower with no bracteoles, giving a four-seriate (quadrate) spike. Subtending bracts ovate lanceolate in shape, with parallel veins. Calyx fifteen-nerved, bilobed the median posterior tooth often distinctly deltoid, the other lanceolate in shape. Corolla bisymmetrical, twice the length of the calyx, five-lobed, the upper two erect and twice as large as the three lateral lobes. Stigma bilobed. Nutlets elliptic, the lateral scar one-third the length of the nutlet, mucilaginous. The important characters for distinguishing the species include leaf shape, indumentum, length of the bract subtending each flower and the shape of the upper middle lobe of the calyx.

The largest section in the genus with at least thirteen species, distributed from the Canary Islands, Cape Verde Islands, Madeira, across North Africa, parts of the Mediterranean basin, the southern Arabian Peninsula and reaching Iran.

1. Median upper calyx lobe narrowly triangular in shape and ± equal in size and shape to the other lobes .... *L. coronopifolia*
   Median upper calyx lobe deltoid in shape, differing in size and shape to all other lobes 2
2. Whole plant with a silver grey indumentum 3
   Plants glabrous or with a sparse to dense indumentum of glandular and/or non-glandular hairs 4
3. Plant a low-growing shrub 20–40 cm. Bracts large, ovate in outline, 1.5 × length of the calyx. Canary Islands and Madeira .... *L. pinnata*
   Plant a robust shrub 50–100 cm. Bracts lanceolate to ovate/lanceolate in outline, 0.5–0.75 × length of the calyx. Canary Islands (Tenerife) .... *L. buchii*
4. Stems glabrous to glabrescent or very sparsely tomentose 5
   Stems with a distinct indumentum of either short hooked stiff hairs, long white stiff hairs, highly branched hairs, or short stalked glandular hairs or combination thereof ... 7
5. Leaves pinnatisect to bipinnatisect. Leaves not fleshy 6
   Leaves irregularly dissected, lobed or dentate. Leaves fleshy .... *L. rotundifolia*
6. Plants woody at the base only, stems herbaceous. Bracts with membraneous wings, orbicular-pentagonal in shape .... *L. maroccana*
   Plants with woody stems, only the seasons growth herbaceous. Bracts without wings, ovate in shape .... *L. canariensis*
7. Stem indumentum of short branched and long white stiff hairs, present at least on the lower stem. Spike with a distinct twist .... *L. multifida*
   Stem indumentum of large branched hairs, short stalked glandular hairs and/or stiff white non-glandular hairs or combination thereof. Spike not twisted 8
8. Stems grey-green indumentum of large branched hairs. Leaves with deep very regular pinnatisect divisions .... *L. minutolii*
   Stems with an indumentum of short stiff hairs, long white stiff hairs and/or short stalked glandular hairs 9
9. Stem indumentum of short stiff hairs often hooked, with long stiff hairs. Glandular hairs few. Bracts ovate with a long acuminate tip with three main nerves 10
   Stem indumentum of short stalked glandular hairs over white stiff hairs. Bracts ovate lanceolate usually with five main nerves 11
10. Stem indumentum of short hooked stiff hairs, with occasional long stiff hairs and short stalked glandular hairs. Algeria and Niger .... *L. antineae*
    Stem indumentum of short stiff simple hairs often interspersed with longer stiff hairs. Glandular hairs absent. Morocco .... *L. tenuisecta*
11. Bracts at least the length of the calyx or exceeding it in length (1–1.5x) .... *L. mairei*
    Bracts 0.6–1 × length of calyx 12
12. Leaves distinctly 2–3 pinnatisect (lobed to midrib). Plant highly aromatic, rather unpleasant aroma .... *L. pubescens*
    Leaves principally pinnatifid (lobed to only half the distance to midrib). Plant aromatic with distinct smell of lemons .... *L. citriodora*

Figure 2.11 L. multifida – close up showing flower spike, Morocco near Oued Laou. (See Color Plate VI.)

7. L. multifida L. *(Figure 2.11)*

Perennial with woody base to 40 cm. Leaves 3–6 cm, deeply dissected pinnatisect to bipinnatisect. Stems with diagnostic short branched hairs and long white simple hairs (sometimes only present near base). Inflorescence stalk usually branched at the base, flower spike 5–8 cm, twisted. Bract elliptic with sharply acute apex, typically with three dark nerves, ± to equal length of calyx. Upper middle lobe of calyx distinctly deltoid in shape. Corolla bicoloured, the lower lobes violet fading to blue-violet on the upper lobes with darker guidelines. Native to South Spain, Italy (Southern tip only), North Africa (Morocco, Algeria, Tunisia and Libya) often associated with open disturbed areas and areas of habitation. Occasionally cultivated as an ornamental, usually treated as an annual.

8. L. canariensis Mill. (L. abrotanoides Lam.; L. multifida L. subsp. canariensis (Mill.)) Pit. and Proust *(Figure 2.12)*

Woody shrub 50–80 cm with glabrous stems. Leaves 4–7 cm, deeply dissected pinnatisect or more often bipinnatisect. Inflorescence spike up to 10 cm, branched at base of main spike. Bract lanceolate to ovate, acute, often with bluish tint, with five main nerves, slightly longer than calyx in flowering state. Upper middle lobe of calyx distinctly broadly triangular in shape. Corolla bicoloured, the lower lobes violet fading to blue-violet on the upper lobes with darker guidelines. Canary Islands (Tenerife, Gomera, Gran Canaria, La Palma, El Hierro). Occasionally cultivated as an ornamental.

Figure 2.12 L. *canariensis* – plant in full bloom, Tenerife near Chio. (See Color Plate VII.)

9. **L. pinnata** L. f.

Shrub 30 (50) cm. Leaves 3–6 cm, deeply dissected pinnatisect, elliptic, each lobe distinctly obovate, pedicel to 1–1.5 cm. Whole plant with a characteristic grey felt of very short branching hairs. Inflorescence spike compact up to 8 cm, branched at the base of the main spike. Bract elliptic, apex acute, with five rather indistinct nerves, exceeding the calyx in length. Upper middle tooth of calyx broadly triangular in shape. Corolla blue mauve. Canary Islands (Lanzarote) and Madeira. Occasionally cultivated as an ornamental.

10. **L. buchii** Webb and Berthel. (**L. pinnata** var. **buchii** (Webb and Berthel.) Benth.)

Shrub 50–70 cm. Leaves 5–8 cm, deeply dissected pinnatisect, ovate, lobes linear to ± obovate, pedicel 0.5–1 cm. Whole plant with a characteristic grey felt. Inflorescence spike up to 10 cm, sometimes branched at the base of the main spike. Bract ovate with acute apex, typically with five rather indistinct nerves, about 0.5–0.75 × the length of calyx. Upper middle tooth of calyx broadly triangular in shape. Corolla tube mauve blue. Canary Islands (Tenerife). Occasionally cultivated as an ornamental.

> *L. buchii* var. *buchii* – flower spikes typically short 4–6 cm, branched many times and clustered at the end of the flower stalk. Leaf pinnatisect the lobes usually obovate, the terminal lobe ± diamond shaped. Northern region of Tenerife, particularly the Anaga Peninsula.
>
> *L. buchii* var. *gracile* M.C. León – flower spike typically long and slender usually over 10 cm, branched only once or twice. Leaf pinnatisect the lobes linear, the terminal lobe linear. Southern region of Tenerife, Teno region.

L. buchii var. *tolpidifolia* (Svent.) M.C. León – leaves linear lanceolate in outline, with irregular shallow to deep lobes. Teno region of Tenerife.

11. L. minutolii Bolle
(L. foliosa Christmann)

Woody shrub 50–80 cm with distinctive very regularly pinnatisect leaves, ovate lanceolate in outline with a greenish/white indumentum of highly branched hairs. Flower stalk usually branched just once. Bracts ± same length as the mature calyx, ovate and long acuminate in shape, the tips tinged dark violet blue. The upper middle tooth of the calyx deltoid in shape. Flowers bicoloured. Canary Islands (Gran Canaria and Tenerife where the population is treated as var. *tenuipinna* Svent.). Occasionally cultivated as an ornamental.

12. L. rotundifolia Benth.

Woody based perennial with fleshy stems and leaves, glabrous. Leaves very distinctive broadly ovate to triangular ovate with margins that are highly variable, usually irregularly dissected and lobed occasionally more dentate or crenate. Bracts 0.75 to ± same length as calyx. Calyx teeth-lanceolate in shape, the middle posterior tooth slightly broader. Corolla red purple in colour, the lobes ± equal in size. Native to the Cape Verde Islands where it grows on cliffs mostly in elevated semi-arid to humid regions. Occasionally cultivated as an ornamental.

13. L. tenuisecta Coss. ex. Ball

Woody based perennial with a characteristic indumentum of short hooked stiff simple hairs with scattered sessile and short stalked glandular hairs, variably interspersed with longer scattered simple hairs. Leaves ovate in outline, usually pinnatisect often with secondary lobing. Bracts distinctive in shape ovate-lanceolate with a long acuminate apex about $0.6 \times$ length of the calyx. Late flowering June to August. Endemic to Morocco, occurring in the High and Middle Atlas at altitudes over 1800 m.

14. L. maroccana Murb.

Woody based perennial with 50–80 cm glabrescent stems and long rather leafless sprawling stems. Spike is short, stout and compact, 1.5–3 cm long. The bracts are diagnostic with large membranous wings giving them an orbicular-pentagonal shape and a short sharp abrupt apex, $c.\ 0.5 \times$ length of the calyx. The upper middle posterior calyx tooth broadly deltoid, the others lanceolate in shape. Corolla violet-blue. Endemic to the High Atlas in Morocco occurring from sea level to 1700 m.

15. L. mairei Humbert var. mairei

A subshrub forming a dome shaped bush 40–50 cm high, the whole plant grey-green in appearance consisting of a distinctive indumentum of long white stiff hairs, variable in density over many short stalked glandular hairs. Leaves ovate-lanceolate in outline, 2–4 pinnatisect. Bracts very large $1.0–1.2 \times$ length of the calyx, rhomboid in shape, the apex acuminate. The upper middle posterior calyx tooth broadly deltoid and shorter than the lateral pair, the lower pair narrowly lanceolate. Corolla dark violet. Endemic to Morocco occurring in the lower regions of SE High Atlas and adjacent ranges on open rocky habitats and desert plains.

var. *antiatlantica* Maire (**L. antiatlantica** Maire, **L. mairei** var. **intermedia**)

It differs in its indumentum of short stalked glandular hairs making the plant viscid with only a few scattered simple or branched non-glandular hairs. Bracts longer than in var. *mairei* 1.2–1.5 × length of the calyx with a long acuminate apex almost becoming spinescent. Endemic to the Anti-Atlas on open rocky habitats and desert plains.

16. **L. antineae** Maire (Figure 2.13)

Woody based perennial with erect and branching annual stems, usually quite leafy. Indumentum of short hooked stiff white hairs and longer simple to once-branched white stiff hairs, variable in density over sparse short stalked glandular hairs. Leaves pinnatisect ovate to ovate-lanceolate in outline. Bracts 0.5–0.75 × length of calyx, ovate-lanceolate with small thin wings, apex long acuminate. The upper middle posterior calyx tooth-deltoid in shape, the upper lateral lobes triangular, the lower lobes narrowly lanceolate. Corolla tube dark violet-blue the lobes becoming bright blue. Endemic to the Central Saharan Massifs of the Hoggar Mountains (Algeria) and Aïr Mountains (Niger) on rocky mountains habitats over 1500 m.

17. **L. coronopifolia** Poir.
(**L. stricta** Delile)

A woody based perennial forming a large bush of annual stems to 80 cm green to grey-green in appearance. Indumentum variable from glabresecent to sparsely pubescent. Leaves pinnatisect. Flowering stems distinctively branched, the common name stagshorn lavender reflecting the many branched peduncle. The flower spike tending to biseriate. The calyx teeth all narrowly

Figure 2.13 L. *antineae* – cultivated at Cambridge University Botanic Garden. (See Color Plate VIII.)

lanceolate in shape, the tips often tinged pink. Corolla sky blue to lilac, the corolla tube narrow and distinctly curved. The most widespread species of lavender occurring from the Cape Verde Islands, across North Africa, West Tropical Africa, NE Tropical Africa, Western Asia (Israel, Jordan and Iran) and the Arabian Peninsula. Found in open rocky and stony habitats and desert plains from ± sea level to *c*. 2000 m.

18. **L. pubescens** Decne.

A woody based perennial with clumps of annual stems 30–60 cm, the whole plant very glandular and highly aromatic. Stems have a distinctive indumentum of long white stiff hairs over numerous short stalked glandular hairs. Leaves ovate in outline, 2–3 pinnatisect. Flower spikes compact and unbranched. Bracts ovate with a long acuminate apex 0.6 to 1 × length of calyx. Corolla tube distinctly curved about 0.5 × along its length, violet-blue. Syria, Jordan, Israel, Egypt, Saudi Arabia and Yemen Arab Republic.

19. **L. citriodora** A.G. Miller

A woody based perennial, strongly aromatic and with a distinctive smell of lemons and a grey–green appearance. Indumentum of simple and branched non-glandular hairs and scattered sessile and stalked glandular hairs. Leaves ovate to broadly triangular in outline, pinnatifid. Spike simple only occasionally branched. Bracts broadly ovate with acute apex 0.75–1.5 × length of calyx. The upper middle posterior calyx tooth broadly triangular the lateral pair narrowly triangular, the lower pair of lobes narrowly deltoid. Corolla bright blue to violet. Endemic to Saudi Arabia and the Yemen Arab Republic found in open rocky habitats between 1900 and 2700 m.

Taxa of uncertain status

L. sublepidota Rech. f.

This species is clearly closely related to *L. coronopifolia* and was described as differing particularly in its indumentum and also characters of the spike, bracts, calyx and leaves. It is native to Iran where *L. coronopifolia* also occurs, but appears to be known from only a single collection. It has not been possible to date to determine whether this is truly a distinct species or a variant that should be included within *L. coronopifolia*.

L. brevidens (Humbert) Maire

This taxon was originally described as a subspecies of *L. coronopifolia* and subsequently recognised as a distinct species. The identity and status of this taxon has seemed to be unclear in the literature and it has probably been a convenient name under which to place some rather aberrant specimens. The identity and status of this taxon is subject to a forthcoming paper by the author.

Hybrids

L. x christiana Gattef. and Maire

A hybrid only known in cultivation between *L. canariensis* and *L. pinnata* which first appeared in the 1930s in garden of J. Gattefossé at Aïn-Seba in Morocco, where he cultivated both parents.

This hybrid is frequently misidentified and sold as either parent, *L. canariensis* or *L. pinnata*. *Lavandula* Sidonie™ originating from Australia is of this parentage.

Forms an attractive large woody shrub to 50–80 cm and now becoming quite widely cultivated. The plants appear to express hybrid vigour in forming such large plants, totally atypical of *L. pinnata* which usually only reaches 30 cm, and exceeding the usual dimensions of *L. canariensis*. The leaves resemble *L. canariensis* in form bearing the typical bipinnate lobing and have the dense white felt of *L pinnata*. The large bracts exceeding the calyx in length are typical of *L. pinnata*. Flowers deep blue mauve.

L. x murbeckiana Emb. and Maire

A hybrid between *L. maroccana* and *L. multifida* first described by the Swedish botanist S. Murbeck (1859–1946) from among seedlings derived from *L. maroccana* growing in the Lund Botanic Garden, Sweden. The plants resemble *L. maroccana* the maternal parent most closely in their general appearance. The leaves are reminiscent of *L. multifida*, resembling their deeply bipinnatisect form closely. The stem indumentum bears sparse long stiff hairs typical of *L. multifida* although lacks the short branched hairs. In *L. maroccana* the stems are almost glabrous. The bracts and spikes are intermediate in character between the two parents. This hybrid appears only to be known from cultivation.

L. x canariensis Mill. × L. buchii Webb and Berthol. var. buchii

This hybrid appears to occur infrequently in the wild where the two parents overlap, in the northern part of Tenerife. Forms a woody shrub to *c.* 100 cm with bipinnatisect leaves (resembling *L. canariensis*) with narrow lobes and usually with a white indumentum (resembling *L. buchii*). The form and branching of the peduncle is very typical of *L. canariensis*, the spikes not being clustered at the apex of the peduncle as in *L. buchii* var. *buchii*. The spikes are slender, 5–10 cm long and not short and robust as in *L. buchii* var. *buchii*. The bracts in the two parent species are very similar, both being ovate acuminate in shape and 0.5–0.75 × the length of the calyx.

Section 5: *Subnudae* Chaytor

Typically woody based perennials or subshrubs with pinnatifid (lobed to only half the distance to midrib) leaves or sometimes leafless. The bracts are distinctly ovate in shape usually with a long spinescent apex and parallel veining except in *L. somaliensis* with reticulate veining. Bracts arranged spirally on the inflorescence axis subtending a single flower in each axil. Calyx 15-nerved, slightly two-lipped, the lobes subequal without an appendage. Nutlets bear a lateral scar 0.25–0.33 × length of the nutlet. The species are morphologically similar and mainly separated on characters of the indumentum, size and shape of the bract.

A section presently of eight species probably more native to the Arabian Peninsula (Saudi Arabia, Oman, the Republic of Yemen), Socotra and NE tropical Africa (Somalia) representing an important centre of diversification in the genus. A more detailed account of the species in this section has been published by Miller (1985). A few species are in cultivation within specialist collections and are all tender and some can be difficult to keep.

Key to species

1 Bracts ovate to obovate, ± abruptly tapering to a point with reticulate veining, leaves regularly pinnatifid (lobed to only half the distance to midrib) with simple segments .... *L. somaliensis*

	Bracts ovate or narrowly triangular, forming a long spinescent tip, parallel veined; leaves present with pinnatifid leaves with toothed segments or leaves absent 2
2	Indumentum of flower spike and lower stem with long soft simple hairs ..... *L. nimmoi*
	Indumentum of flower spike with fine short hairs, short woolly hairs or dense long woolly branched hairs 3
3	All stems 6–8 angled, bracts narrowed above to a fine bristle like tip .... *L. setifera*
	All stems four-angled, bracts with stiff spinescent tip 4
4	Bracts 6.5–12 mm long, greatly exceeding calyx, calyx purple when in flower .... *L. aristibracteata*
	Bracts 2–7 mm long, calyx green to grey-green or if purplish then bract not exceeding subtending bract in length 5
5	Plants ± leafless or with one or two leaves on lower part of stem 6
	Plants leafy 7
6	Spike lax in fruit, plant branched from nodes and often tufted at upper nodes. Sessile glands numerous in nerves between calyx .... *L. macra*
	Spike condensed in fruit, plant branched mainly from the base. Few sessile glands between nerves on calyx .... *L. subnuda*
7	Bracts longer than calyx, calyx teeth ± reflexed in fruit; bracts and calyx with sparse indumentum of long woolly hairs .... *L. galgalloensis*
	Bracts shorter than calyx rarely longer, calyx teeth erect in fruit; bracts and calyx with indumentum of fine short hairs or short woolly hairs 8
8	Stems and leaves densely hairy, upper leaves pinnatifid .... *L. dhofarensis*
	Stem and leaves glabrescent or sparsely hairy, uppermost leaves typically entire with one or two teeth only .... *L. subnuda*

20. **L. subnuda** Benth. *(Figure 2.14)*

A sprawling almost leafless bush to 2 m, branching from the base. Stem indumentum variable glabrescent or with sparse simple and branched hairs. Lower leaves ovate in outline and pinnatifid the upper ones becoming elliptic to rhombic and entire. Spike (1.5–)2–4(–8) cm long. Bracts ovate with a spinescent or narrowly triangular tip, 0.3–0.6 × as long as calyx (occasionally just exceeding calyx in length). Corolla variable in colour from shades of pale to dark blue. Endemic to Oman on rocky slopes, stony places and cliff crevices to 800 m.

21. **L. macra** Baker

A species very similar to *L. subnuda* but usually totally leafless or with very sparse leaves, tending to branch from nodes above the base, with several branches arising from a node giving a tufted appearance and the leaves which when present are always pinnatifid. Stem indumentum variable from glabrous or with simple or branched hairs. Spike 2–7 cm long elongating and becoming lax in fruit. Bracts ovate with a sharp to long spinescent tip, 0.25–0.33(–0.6) × length of calyx. Calyx with conspicuous sessile glandular hairs between nerves. Corolla pale blue to violet. Native to Oman, Republic of Yemen and Somalia among limestone hills and wadi beds to 1600 m.

22. **L. dhofarensis** A.G. Miller

Woody based perennial forming either dense or open straggly clumps with an indumentum of branched hairs throughout. Stems leafy to 40 cm. Leaves ovate to lanceolate in outline and

28 *Tim Upson*

Figure 2.14 L. *subnuda* – cultivated at Cambridge University Botanic Garden. Section Subnudae. (See Color Plate IX.)

pinnatifid. Spike 1.5–6 cm. Bracts ovate with spinescent tip, (0.5–)0.75–1 × length of calyx. Corolla lilac, mauve or pale purple. Two subspecies are recognised: subsp. *ayunensis* – a plant forming a dense clump with erect stems, the upper internodes 2–4 cm and a dense white indumentum covering stem surface; subsp. *dhofarensis* – forming a more open straggling clump, the upper internodes (2–)4–9 cm and a sparse white indumentum not covering stem surface. Endemic to the Dhofar region of southern Oman. The subsp. *dhofarensis* occurs on escarpment mountains, subsp. *ayunensis* on the drier northern slopes of mountains.

23. **L. setifera** T. Anderson

A bushy often leafless perennial with slender wiry stems to 50 cm, 6–8 angled, glabrous or with an indumentum of simple or branched hairs. Leaves oblong-lanceolate in outline and pinnatifid. Spike 1.5–3 cm long. Bracts ovate or narrowly triangular narrowing gradually into a long bristle like tip, 1.25–1.5 × length of calyx. Corolla white or pale lilac. Native to the Republic of Yemen near Aden and Somalia, sea level to 30 m.

24. **L. nimmoi** Benth.

A perennial usually with a woody base. Stems slender, wiry and leafy on its lower parts 30–60 cm long with an indumentum of long simple hairs becoming glabrous above. Leaves ovate to oblong-ovate, pinnatifid to bi-pinnatisect. Spike 2–5 cm. Bracts ovate tapering to a sharp tip, 0.33–0.75 × length of calyx. Calyx with a distinctive indumentum of long simple hairs and sessile glands. Corolla clear blue. Endemic to Socotra.

25. **L. galgalloensis** A.G. Miller

Woody based aromatic perennial with twiggy stems leafy on the lower part, 15–60 cm. Leaves ovate in outline, pinnatifid. Spike 2–4 cm. Bracts ovate with long spinescent tip, *c.* 1.25 × length of calyx. Calyx with an indumentum of long woolly hairs, the teeth reflexing in fruit. Corolla lilac. Endemic to northern Somalia growing on limestone escarpments, 1600–1800 m.

26. **L. aristibracteata** A.G. Miller *(Figure 2.15)*

Woody based aromatic perennial with leafy stems to 35 cm. Indumentum of simple and sessile glandular hairs. Leaves variable, elliptic, ovate, obovate to triangular in outline, simple or pinnatifid. Spike 2–5(–8) cm. Bracts often purplish tinged, ovate with a very long spinescent tip, 1.25–2 × length of calyx. Calyx distinctly purple-tinged, with fine short hairs, the teeth reflexed in fruit. Corolla blue or purplish blue. Endemic to the Surundi Hills in northern Somalia, 1500–2000 m.

27. **L. somaliensis** Chaytor

Plant with numerous stems rising from a woody rootstock, leafy, simple or branched to 20–40 cm, with an indumentum of long woolly branched hairs. Leaves ovate to ovate-oblong in

Figure 2.15 L. *aristibracteata* – cultivated at Cambridge University Botanic Garden. Section Subnudae. (See Color Plate X.)

outline and regularly pinnatifid. Spike 1.5–3.5 cm. Bracts ovate to obovate, abruptly tapering to a sharp tip (acuminate), 0.3–0.75 × length of calyx, papery and with characteristic reticulate veining. Calyx with long woolly branched hairs and many sessile glandular hairs. Corolla blue. Endemic to the mountains of northern Somalia, 1600–1700 m.

Section 6: *Chaetostachys* (Benth.) Benth.

Herbaceous plants with green fleshy stems, distinctly rectangular with long internodes. Leaves numerous, usually large 7–10 cm and distinctly bipinnatisect. Flower stalks and spikes often branched. The single-flowered cymes are arranged spirally on the axis. Bracts ovate spinescent with parallel-veins. Calyx 15-nerved, the lobes ± subequal, slightly bilabiatae. Corolla twice the length of the calyx, lobes small, the lower middle lobe *c.* 2 × the size of the lateral lobes. Nutlets elliptic, black to dark brown in colour. Lateral scar present, over 0.75 × length of nutlet.

This section comprises the two native Indian species. It is characterised by their herbaceous nature, the large deeply bipinnatisect leaves and the nutlets which bear a lateral scar usually 0.75 × length of the nutlet. The corolla differs from other species in the lower median corolla lobe which is distinctly larger than the lateral lobes. The single-flowered cymes arranged spirally, the spinescent bracts and the subequal corolla lobes suggests its closest affinities are with section *Subnudae*. Not in general cultivation apart from a few specialist collections, tender and difficult to overwinter.

Key to species

Bracts concealing calyx .... *L. gibsoni*
Bracts only covering basal part of calyx .... *L. bipinnata*

28. L. gibsonii Grah.

Herbaceous plant up to 1 m with an indumentum of short soft hairs (pilose). Leaves with very short petiole, pinnate to bipinnate upto 13 cm long. Spike to 5.5 cm usually branched at base of spike. Bracts broadly lanceolate with acute apex, equal or slightly longer than calyx and concealing it. Corolla with rather small lobes. India, Deccan Peninsula.

29. L. bipinnata (Roth) Kuntze *(Figure 2.16)*

A highly variable herbaceous plant 15–100 cm. Indumentum of short hairs, particularly dense on stems and inflorescence. Leaves sessile or shortly petiolate, 2–12 cm long and usually distinctly bipinnatisect. Spike 4–7.5 cm, branching from base of spike. Bracts ovate tapering to a long sharp tip, 0.3 × length of calyx and only concealing the base of calyx. Corolla tube twice length of calyx, very pale blue in colour. Central and south India.

A number of varieties have been recognised based on shape and length of bract. Further study is desirable to determine if these are distinct taxa or represent natural variation in this species and hence have not been included in this treatment.

Other species

The following three species have not formally been classified within the present sectional classification and are subject to further work to clarify their position.

Figure 2.16 L. *bipinnata* – close up of flower, cultivated at Cambridge University Botanic Garden. Section Chaetostachys. (See Color Plate XI.)

30. L. hasikensis A.G. Miller

Woody shrub to *c.* 30 cm, with a characteristic dense white tomentose indumentum. Leaves oblong-ovate, small 0.2–1.5 cm long, with 1–3 pairs of rounded or ± triangular lobes. Spike capitate, *c.* 1 cm long, with single-flowered whorls, spirally arranged on axis, which lengthens in fruit to *c.* 4–5 cm. Bracts with a short sharp tip arising from between the two wing-like and membraneous ± orbicular lateral lobes, parallel-veined. Calyx 15-nerved, all five lobes ± equal in size and shape. Corolla tube *c.* 2 × length of the calyx, lilac in colour. Endemic to Dhofar.

An extremely distinct species of *Lavandula* with no clear affinities within the genus. The single-flowered cymes borne in a spiral arrangement, the subequal calyx teeth, form of the corolla and the biogeography of this species would suggest some affinity to section *Subnudae*. However, the capitate inflorescence with the rachis extending in fruit, the bracts which are parallel-veined and extremely broad with two wing like lateral orbicular lobes and the leaves which are oblong ovate with rounded or triangular lobes are all unique to this species and hence would be a rather anomalous in any of the presently described sections.

L. atriplicifolia Benth. and L. erythreae (Chiov.) Cufud.

The following two species which are native to the southern Arabian Peninsula and Ethiopia are evidently closely related. They are distinct and instantly recognisable by: their compact drooping spikes; the single-flowered cymes with bracteoles (none of the other single flowered species bear bracteoles) in a spiral arrangement on the inflorescence axis; the papery bracts ovate in shape with reticulate venation; the regular calyx with five lobes all equal in size and fifteen-nerved; the subequal stamens (unequal in all other *Lavandula* species) included within the corolla tube; and particularly the corolla which has five equal lobes (often described as star shaped) and are yellow brown in colour. The flowers are also described as having a musty odour, which together with their colour suggest they maybe fly pollinated.

These species have previously been classified within both *Lavandula* and a separate genus *Sabaudia* Buscal & Musch. They have been included within *Lavandula* in this treatment on the basis that these taxa bear the nectary lobes opposite the ovary lobes, the compact spike and stamen arrangement which are the main characters defining the genus *Lavandula*.

Key to species

Spike (1–)1.5–3 cm long, bracts 0.75–1 × length of calyx, interior of
corolla tube glabrous, Saudi Arabia and Republic of Yemen . . . . *L. atriplicifolia*
Spike 1–1.5 cm long, bracts 0.5 × length of calyx, interior of corolla
tube hairy, Eritrea . . . . *L. erythreae*

31. L. atriplicifolia Benth. (Sabaudia atriplicifolia (Benth.) Chiov.)

Woody perennial forming a bushy, weak stemmed aromatic shrub, 30–100 cm, the whole plant with a dense white indumentum. Leaves simple, entire and sessile, linear lanceolate in outline. Peduncle branched and bearing a compact drooping spike, 1.5–3 cm long. Bracts persistent, almost papery, obovate to circular in outline, 0.75–1 × length of calyx. Corolla *c.* 3 × the length of the calyx, the inside of the corolla tube glabrous, the five lobes all equal, narrowly lanceolate to triangular in outline, yellow brown in colour. Native to Saudi Arabia and the Republic of Yemen on dry rocky areas over 2500 m.

32. L. erythreae (Chiov.) Cufud. (Sabaudia erythreae Chiov.)

Woody perennial to 100 cm, the whole plant with a dense white indumentum. Leaves simple very rarely with 1–2 lobes, entire and sessile, lanceolate to narrowly elliptic in outline. Peduncle highly branched and bearing a compact drooping spike, 1–1.5 cm long. Bracts persistent, almost papery, obovate to circular in outline, 0.5 × length of calyx. Corolla *c.* 3 × the length of the calyx, the inside of the corolla tube hairy, the five lobes all equal, narrowly triangular to lanceolate in outline, the margins reflexed, pale yellow to yellow brown in colour. Native to Eritrea.

Intersectional hybrids

Several intersectional hybrids are recorded and are becoming widely grown as ornamentals particularly in Australia, New Zealand, South Africa and the Mediterranean.

L. x allardii Hy

Large woody shrub to 1.5–2 m tall (sometimes to 4 m), the stems with long internodes. Leaves sessile, linear to lanceolate, tending to obovate, with two distinct leaf types, either mostly entire or partially toothed with 1–7 teeth on each side above the centre and towards the apex. Both leaves and stems grey-green. Flower spikes long and narrow, 7–12 cm, often interrupted. Each cyme has 7–9(–11) flowers, bracteoles present. Bracts broadly ovate-lanceolate, tapering to a sharp apex and with no coloured bracts borne at apex of spike as in one parent *L. dentata*. Corolla shades of blue.

A sterile hybrid that originated in cultivation between *L. dentata* and *L. latifolia*. Flowers over a long period June–September in the United Kingdom. Not fully hardy only surviving a few degrees of frost. In frost-free areas it can make a good hedging plant and imposing specimen. A number of different clones appear to be in cultivation some of which have been named: 'African Pride' – common in South Africa with occasionally toothed foliage; Clone B – refers to material from Australia with mainly toothed foliage.

L. x heterophylla Poir.

A hybrid originating in cultivation between *L. dentata* and *L. angustifolia*. Similar in general appearance to *L. x allardii*, but it would appear to be a less vigorous plant and with much smaller bracteoles which are obvious as in *L. x allardii*. Many plants grown under this name are probably *L. x allardii* and this cross appears to be rare in cultivation. Further work is required to confirm the identity of *L. x heterophylla*.

L. dentata L. × L. lanata Boiss.

A bushy woody shrub to 75 cm. Leaves lanceolate to obovate with a dense silver grey indumentum, the leaves entire or often toothed. Flower spikes conical in shape 6–10 cm with purple mauve flowers. Bracts diamond shape, violet coloured at least when young, more or less becoming sterile near apex of the spike and forming a small tuft of bracts (a coma) present, bracteoles 2–3 mm. Corolla deep violet-blue.

The cultivar 'Goodwin Creek Grey' selected and bred in the United States appears to be the only cultivar available with this parentage.

References

Andrew, S. (1994). Lavenders in Cultivation. *The Lavender Bag*, 1: 1–12.
Brummitt, R.K. and Powell, C.E. (1992). Authors of Plant Names. Royal Botanic Gardens, Kew.
Cantino, P.D., Harley, R.M. and Wagstaff, S.J. (1992). Genera of Labiatae: Status and Classification. In R.M. Harley and T. Reynolds (eds), *Advances in Labiate Science*: 511–22. Royal Botanic Gardens, Kew.
Chaytor, D.A. (1937). A Taxonomic Study of the Genus *Lavandula*. *Journal of the Linnean Society – Botany* 51: 153–204.
Gingins De La Sarraz, F.C.J. (1826). *Histoire Naturelle des Lavandes*. Paris.
Green, M.L. (1932). Botanical Names of Lavender and Spike. *Kew Bulletin*: 295–7.
Lundmark, J.D. (1780). *De Lavandula*. Dissertatio Academica. Uppsala.
McNaughton, V. (2000). *Lavender, the Grower's Guide*. Garden Art Press. Woodbridge.
Miller, A.G. (1985). The Genus *Lavandula* in Arabia and Tropical NE Africa. *Notes of the Royal Botanic Garden, Edinburgh* 42(3): 503–28.
Rozeira, A. (1949). A Secciao *Stoechas* Gingins do genero *Lavandula* L. *Brotéria Série de Ciências Naturais* 18(46): 1–82.

Suarez-Cervera, M. and Seoane-Camba, J.A. (1986). Taxonomia numerica de algunas especies de *Lavandula* L., basada en caracteres morfologicos, cariologicos y palinologicos. *Anales del Jardin Botanico de Madrid.* 42(2): 395–409.

Suarez-Cervera, M. and Seoane-Camba, J.A. (1989). Estudio morfologico del genero *Lavandula* de la Peninsula Iberica. *Biocosme Mésogéen, Nice* 6(1–2): 21–47.

Tucker, A.O. and Hensen, K.J.W. (1985). The cultivars of Lavender and Lavandin (Labiatae). *Baileya* 22(4): 168–77.

3 History of usage of *Lavandula* species

Jo Castle and Maria Lis-Balchin

Introduction

The term lavender is considered to come from the Latin 'lavando' part of the verb 'lavare' to bathe, the Romans having used many plants to perfume their baths. The Greeks and Romans also referred to lavender as nard, from the Latin Nardus Italica, after the Syrian town Naarda. This was the beginning of much confusion as to which plant was being referred to in classical and medieval times. *Lavandula* is obvious, however nard and spike can refer to spike lavender or to spikenard (a plant imported from India during the Middle Ages and equally popular then for its aromatic properties). Despite much learned investigation into the identification of lavender in the writings of classical authors; it has remained impossible to unquestionably identify *L. vera* or *L. spica*. *L. stoechas* is, however, distinctly referred to by both Dioscorides and Pliny (Gingins-Lassaraz 1826 in Fluckiger and Hanbury).

An alternative, but less likely explanation from Victorian times connected the name to the Latin 'livere' meaning to be livid or bluish (Festing, 1989).

Historical review of the use of lavender

The classical physicians

Lavender has been used as a healing plant and was first mentioned by Dioscorides (*c.* 40–90 AD) who found what was probably *L. stoechas* growing on the islands of Stoechades (now known as Hyeres); this was used in Roman communal baths (Festing, 1989). Dioscorides attributed to the plant some laxative and invigorating properties and advised its use in a tea-like preparation for chest complaints (Festing, 1989). The author also recounts that Galen (129–99 AD) added lavender to his list of ancient antidotes for poison and bites and thus Nero's physician used it in anti-poison pills and for uterine disorders. Lavender in wine was taken for snake bites stings, stomach aches, liver, renal and gall disorders, jaundice and dropsy.

Pliny differentiated between *L. stoechas and L. vera*, the latter was apparently used only for diluting expensive perfumes. Pliny the Elder advocated lavender for bereavement as well as promoting menstruation.

Abbess Hildegard

The Abbess Hildegard (1098–1179) of Bingen near the Rhine in what is now Germany, was the first person in the Middle Ages to clearly distinguish between *L. vera* and *L. spica* (Fluckiger and Hanbury, 1885; Throop, 1994):

> On Palsy one who is tormented should take galangale, with half as much nutmeg, and half as much Spike lavender as nutmeg, and equal weights of githerut (probably Gith or Black

Cumin) and lovage – but of each one, more than the spike lavender. To these he should add equal weights of female fern and saxifrage (these two together should be equal to the five precious ingredients). Pulverise this. If one is well, he should eat this powder on bread, if ill, he should eat an electuary (soft pill) made form it.

In a chapter on lavender she alluded to its strong odour and many virtues (Throop, 1994: chapter XXXV):

> Lavender (*Lavandula*) is hot and dry, having very little moisture. It is not effective for a person to eat, but it does have a strong odour. If a person with many lice frequently smells lavender, the lice will die. Its odour clears the eyes (since it posses the power of the strongest aromas and the usefulness of the most bitter ones. It curbs very many evil things and, because of it, malign spirits are terrified).

Lavender continued to be used for de-lousing until about 1870s, blotting paper being soaked in the oil and applied to children's heads.

Hildegard also recommends this decoction of lavender for pulmonary congestion, translated from the French (Hertzka and Strehlow, 1994):

> To cook lavender of spic (spike) with wine, or if one has no wine, with honey and water, put it in case to cool often, soften the suffering in the liver, and in the lungs and the vapour in the chest (pulmonary congestion), and the wine of lavender I assure you is a science pure and clean'.
>
> (PL 1140 C)

Hildegard distinguishes expressly between the 'lavande aspic sauvage' (*L. spica*) and the noble 'lavande de jardin' (*L. vera*). Furthermore, in her descriptions of the different types of rest and sleep she states that to prepare the nervous system for sleep, a walk followed by a bath steeped in lavender is beneficial (Hertzka and Strehlow, 1994).

Summary of uses recommended by Hildegard

For palsy (a powder with other ingredients), for head lice, to clear the eyes when smelt, to curb malign spirits and for pulmonary congestion in wine or honey. To ensure a restful nights sleep she recommends a bath with lavender after a walk.

Middle Ages

During the Middle Ages, 'strewing herbs' in churches and houses incorporated lavender. Lavender was used in medicines in medieval Wales and England in conjunction with numerous other herbs, including herb robert, valerian, wormwood, elecampagne, parsley, fennel etc.

A poem of the school of medicine in Salerno around 1020 AD entitled 'Flos Medicinae' (de Renzi) gives the following lines:

> Salvia, castoreum, lavandula, primula veris,
> Nasturtiom, athanas haec sanat paralytica membra
>
> (Sage, Beaver gland excretion, lavender, primrose,
> Nasturtium, are cleansing and soothing for paralytic limbs)

William Turner (1508–68)

He stated that 'because wyse men founde by experience that it was good to washe mennis heades with, which had anye deceses therein'. Indeed Turner was a passionate gardener, creating gardens wherever he was living, he writes of growing 'Stechas or Lavender Gentle (*L. Stoechas*), a variety not seen in England, growing in my gardens in Germany'. And 'Stechas groweth in the islands of France over against Marseilles which are called Stechades, whereupon the herb got its name'.

John Gerard (1545–1612)

First, he writes of common lavender – *L. flore caeruleo* (or most probably *L. vera*) the drawing of which has round tips to the leaves and slightly drooping flowers. White floured lavender – *L. flore albo*. And lavender spike or in Spanish spica – *Lavandula* minor sive spicae which he describes as having pointed tips to the leaves and a more upright habit (if lavender can ever be described as upright in its behaviour!). He then says that:

> We have growing in our English gardens and being of a small kind, altogether lesser than the other, and the floures of a more purple colour and grow much less and shorter heads, yet have a far more grateful smell. The leaves are less and whiter than those of the ordinary sort. This doth grow in plentie in His Majesties Private Garden at Whitehall. And this is called Spike, with out addition and sometimes Lavender Spike and this by distillation is made that vulgarly known and used oile which is termed Oleum Spicae, or oile of Spike. In Spain and Languedocke in France, most of the mountains and desert fields, are as it were covered over with Lavender. In these cold countries they are planted in gardens.

He reminds us that some think it is the sweet herb cassia which Virgil mentions, but states wisely that here is another type of cassia sold in the shops called cassia lignea, and also cassia nigra or cassia fistula.

He writes: 'Lavender is hot and dry in the third degree, and of a thin substance, consisting mainly of airy and spiritual parts, good for cold diseases of the head'. He advocates: 'The distilled water of lavender smelt unto, or the temples and forehead bathed therewith is refreshing for those with Catalepsie, a light Megrim (migraine) and to them that have the falling sickness (epilepsy) and that swoune (faint) much'. And continues, 'the floures of lavender picked from the knaps, I mean the blew part and not the huske, mixed with Cinnamon, Nutmegs and Cloves made into a pouder and given to drinke in distilled lavender water, doth help the panting and passions of the heart; prevaleth against giddiness or swimming of the braine, and palsie'. In other words a decoction of lavender distillate with powdered lavender flowers, cloves, nutmeg and cinnamon, is beneficial to what may be panic attacks, palpitations, giddiness and the shakes associated with Parkinson's.

He cautions against taking lavender 'when there is an abundance of humours' and he advised against the use of lavender 'taken in distilled wine: in which such kinds of herbes, floures, or seeds, and certain spices are infused and steeped, though most men do so rashly'. He continues 'for by using such hot things that fill and stuffe the head, both the disease is made greater, and the sick man also brought into danger'. This is probably referring to the distillation of herbs, spices and wines, which produced very potent spirits in the stills which were abundant in big households of the period.

Gerard also suggests that a conserve made with lavender flowers and sugar is also very good for the diseases previously mentioned, taken in the amount of a bean in the morning fasting and advises washing those with the palsie with either lavender distillate or lavender oil and olive oil.

Gerard also admonishes the 'unlearned physitians and diverse and over-rash Apothecaries and other foolish women' who treat people with such mixtures regardless of their condition, for example, those with 'Catuche or Catalepsis with a fever; to whom they can give nothing worse, seeing those things do very much hurt and often times bring death it selfe'.

L. stoechas *after Gerard*

He describes French lavender or sticados also known as stickedoue and sticadoue, which has spiky heads out of which the flowers grow, Gerard calls this 'Stoechas sive spica hortulana'. Jagged sticados or lavender with the divided leaf he calls 'Stoechas multisida'. Toothed sticados, with nicked or toothed leaves like a saw for which he gives 'Stoechas folio serrato', and naked stoechas have long naked stems on which the spike of flowers grow, this he calls 'Stoechas summis cauliculus nudis'. He gives clear descriptions of each variety and again these are illustrated, but his Latin names have no real botanical significance.

He continues, 'These herbs do grow wilde in Spaine, in Languedocke in France, and in the islands called Stoechas over against Manilla, we have them in our gardens and keep them with great diligence from the injurie of the cold', in other words considered very tender. Gerard cites Dioscorides and Galen and gives the names in Latin (stoechas), High Dutch (stichas kraut), Spanish (thomani and cantuesso) and in English (French lavender, steckado, stickadoue, cassidonie, and by some simple people cast me down).

For medicinal use he cites Dioscorides as teaching that a decoction of French lavender helps diseases of the chest, and is with good success mixed with counter poisons. The later physicians are not named but cited as writing that the flowers are 'most effectual against paines in the head, and all diseases proceeding from cold causes, and therefore they be mixed in all compositions which are made against head-ache of long continuance, the Apoplexie, the Falling Sickness, and such like diseases'.

Lastly, Gerard states that the 'decoction of the husks and floures drunk, openeth the stoppings of the liver, the lungs, the milt (melts), the mother (womb), the bladder and in one word all other inward parts, cleansing and driving forth all evill and corrupt humours, and provoking urine'.

A summary of uses suggested by Gerard

L. vera was used to treat catalepsie (?), megrims (migraines), epilepsy, fainting and panting and passions of the heart (the latter may be panic attacks or palpitations and heart problems). He also includes giddiness, and palsy (Parkinson's etc.), and lastly a conserve of lavender as being good for all these diseases.

L. stoechas he recommends as good for diseases of the chest (lungs), in counter poisons (theriac and hiera picra), pains in the head, diseases of cold cause and in compositions (compounds) for headaches of long history. Also for apoplexy, epilepsy and similar diseases and lastly a decoction to open all internal organs and provoke urine.

Shakespeare (1564–1616)

He mentioned lavender only once, although it was grown in herb gardens, especially knot gardens. It was probably not a common garden plant in his time, though mentioned by Spencer as 'the lavender still gray' and by Gerard as growing in his garden and the King's. Two centuries later, Leyel recommended 'a tissane or even a spray of lavender to cure nervous headaches, especially if worn under the hats of harvesters'! (Grieve, 1937).

John Parkinson

Herbal Apothecary to King James I (1603–25) and author of the 'Theater of Plants and Herball' and 'The Garden of Plants', he wrote that lavender was useful for the senses and should be used for scenting linen, clothes, gloves, leather and also that dried flowers used to soak up moisture in a cold brain. He says of 'Sage and lavender, both the purple and the rare white (there is a kinde herof that beareth white flowers and somewhat broader leaves, but it is very rare and seene in but a few places with us, because it is more tender and will not so well endure in our cold winters)'. He continues to say that it is put in bathes, ointment and things for cold causes. The seed is much used for worms (back to Hildegard again, but this time the seed), he also recommends it for pains of the head and brain (Grieve, 1937: 469).

Nicholas Culpeper (1616–54)

He called himself an astrological doctor, physician who was trained as an apothecary and he proclaimed that lavender 'was owned by Mercury, and carries his effects very potently'. 'Lavender is of special good use for all the grief's and pains of the head and brain that proceed from a cold cause, as the apoplexy, falling sickness, dropsy, sluggish malady, cramps, convulsions, palsies and faintings'. 'It strengthens the stomach and frees the liver and spleen from obstructions, provokes women's courses and expels the dead child and after birth'. So far he has repeated what was said by Gerard, but has added the mysterious sluggish malady and cramps and an improbable faith in Lavender's ability to expel the dead child and afterbirth from the mother's body. The same words are used for pennyroyal, considered to be an abortifacient, and this is perhaps the only reference to an abortifacient quality attached to lavender.

'The flowers of lavender steeped in wine, helps them to make water that are stopped, or are troubled with the wind and colic, if the place be bathed therewith. A decoction made with the flowers of lavender, hore-hound, fennel and asparagus root, and a little cinnamon, is very profitable used to help the falling sickness, and the giddiness or turning of the brain: to gargle the mouth with the decoction thereof is good against the toothache.' The theme of falling sickness continues but here we are moving away form Hildegard and Gerard: Culpeper is advocating its use in wind, colic, and toothache.

He returns to the plot however and says that 'Two spoonsful of the distilled water of the flowers taken, helps them that have lost their voice, as also the tremblings and passions of the heart, and the faintings and swoonings, not only being drunk but applied to the temples or nostrils to be smelled unto', now we have returned to almost exactly what Gerard said, with a dash of Hildegard.

Oil of spike

Culpeper particularly warned against 'The chemical oil (essential oil) drawn from lavender, usually called oil of spike, is of so fierce and piercing a quality that it is cautiously to be used. Some few drops being sufficient to be given with other things, for inward or outward griefs (troubles)'. Finally, he says that 'it is not safe to use it where the body is replete with blood and humours, because of the hot and subtle spirits wherewith it is possessed'.

He repeats all this information in his 'the English Physician and Family Dispensatory' in a briefer format. Under 'Simple Waters Distilled' he cites Lavender, and under 'Compounds, Spirit and Compound Distilled Waters' he warns 'Let all young people forbear them whilst they are in health, for their blood is usually hot enough without them' and again ' ... not to be meddled with by people of hot constitutions, when they are in health... If they drink of them moderately now and then for recreation ... they may do them good', we must remember that a great debate had already begun a hundred years after spirits of alcohol had become commonplace as to its use in medicine and recreationally.

He then gives a compound spirit of lavender ascribed to Matthias:

> Take of Lavender flowers one gallon, to which pour three gallons of the best Spirits of Wine (Aqua Vitae), let them stand together in the sun six days, then distil them with an Alembick with this refrigeratory: Take flowers of Sage, Rosemary and Bettony of each a handful. The flowers of Borage, Bugloss, Lilies of the Valley and Cowslips of each two handfuls. Let the flowers be infused in one gallon of the best spirits of wine. And mingled with the forgoing Spirit of Lavender flowers, adding the leaves of Bawm (Melissa), Featherfew (Feverfew), and Orange tree freshly gathered. The flowers of Stoechas (the other type of Lavender) and Orange tree, May berries (Hawthorn), of each one ounce. After convenient digestion distil it again, after which add Citron pills the outward bark, Peony seed husked, of each six drams, Cinnamon, Mace, Nutmegs, Cardamoms, Cubebs, and Yellow Saunders of each half an ounce. Wood of Aloes one dram, the best Jujubes the stones being taken out half a pound; digest them six weeks, then strain it and filter, and add to it prepared Pearls two drams, Emeralds prepared a scruple, Ambergrease, Musk, Saffron of each half a scruple, Red Roses dryed, Red Saunders of each half an ounce yellow saunders, Citron pills dryed, of each one dram. Let the species being tyed up in a rag be hung into the afore mentioned spirit.

Culpeper then gives a list of complaints as to why the College of Physicians has not clarified this recipe, making clear that he realises the difficulties in getting fresh orange leaves and flowers, and borage, bugloss and cowslips flowering together and thus fresh. The very lengthy list of ingredients including tiny amounts of precious stones and pearls is a common format for physicians of the late sixteenth century.

He lists the 'Simple Oils by infusion and Decoction', but gives neither oil of lavender nor oil of spike lavender, but only oil of nard or spikenard from India (making it clear which he means).

He does however end this chapter with a statement of interest to modern aromatherapists: 'That most of these oils, if not all of them, are used only externally, is certain'.

> And it is certain that they retain the virtues of the simples wherof they are made, therefore the ingenious might help themselves.
>
> (Culpeper, 1653)

Culpeper in summary

Lavender for pains in the head and brain from cold cause, apoplexy, epilepsy, dropsy (Oedema or severe water retention), the sluggish malady, cramps, convulsions, palsies and fainting. It is strengthening for the stomach and opening for the liver and spleen. It provokes women's periods, and expels dead foetuses and afterbirth. Finally, bathing the skin with a decoction of lavender is good for wind and colic, gargling for toothache and for voice loss.

Thomas Palmer (d. 1696)

He was an American physician practising in the harsh world of the newly founded colonies, and he also preached, but not very well. His medical practise was better, but so hard were these times he could not afford a horse to get him around his practise. In 1696 he wrote 'The Admirable Secrets of Physick and Chyrurgery', written in New England, giving us a window into the work of a colonial practitioner at the end of the seventeenth century. In Palmer's world there was little room for unwarranted frills in medicine, despite this many drugs considered essential were transported from the Old World because they were believed to be irreplaceable.

Only under 'Cold Distempers of the Heart' do we find lavender:

> The weakness, slowing and thinnes of the pulse & such a breath know this distemper. The air breathed out appears cold and sometimes the whole body is cold. Medicines that heat the heart are: Balm, Rosemary, Cardus Benedictus (Blessed Thistle), Calamint, Angelica, Rosemary flowers (an ancient favourite), Lavender, Lily of the Valley, Citron seeds, grains of Alchornes (Kermes or Alkermes a scarlet red colouring), Lignum Aloes (Aloes wood), Cinnamon, Cloves, Zeadary (Zedoary), Mace, Nutmegs, Amber, Musk. Of these there are divers compositions, waters & spirits, conserves, syrups, preserves, oyls, species; confections of Alkermes, Treacle & Mithradite, all sorts of Aqua Vitae and electuarys.

Of these only balm, rosemary, blessed thistle, calamint, angelica, lavender and lily of the valley could have been grown in New England, the rest are very expensive Mediterranean or tropical imports.

He also suggests as treatment for palpitations or passion or beating of the heart in addition to medicines: external applications of bags with melissa flowers, lavender, rosin (pine resin), borage, bugloss or white wine and water of musk roses. Under none of the other infirmities does he mention lavender, suggesting that it was probably also an import or difficult to grow.

William Salmon (1644–1713)

A seventeenth century commentator and the writer of a Herbal, he states that lavender was 'good against the biting of serpents, mad dogs and other venomous creatures (a dramatic way of describing bites), being given inwardly and applied poultice-wise. The spirituous tincture of the dried seeds or leaves, if given prudently, cures the hysterick fits though vehement and of long standing'. And it is 'abstersive (detergent), aperitive (laxative), astringent, cephalic (for megraine), discursive (dispels), diuretick (stimulates urine), incisive, neurotick (nervine), stomatick (stomachic), cordial (gentle or used to make a cordial), nephretick (kidney-action), hysterick, alexipharmick (?) and analeptick (?) and antiparalitick (for paralysis), being of very subtil and thin parts.' He approved it for 'convulsions, epilepsy's, palsies, tremblings, vertigoes, lethargies, swoonings, hysteric fits, other diseases of the head, brain, nerves, womb, also bites of mad dogs as well as snake bites'.

He also wrote of Hiera Picra, which is an antidote and cure all with multiple ingredients very similar to Treacle or Theriac made under seal by various cities including London, the origins of which go back the ancient Greeks.

The growing of lavender in history

In the nineteenth century *L. vera* was considered indigenous to the mountainous regions of the countries bordering the western half of the Mediterranean basin. Occurring in Eastern Spain,

Southern France (extending Northward to Lyons and Dauphiny), Upper Italy, Corsica, Calabria and Northern Africa. In cultivation it grows very well in the open air throughout the greater part of Germany and as far North as Norway. Dried lavender flowers were the object of some trade in the South of Europe. Lavender and orange flowers were exported form France in 1870 to the extent of 110,958 kg (244,741 lbs), chiefly to the Barbary states (N. Africa), Turkey and America. There is no data given for the amount of volatile lavender oil imported into England, where it was much used as a perfume and was considered to have stimulant properties (Fluckiger and Hanbury, 1885).

Mrs Grieve (1937)

L. vera

Describing English lavender (*L. vera*), she clarifies the *L. officinalis* of the sixteenth century botanists as being two distinct plants, known and named by the French botanist Jordan as lavender of dauphine (*L. delphinesis*) and *L. fragrans* from which the French distil lavender oil.

L. spica

She continues by describing spike lavender (*L. spica* DC or *latifolia* Vill.) which also grows in the mountainous districts of France, the flowers which yield three times as much essential oil, but of a second rate quality, less fragrant than that of true lavender. It is called by Parkinson the lesser or minor lavender and by some 'Nardus Italica' (some believing it is the spikenard of the bible).

Grieves continues by explaining that *L. spica* and *L. fragrans* often form hybrids known as bastard lavender, care having to be taken that neither the 'bastard' nor the spike lavender are among the true lavender during distillation as both will injure the quality of the essential oil. White lavender, which is sometimes found in the Alps and is described by Gerard, is probably a form of *L. delphinesis*.

L. stoechas

Finally, in her description of the varieties of lavender she comes to *L. stoechas* also known as French lavender, for which Gerard gives four varieties. Mrs Grieves ascribes this kind of lavender the classical Romans and Libyans who used it as a perfume for the bath. In Spain and Portugal in the 1930s it was used to strew the floors of churches and houses on festive occasions. The flowers were used in England medicinally until the middle of the eighteenth century, being called 'Sticadore' in has even more ancient pedigree as one of the ingredients of the 'Vinegar of the Four Thieves' famous in the Middle Ages. It is generally not used in the distillation of oil, though in France and Spain the country people extract the oil in a simple manner and use it to dress wounds, this is done by hanging the flowers downwards in a closed bottle in the shade. The Arabs use the flowers as an expectorant and anti-spasmodic.

Mrs Grieve ascribes lavender as being formerly used as a condiment and for flavouring 'dishes to comfort the stomach', and conserves for the table. She states that it has aromatic, carminative and nervine properties and though largely used in perfumery it was in her day employed as a flavouring agent, in pharmacy to cover the disagreeable odours in ointments and other compounds. Red lavender lozenges were employed as both a mild stimulant and for their agreeable taste.

The essential oil (or a spirit of lavender made from it), however, had much wider applications: according to Grieve it is an admirable restorative, and tonic against faintness, nervous palpitations, weak giddiness, spasms and colic. Its pleasant smell provokes the appetite, it raises the spirits, and dispels flatulence taken on sugar with a dosage of 1–4 drops. She also recommends a few drops in a footbath to relieve fatigue and external application for toothache, neuralgia, sprains and rheumatism; and that it is a powerful stimulant in treating hysteria, palsy and similar disorders of debility and lack of nerve power; quoting Gerard on its external application for this purpose. However, later in her chapter on the subject she says that an infusion of lavender tops made in moderate strength is excellent for headaches from fatigue; 'an infusion taken too freely, however, will cause griping and colic and lavender oil in too large doses is a narcotic poison and causes death by convulsions'. She quotes Culpeper's warning regarding the taking of oil of spike (disregarding the difference between oil and an infusion) in support of this statement.

With regard to usage in the 1930s Grieve affirms that oil of lavender could help in some cases of mental depression and delusions, and nervous headaches, if rubbed into the temples; faintness could be cured by compound tincture of lavender (red lavender). This tincture, which contains lavender, rosemary, cinnamon bark, nutmeg and red sandalwood macerated in spirits of wine for seven days, had remained in the British Pharmacopoeia for 250 years and was known as 'Palsy drops' in the eighteenth century and when it first appeared, and in the seventeenth century contained over thirty ingredients (very typical of compounds of the period). Statements for its efficacy when first made 'Official' included all those ascribed to Hildegard, Gerard and Culpeper for lavender, plus loss of memory, dimness of sight and bareness of women.

To summarise Mrs Grieve's advice

Lavender oil is of service when used to anoint the temples and forehead for headaches, as an external massage for paralysed limbs (at a time when polio was still widespread). Hot fomentations of lavender in bags, applied hot, will aid the recovery of local pain. Distilled water of lavender is a gargle for hoarseness and loss of voice. The French Academy of Medicine used oil of lavender for swabbing wounds and other antiseptic purposes during the war and the oil has been subsequently used in the treatment of sores, varicose ulcers and burns and scalds. In veterinary practice lavender oil is used in the elimination of lice and other parasites, and finally the oil is increasingly used in the embalming of corpses.

Note on use as antiseptic

The antiseptic power of lavender oil is not regarded as high, as *in vitro* antimicrobial work by Lis-Balchin *et al.* (1998), showed that lavender has relatively low antibacterial activity and that it is very variable, from batch to batch of commercial lavender oil. It would, however, make the septic wound and ward smell somewhat better.

Main functions of lavender in the past

There is a mystery surrounding the actual appearance or reappearance of lavender in Britain after Roman times (Festing, 1989). The Huguenots have been suggested as possibly bringing it over from France after 1685, however, a poem written by Master Jon Gardener in 1440 suggests that lavender was already growing in Britain by then as do many other references to its medical usage. Many rhymes pertaining to lavender were printed and recited around 1672–85, including the children's rhyme: 'lavender green, lavender blue, I shall be king and you shall be queen', (with and without 'diddle-diddle'), suggesting that lavender was well established for centuries.

Cleanliness

From earliest times lavender has been associated with cleanliness and purity, since antiquity bathing was included in the Regimen Sanitatis or writings on the care of the body. Medical, literary and ecclesiastical documents all reveal that bathing played a significant part in medieval life. By the twelfth century there were baths (very large wooden tubs) in the houses of the richer classes, in monasteries, and public baths in towns and villages often with rooms set aside for resting after therapy (Berger, 1999). Bath houses were places for socialisation and intimacy, as we find from the concerns in ecclesiastical quarters, Burchard of Worms (1008–12) in (McNeil and Garner, 1990):

> Hast thou washed thyself in the bath with thy wife and other women and seen them nude, as they thee? If thou hast, thou should fast for three days on bread and water.

The type of water best used in a bath was specified, hard or soft, river, rain or snow and should the patient take a steam bath or a water bath, for which Hildegard of Bingen gives much detail regarding thermal springs. For the former, plant extracts were thrown on heated stones in a confined environment, for the latter, if they were medicinal plant materials, they were added to the water (Berger, 1999).

The Liber Niger or Black Book of Edward VI (1547–53), (Leyel, 1937) gives a reference to a 'lavender man' authorised to obtain from the spicery enough soap for the King's personal washing.

Lavender was boiled in water and this was used for washing clothes. Shirts and sheets smelling of lavender were recognised as especially clean and thus those hospitals (some of which were hotels for travellers) and inns with linen smelling of lavender and also pots of lavender on the sills, were frequented in preference to others (Festing, 1989). Lavender was also associated with bridal beds.

Occult

Occult properties have also been associated with lavender, as it was among the mint and Feverfew, which were consecrated to the patronesses of witches and sorcerers namely Hecate (goddess of the infernal regions) and her daughters Medea and Circe.

Use of different *Lavandula* species

Spike lavender oil

Spike lavender was said by some authors to have been mainly used in veterinary practice, as a prophylactic in cases of impending paralysis (Festing, 1989). It was too camphoraceous and never worth more than one-fifth of true lavender oil, but Gerard (1633) lists its uses as the main medicinal lavender. Spike oil was also used in the manufacture of fine varnishes and lacquers with oil of turpentine and used for painting on porcelain. Its one medicinal value was for promotion of hair growth.

L. stoechas L.

Used by Muslim physicians who consider it to be 'cephalic (tonic), resolvent, deobstruent, and carminative' and prescribe it in 'chest infections and for expelling bilous and phlegmatic humours' (Said, 1996). He suggests that the vernacular 'Ustukhuddus' come from the Roman 'satukhas'. It appears to be the 'Astadus' or 'Astiqus' of Ibn Sina. Other folk-medicinal Indian writers have accredited *L. stoechas* with cephalic virtues and called it the 'broom of the brain' because it sweeps away all phlegmatic impurities, removes obstructions, strengthens its powers of expelling waste crudities and clarifies intellect'.

External uses of lavender oil

It was rubbed externally for stimulating paralysed limbs, mixed with three-fourth spirits of turpentine it gave the Oleum spicae, used to massage stiff joints and cure sprains (Grieve, 1937). Local pains were relieved by bags of lavender applied hot. Lavender oil was used in veterinary practice to kill lice and other parasites. It was also used as a vermifuge. The dried flowers were still used for perfuming of linen and keeping insects at bay. Apparently, in the United States, lavender was used to keep away mosquitoes, proof of the effectiveness of this however has not been forthcoming.

Lavender oil was also used in embalming corpses, as early as Tudor times!

Distillation of lavender

Historical survey

Distillation is thought to have originally occurred before the sixteenth century and there is some evidence to suggest that it was already used in ancient times (perhaps even Ancient Egyptian times) and was practised by the Chinese and Arabic nations 4,000–5,000 years ago. The first alembic stills were crudely represented in a few drawings from the pre-Christian period, and distillation was subsequently re-introduced to Europe around the fourteenth century due to the influence of Arabian physicians, and became all the rage in the fifteenth century. There is an early account of how Louis XIV (or his courtiers) had noticed that the rose petals liberally sprinkled on top of the water channels and in fountains would form an oily substance on top of the water. However, there was no correlation at that time between chemical oils (essential oils) or perfumes and the floating oil. In the Middle Ages, perfumes, including lavender extracts, were not accepted by the church and authorities and women wearing such products were condemned as whores by ecclesiasticals. It was acceptable for men to wear these 'dreadful odours' and there is evidence that perfumes were brought back form the crusades and were available to their wives, but mostly to be used to scent chamber and bed linen.

Sixteenth century England: distillation takes off

Lavender oil was first distilled in conjunction with other herbs and spices in the Still rooms of sixteenth century England. Lavender was a component of numerous distillation recipes and was also used as a medicine in extracts, but was not distilled alone. The herbs and spices were mixed with wine before distillation and hence the resulting distillate had a very strong alcoholic composition. This fact suggests that many of the recipes were actually used as alcoholic beverages in the same tradition as the numerous scented and flavoured alcoholic beverages produced in Europe mainly by monks in their many monasteries, resulting in the well-known Benedictine beverages.

In Elizabethan times, perfumes consisted of numerous Damask waters, which also included orris, the animalistic musk, civet and castoreum as well as numerous herbs and spices (like cloves or cinnamon).

Victorian England

It was in the Victorian times that lavender became one of the three most popular scents, together with rose and violet flowers. The famous Eau de Cologne was based on lavender oil. A recipe from 1834 gives a splendid, if very expensive, cologne:

Bergamot oil 6.2 kg, neroli oil 0.8 kg, lavender oil 1.2 kg, lemon oil 3.1 kg, clove oil 1.6 kg, rosemary oil 0.8 kg, with alcohol 90° added to make up to 100 l.

Yardley, best known for its lavender products, first became active under Charles I, before the Great Fire of London, and rose to prominence as the House of Yardley in 1770. After a period of bad business in the nineteenth century and early twentieth century, advertising in the 1930s made perfumery, as well as Yardley, popular once again.

Lavender and lavender oil today

Nowadays, lavender is used mainly as the essential oil in manufactured products like cosmetics, perfumes, soaps etc, but there has been a general reversion to the use of the dried lavender plant itself in the homespun industry of making products like herbal pillows and lavender bags (see Chapter 18).

Spike lavender is included in some veterinary shampoos and other products as an insect repellent especially for fleas (Potter's, 1988). Lavender, usually the synthetic type, is used in household products like furniture polish, general cleaners, odour-repellent sprays. It is also used in cooking by some enthusiasts and in the Food Industry (Fenaroli, 1998).

Uses in aromatherapy as a cure for mind and body

Nowadays, lavender oil has once again being put into the front-line as a cure-all for mind, body and spirit and used by aromatherapists for every possible malady, irrespective of the fact that the few clinical trials have not offered satisfactory evidence of efficacy. Second, many people these days actually hate the smell of lavender and therefore any possible psychological effect could be completely over-ruled by their dislike of the odour.

References

Berger, M. (1999) *Hildegard of Bingen—On Natural Philosophy & Medicine*, selections from *Cause et Cure*, Brewer: Cambridge.
Chaytor, D. A. (1937) A taxonomic study of the genus *Lavandula*. *Linn. Soc.*, 51, 153–204.
Chevallier, A. (1996) *Encyclopaedia of Medicinal Plants*, Dorling Kindersley: London.
Culpeper, N. (1653a) *Culpeper's Complete Herbal & The English Physician and Family Dispensatory*, Facsimile: Wordsworth editions, 1995.
Culpeper, N. (1653b) *Culpeper's Complete Herbal and English Physician*, 1st edn 1653, Facsimile Harvey Sales: London, 1981.
Fenaroli, G. (1998) *Fenaroli's Handbook of Flavour Ingredients*, Vol. 1, 3rd edn, CRC Press: Boca Raton.
Festing, S. (1989) *The Story of Lavender*, 2nd edn, Heritage in Sutton Leisure: London.
Flanders, A. (1995) *Aromatics*, Mitchell Beazley: London.
Fluckiger, F. and Hanbury, D. (1885) Pharmacographia – the history of the Principle Drugs of Vegetable Origin.
Gerard, J. (1633) *The Herbal or General History of Plants*, the complete 1633 edition as revised and enlarged by Thomas Johnson, Facsimile edition, Dover Publications Inc.: New York, 1975.
de Gingins-Lassaraz, F. (1826) *Histoire des Lavande*, Genevre: Paris.
Grieve, M. (1937) *A Modern Herbal*. Reprinted in 1992. Tiger Books International, London.

Hertzka, G. and Strehlow, W. (1994) *Manuel de la Médicine de Sainte Hildegarde*, 2nd edn, Edition Resiac, Monsurs.
Leyel, C. F. (1937) *Herbal Delights*, Faber and Faber: London.
Lis-Balchin, M., Deans, S. G. and Eaglesham, E. (1998) Relationship between the bioactivity and chemical composition of commercial plant essential oils, *Flav. Fragr. J.*, 13, 98–104.
McNeil, J. and Garner, H. M. (1990) *Medieval Handbooks of Penance*, Columbia University Press: Columbia, USA.
Palmer, T. (1984) *The Admirable Secrets of Physick and Chyrurgery*, Forbes T. R. (ed.), Yale University Press: USA.
de Renzi, S. (1857) *Collectio Salernotano, Storia documentata della Scuola medica di Salerno*. 2a. ed., Napoli, I – 417–516.
Rose, J. (1982) *Herbal Body Book*, GD – Perigee Book: New York.
Ryman, D. (1991) *Aromatherapy*, Piatkus: London.
Said, H. M. ed. (1996) *Medicinal Herbs*, Vol. 1., Bait al-Hikmah, Madinat al-Hikmah: Pakistan.
Salmon, W. (1710) Quoted by Festing (1989).
Vickers, L. (1991) *The Scented Lavender Book*, Ebury Press: London.
Williamson, E. M. and Evans, F. J. and C. W. (eds), (1988) *Potter's New Cyclopaedia of Botanical Drugs and Preparations*, Revised edn, Daniel Co. Ltd.: Saffron Walden.

Appendix: transcriptions of texts in historical section

Abbess Hildegard

When a person with palsy (possibly Parkinson's disease) is afflicted they should take galangale (a rhizome with similar properties to ginger), with half as much nutmeg (50 per cent of the amount of galangale), and half as much of spike lavender as nutmeg, plus an equal amount of githrut (probably gith or black cumin) and lovage. To these he should add equal weights (amounts) of female fern and saxifrage (these two together should be equal to the five precious ingredients). Pulverise these in a pestle and mortar. If the patient is (well) strong, he should eat this powder on bread, if (ill) weak he should eat an electuary (soft pill made with honey) made from it.

So today we might say, for example, the five precious ingredients: 100 gms of galangale; 50 gms of nutmeg; 25 gms spike lavender; 12.5 gms each of githrut and lovage. To this add: 100 gms each of female fern and saxifrage.

The second recipe quoted is easier to understand, but less obviously effective.

Lavender is hot and dry (referring to its properties under the Galenic system of medicine), having very little moisture (it is indeed a dry herb). It is not pleasant to eat, but does have a strong smell. If a person with many lice frequently smells lavender, the lice will die (dubious, but this may mean fumigation with lavender). Its odour clears the eyes, since it possesses the power of the strongest aromas and the usefulness of the most bitter ones. It certainly has an aroma that ranks in nature among the strongest of the pleasing smells, and the effectiveness of the most bitter plants (rue and wormwood), she is saying here that it is as effective as these. As to its curbing evil things and terrifying malign spirits we can make no comment.

Hildegard's third recipe is for congestion of the lungs: to cook spike lavender with wine, or if one has no wine, with honey and water. Cool it often (in a case) and it will soften hardness of the liver, and congestion of the lungs. Wine of lavender is a science pure and clear (a true remedy).

John Gerard

We grow a smaller kind of lavender in our English gardens, much smaller than other varieties, the flowers are of a deeper purple and grow smaller shorter heads with a more pleasant smell. The leaves are smaller and whiter than ordinary lavender. This grows in his Majesties Private

Garden at Whitehall. This variety is called simply spike, and sometimes spike lavender and by distillation an oil is made what is vulgarly (by common people) called oile of spike or oleum spicae. In Spain, and Languedoc in Southern France, most of the mountains and scrub fields are covered with this lavender. In cold countries we plant it in our gardens.

On L. vera

Lavender (vera) is hot and dry in the third degree, and of thin substance, consisting mainly of airy and spiritual parts (referring also to its properties under Galen's system, stating that lavender is a hot dry, airy plant, with the relevant parts growing above the ground).

It is good for diseases of the head, distilled lavender water inhaled, or for bathing the temples and forehead is good for catalepsie (no translation), mild migraines, for those who have epilepsy and those that faint a lot. The flowers of lavender picked from the heads, I mean the blue part not the husk, mixed with cinnamon, nutmeg and cloves, is made into a powder and given as a drink in distilled lavender water. This helps palpitations and passions of the heart (possibly panic attacks or more probably heart disease) and is excellent for giddiness and palsy (the shakes, possibly Parkinson's disease).

Gerard also cautions against taking lavender 'when there is an abundance of humours', in other words when the body is out of its humoural balance. He advises against taking lavender and other herbs, seeds and spices in distilled wine or in other words brandy. This he states will make disease greater, and thus put a sick man in danger. He also suggests that unqualified doctors, over keen apothecaries and foolish women who treat people with these mixtures without regard to their diseases are doing more harm than good and indeed often do much harm, even bringing on death itself.

Gerard on L. stoechas *or French lavender*

He describes it in various terms common at the time (sticados, stickedoue and sticadoue and its names in European languages). He gives subspecies as jagged sticados or 'Stoechas multisida', toothed stoechas or 'Stoechas folio serrato' and naked stoechas which he calls 'Stoechas summis cauliculus nudis'. These are his Latin names and have no botanical significance. He describes their growing wild in Spain, Southern France and the island near Manilla called Stoechas, stating that they were grown in English gardens, great care being taken to protect them from frost damage.

French lavender he says is effective against pains in the head and all diseases, which result from cold cause. It can therefore be mixed in medicines for headaches, which have persisted for some time, apoplexy, epilepsy and similar diseases.

He concludes by recommending that drinking a decoction of the flowers and husks will open stopping or obstructions of the liver, lungs, melts, womb and bladder. In other words all the internal organs, cleaning and driving out all unpleasant matter and provoking urine. Bearing in mind that it was common practice to add large numbers of drugs to a decoction or compound in Gerard's day. He also recommends its use in counter poisons or antidotes, such as theriac or treacle and hiera picra, which were made in various places around Europe, under strict license and were often included in treatments where poison or what we would call septicaemia were a risk.

Nicolas Culpeper

Culpeper says that lavender (vera) is particularly good for all illnesses and pain of the head and brain with cold causes. Such as apoplexy, epilepsy, dropsy (swelling of the limbs caused by a failing heart), the sluggish malady, cramps, fits, shakes and fainting. He continues that it strengthens the stomach, opens obstructions of the liver and spleen, provokes menses of women, and expels both a stillborn child and afterbirth (the only reference to lavender as an abortifacient).

For those who have trouble passing urine he suggest lavender flowers steeped in wine, saying that this is also good for wind and colic if the stomach is bathed with the decoction. For epilepsy and giddiness he suggests a decoction of lavender flowers, hore-hound, fennel, asparagus root and a little cinnamon and adds that if one has toothache this decoction is good as a gargle.

He ends his discussion of *Lavandula vera* by suggesting two spoonfuls of distilled water of lavender flowers, helps those who have lost their voice, also those with tremblings and passions of the heart and those who faint. This cannot only be drunk but also applied to the temples or the nostrils.

Culpeper on spike lavender

In Culpeper's day the art of distilling essential oils was being perfected, a new process, they were known as chemical oils. He warns against the use of the chemical oil drawn from spike lavender, called 'Oil of Spike' as being so fierce and piercing in its qualities that it must be used cautiously. Just a few drops are sufficient, given with other things, for internal and external troubles. Lastly, he warns that the oil of spike is not safe to use when the body is full of illnesses (blood and tumours) because it is too heating.

Elsewhere he warns young people who are in good health from taking distillates of single plants and against distillates of compounds and spirits. Saying that these are not to be meddled with by those of hot constitutions (the young) when they are healthy, for their blood is hot enough without them. If they drink them in moderation occasionally for recreation, when in good health, that may do them some good.

Culpeper then gives a complex recipe for compound spirit of lavender according to Matthias: Take one gallon of lavender flowers, to which add three gallons of the best brandy (spirits of wine) and leave to stand in the sun for six days. Then distil them in an alembic adding the following: flowers of sage, rosemary and bettony (a handful each), flowers of borage, bugloss, lilies of the valley and cowslips (two handfuls of each). Let these flowers infuse in one gallon of the best brandy and mix them with the above spirit of lavender flowers, adding melissa leaves, feverfew, and orange flowers freshly gathered. The flowers of lavender stoechas, orange and hawthorn add (of each one-ounce). Leave it to stand (convenient digestion) and distil again, after which add lemon peel and husked Peony seed (six drams each). Then add cinnamon, mace, nutmegs, cardamoms, cubebs, and yellow sandalwood (half an ounce each), aloes wood (one dram) and half a pound of the best jujubes with the stones taken out. Leave it to stand (digest) for six weeks, then strain and filter it, adding two drams of prepared pearls, a scruple of prepared emeralds, half a scruple of ambergris, musk, and saffron, half an ounce of dried red roses and half an ounce of red sandalwood, and a dram each of lemon peel and yellow sandalwood. Let the species (the last sets of ingredients) be tied up in a piece of cloth and suspended in the spirits. Culpeper then points out that this recipe is impossible to follow. As the flowers required never flower at the same time, and thus can never be gathered fresh. The addition of precious stones and animal secretions to medicines is a common method of the time (especially if the patient was wealthy).

Thomas Palmer

Recommends lavender for the treatment of 'cold distempers of the heart' where the pulse is weak, slow and thin, the breath is cold and sometimes the whole body is cold. He includes lavender in a list of plants that are helpful to this condition made up into the form of distilled waters and spirits, conserves, syrups, oils, compounds, soft pills made with honey and all sorts of aqua vitae.

William Salmon

Recommends lavender for snakebite, the bites of mad dogs and other poisonous creatures, applied internally and as poultices. A spirit-based tincture of the dried seeds or leaves, given prudently, cures hysterical fits of long standing. This tincture is also detergent, laxative, astringent, good for migraine, dispels, stimulates urine, is nervine, stomachic, used to make cordial, good for the kidneys, and for paralysis. He approved of it for fits, epilepsy, the shakes, vertigo, lethargy, fainting, hysteric fits etc.

It seems that not only does the number of individual plants used in a compound increase as the centuries pass, but the number of diseases that each plant cures also increases.

4 History of nomenclature of *Lavandula* species, hybrids and cultivars

Maria Lis-Balchin

Classification of lavender through the ages

Different kinds of lavender have been classified and reclassified at will over the years. In 1937, Mrs Chaytor from Kew reclassified the genus on the basis of combination of two previous classifications. The new scheme gave five sections based on differences in habit and anatomy, but had many intermediates which could not be properly segregated.

All the common garden and commercial plants belong to two main sections: *Stoechas* and *Spica*. *Stoechas* has four species: *Lavandula stoechas, L. dentata, L. viridis* and *L. pedunculata*. *Spica* has three: *L. officinalis, L. latifolia* and *L. lanata*. Chaytor (1937) believed that most garden plants were hybrids between true lavender, *L. angustifolia* and spike, *L. latifolia*, the silvery leaves and camphoric smell been reminiscent of the latter parent.

Cultivated *Lavandula*

DeWolf (1955) gave a brief history of the cultivated *Lavandula* species including the fact that *L. stoechas, L. pedunculata* and *L. dentata* were mentioned in Roman times, but not *L. spica* and *L. latifolia*. In the twelfth century, St Hildegard first mentioned the present cultigens.

Throughout the Middle Ages, the European plants in the genus *Lavandula* were separated into two genera: *Stoechas* (*L. stoechas, L. pedunculata* and *L. dentata*) readily recognized by the development of some of the uppermost floral bracts of the inflorescence into corona, and *Lavandula* which included *L. spica* and *L. latifolia*. It was Linnaeus who united the two genera. Linnaeus (the younger) referred to the name Lavender being derived from the Latin 'Lavare', presumably due to the use of the aqueous or oleaginous infusions of the plants. In the 'Species Plantarum' of 1753, he recognized but five species in the genus: *L. multifida, L. dentata*, from Spain, and *L. stoechas* (which *included L. pedunculata*) and *L. spica* from Southern Europe. In 1780, Linnaeus' son published a short monograph on the genus which was reprinted in 1790 by Schreiber. He added *L. pinnata* and also *L. carnosa*, which is now placed in the genus *Anisochilus*.

Baron Frederic Charles Jean Gingins de Lassaras (1826) recognized twelve species, divided into three sections. Bentham in his 'Labiatarum Genera et Species' (1832–36) recognized a fourth section and in the 'Prodromus' for De Candolle (1848), there were eighteen species. Eighty-nine years later Chaytor recognized twenty-eight species, and had six sections.

De Wolf (1955), in his discussions of the section *Spica*, states that Linnaeus in the 'Species Plantarum' had the two different taxa, but due to carelessness and misapplication of names, synonymy of the two species had become rather involved. The taxon called variously *L. vera* DeCandolle or *L. officinalis* Chaix, should be called *L. spica* Linnaeus; the taxon in the past called *L. spica* by many authors, should be *L. latifolia* Villars.

Description of the species (Chaytor, 1937)

The stoechas *group*

L. dentata originates from arid Spain, Gibraltar, and other parts of the Mediterranean areas, and some Atlantic islands. It has rounded teeth on its dark green leaves and bullrush-like flower spikes.

L. stoechas has square flower heads with plumes of plump purple brachts with a heavy pine-like smell. The oil was extracted by hanging it upside down in a closed bottle in the hot sun. Grows along Mediterranean area to Greece.

L. viridis grows in South Portugal and Madeira. It is intermediate between *L. stoechas* and *L. pedunculata*. The latter is mainly found in Spain, and has violet bracts and a shorter and fatter head than *L. stoechas*.

The spica *group*

L. angustifolia is a subdivision of *L. officinalis*, the latter is native to west Mediterranean areas extending up to 6000 feet. This is the source of true lavender oil but it is not as productive as *L. intermedia*.

Lavandin is known in the United States as *L. hortensis*: it is a hybrid between true lavender *L. angustifolia* and spike lavender, *L. latifolia*. This is a long-lasting, tall-growing plant with a good oil yield, about five times that of true lavender, though not always as sweet-smelling.

L. latifolia is the spike lavender of perfumery, similar to true lavender in appearance but flowering later. It is found growing naturally in Spain, with some plants growing high up in the mountains and also at low altitude.

L. lanata grows in Spain in mountains, for example, Sierra Nevada; it is thickly woolly with bright violet flowers and a mentholic odour.

Genus *Lavandula* in Arabia and tropical NE Africa

In 1985, Miller wrote a monograph on the fourteen species in the area of the Arabian peninsula, Socotra and Somalia. Five of these species were completely new: *L. aristibracteata* A. G. Miller (Somalia); *L. citriodora* A. G. Miller (N. Yemen and Saudi Arabia); *L. dhofarensis* A. G. Miller (Oman); *L. galgalloensis* A. G. Miller (Somalia) and *L. hasikensis* A. G. Miller (Oman).

Section *Subnuda* Chaytor (*J. Linn. Soc. Bot.* 51: 200 (1937))
Originally included just four species, to which were added three of the above:
– *L. aristibracteata* A. G. Miller (Somalia)
– *L. dhofarensis* A. G. Miller (Oman)
– *L. galgalloensis* A. G. Miller (Somalia)

Chaytor (1937) had originally placed *L. somaliensis* Chaytor in section *Pterostoechas*, 'as the opposite paired arrangement of the bracts excludes it from section *Subnuda*; however, Miller (1985) found that all the samples he examined had spirally-arranged bracts and includes this with those above in section *Subnuda*. All the species, in this particular area, are closely related and separated mainly on differences in the indumentum and size and shape of bracts.

Section *Pterostoechas* Gingins (*Hist. Nat. Lavand.* 120: 158 (1826))
This is represented in Arabia by four species restricted to the mountains of the western escarpment and comprises:
– *L. coronopifolia* Poiret
– *L. pubescens* Decaisne
– *L. atriplicifolia* Bentham
and the new species:
– *L. citriodora* A. G. Miller

All except *L. atriplicifolia* Bentham are closely related.

Hybrids

The subject was researched by Vinot and Bouscary (1971), who wrote theoretically on a fertile lavandin hybrid and generally about the very high genetic variability within the species, because the frequently isolated original populations have undergone transformations independently from one another and each character possesses different alleles. The authors continued that a return to more stable genotypes is fairly readily accomplished by self-fertilization. Their work on hybrids indicated that after the fourth generation, the seedlings were often true to form. The authors also discussed convergence and back-crossings with either parent which give more fertile products.

Lavender and lavandin cultivars

DeWolf (1955) stated that the true lavender, *L. spica* Linnaeus (or *L. vera* DeCandolle or *L. officinalis* Chaix is the lavender of the perfume trade and of the gardens. It has produced many cultivars by hybridization with *L. latifolia*. De Wolf states further that the variety *L. spica* var. *angustifolia* Rouy and Foucaud is distinguished on the basis of the shorter peduncles (5–8 cm) long compared with a typical variety (15–20 cm long); it also has narrower leaves. He thought that it included all the cultivars with 'nana', 'compacta' or 'dwarf' in the names. He also recognized two other varieties: *L. spica* var. *delphinensis* Rouy and Foucaud and *L. spica* var. *pyrenaica* Bentham.

DeWolf discussed *L. latifolia* Villars (or spike lavender) which he says produces the oil of spike which is inferior to lavender oil for perfumery and is used for soaps.

The chemical compositions also differ as does their habitat, with *L. spica* growing at 700–1000 m and *L. latifolia* below 700 m. The hybrid *L. x intermadia* Emeric ex Loiseleur, is intermediate between the two parents.

Initial hybridizations

The cultivars of lavandin probably arose initially from hybridisations between *L. angustifolia* Miller and spike *L. latifolia* Medicus (Hy, 1898; Chaytor, 1937). This is also shown in horticultural manuals (Bean, 1973; Hillier and Son, 1972). The fancy-named varieties were said by DeWolf (1955) to be lavandin.

Lavandula cultivars

Tucker and Hensen (1985) state that all the cultivars of *Lavandula* examined by them were either *L. angustifolia* Miller subsp. *angustifolia* (lavender) or *L. x intermedia* Emeric ex Loiseleur (lavandin), with none from spike *L. latifolia*. The authors discuss the different cultivars according to their colours.

The example given for the white variety was *L. angustifolia* Miller 'Alba' which was first described as:

L. vera deCondolle b. alba (Gingins de Lassaraz, *Hist. Nat. Lavandes*, 147 (1826)) and then as:

- *L. spica* Linnaeus b. alba (*Sweet Hort. Brit*. 316, non Weston (1770)).
- *L. officinalis* Chaix f. albiflora (Rehder, in *J. Arnold Arb*, 20: 428 (1939)).
- *L. officinalis* Chaix f. alba (Gingins de Lassaraz) (Rehder, in *J. Arnold Arb*. 26: 77 (1945)).

It is also confused with 'Nana Alba' and *L. x intermedia* 'Alba'.

There is another *L. angustifolia* Miller 'Nana Alba', which was introduced before 1938, and could cause confusion!

The pink variety

The pink variety of *L. angustifolia* Miller 'Rosea' was again introduced before 1937, and is almost identical to the 'Hidcote Pink' (before 1958).

Dark violet-flowered cultivars of lavender

These include:

L. angustifolia Miller 'Dwarf Blue' (before 1911), with several synonyms, for example, 'Blue Dwarf', New Dwarf Blue', 'Baby Blue', 'Hardy Dwarf' and 'Nana'.

Other varieties

These include:

- 'Hidcote' (1950), synonymous with 'Hidcote Blue', 'Hidcote Purple' and confused with 'Nana Atropurpurea'.
- 'Loddon Blue' (before 1963).
- 'Mitcham Grey' is probably a selection of 'Nana Atropurpurea'.
- 'Munstead' was raised by Miss Gertrude Jekyll and introduced by Barr in 1916. Synonyms are 'Munstead variety', 'Munstead Dwarf', 'Munstead Blue'.

Lavender-blue flowered cultivars

These include:

- *L. angustifolia* Miller' Bowles Early' (1914), syn. 'Bowles variety', 'Munstead', 'Miss Dunnington'.
- 'Compacta' (before 1901), also sold as 'Munstead'.
- 'Folgate' (before 1933), developed by Linn Chilvers of Norfolk Lavender, syn. 'Folgate Blue', 'Folgate variety'.

- 'Maillette' introduced by Pierre Grosso, and confused with lavandin 'Grosso'.
- 'Twickle Purple' (before 1922), syn. 'Twickle', 'Twinkle Purple', 'Twickes Purple'.

Other cultivars include:

- 'Graves', 'Gwendolyn Ashley', Irene Doyle'.

Lavandin cultivars (*L. x intermedia*)

White-flowered

L. x intermedia Emeric ex Loiseleur 'Alba'

Lavender-blue-flowered

L. x intermedia Emeric ex Loiseleur 'Dutch' (before 1923), syn. *L. vera* Hort. Angl, *L. spica* Hort. Neerl, 'Early Dutch'.
  Other cv's include:

- 'Grappenhall' (1902).
- 'Grosso' (1972) in Vaucluse District of France, is resistant to the mycoplasm which decimated 'Abrialii' in France.
- 'Hidcote Giant' (before 1958).

'Old English' (before 1958), syn. *L. spica* Hort. Angl. (Anonymous, 1963).
  Other cv's include: 'Provence', 'Seal', 'Silver grey', 'Abrialii", Waltham Giant' etc.

Lavender grown for oil production

L. angustifolia is mainly propagated by seed, sown in spring or autumn, depending on the severity of the winters in the region (Weiss). Sowing can be done directly into fields but more often the first stage is in nursery beds, where the plants remain for about a year. Clonal plants are made via cuttings. Healthy mother plants are cut down near ground level and the branches can be stored for months before preparing the cuttings of 10–15 cm with 1–2 branchlets. These are also planted in a nursery, usually in the spring, for a year. Green cuttings can be used but these require tender care, growth hormones and misting.

The plants are planted out in rows 1.5 m apart with 0.4–0.4 m in between rows; giving 10,000 plants per ha for *L. angustifolia* and about half for the hybrids (Weiss, 1997).

Husbandry has now improved the lavender crops and include fertilizers, often as ash. The soil is loosened superficially 2–3 times a year to remove weeds or else weed killers are used. There are many lavender pests and diseases which have attacked crops since hundreds of years, and this reduces a possible 15–20 year lifespan to 3 years. The lavender fields in France and elsewhere are therefore declining as they become uneconomical. Root rot due to *Armillaria mellea* is very serious as a fungal disease; the insect *Thomasiniana lavandulae* (Diptera) is the most serious as its larvae feed under the bark, causing damage to the tops of branches. Other diseases are due to the fungus *Rosellinia necatrix*, the homoptera *Hyalesthes obsoletus, Cechenotettix martini, Eucarazza elegans*; coleoptera include *Arima marginata, Chrysolina americana* and *Meligethes subfumatus*; Lepidoptera include *Sophronia humerella; Argyrotaenia pulchellana, Pyterophorus spicidactyla*, and many others (Chaisse and Blanc (1990)). The decline in lavender has been going on for over twenty-five years.

Harvesting was done by hand, especially in the mountains, using a sickle, but mechanical harvesters are now fully developed, cutting 7500 kg per day compared to hand harvesters cutting 500 kg. The yield of lavender oil is 40 kg/ha and lavandin is up to 120 kg. Spike lavender yields 50 kg/ha. The harvested lavender is left in the fields for a few days then steam distilled or extracted with CO_2 or other solvent.

References

Andrews, S. (1994) Lavandula hybrids, *Lavender Bag*, 1, 4–5.
Chaytor, D. A. (1937) A taxonomic study of the genus *Lavandula, J. Linn. Soc. Bot.*, 51, 153–204.
Chaisse, E. and Blanc, M. (1990) Les ravageurs de la lavande et du lavandin, *Phytoma*, 419, 45–6.
DeWolf, G. P. (1955) Notes on cultivated Labiates 5, *Lavandula, Baileya*, 3, 47–57.
Germer, R. (1985) *Flora des pharaonische Agypten*, Mainz.
Hohmann, J., Zupko, I., Redei, D., Csanyi, M., Falkay, G., Mathe, I. and Janicsak, G. (1999) Protective effects of the aerial parts of Salvia officinalis, Melissa officinalis and *Lavandula angustifolia* and their constituents against enzyme-dependent and enzyme-independent lipid peroxidation, *Planta Med.*, 65, 576–78.
Kourik, R. (1998) *The Lavender Garden*, Chronicle Books, San Francisco.
Kuntze, O. (1981) Revisio generum Plantarum, Leipzig.
Manniche, L. (1989) *An Ancient Egyptian Herbal*, British Museum Publications Ltd., London.
Miller, A. G. (1985) The genus *Lavandula* in Arabia and tropical NE Africa, *Notes Royal Botanic Gardens Edinburgh*, 42, 503–28.
Pitard, C. J. M. and Prost, L. (1908) Les Iles Canaries, Flores-de L' Archipelago, Paris.
Tucker, A. O. and Hensen, K. J. W. (1985) The cultivars of lavender and lavandin (Labiatae), *Baileya*, 22, 168–77.
Tustin, M. (1996) Handy hints for successful growing, *Lavender Bag*, 5, 22–3.
Vinot, M. and Bouscary, A. (1971) Studies on lavender (VI) The hybrids, *Recherches*, 18, 29–44.
Weiss, E. A. (1997) *Essential Oil Crops*, CAB International, Oxon.
Wilkınson, A. (1998) *The Garden in Ancient Egypt*, The Rubicon Press, London.

5 Lavender growing in England for essential oil production

Maria Lis-Balchin and Henry Head

England's modern lavender industry: Norfolk Lavender

History of the business

The family-run business originated in 1932, when Linn Chilvers founded Norfolk Lavender and planted the first six acres in Heacham, Norfolk. The demand for lavender during the First World War had been intense as it was used as a disinfectant for wounds when mixed with sphagnum moss. Most of the other English lavender companies around London had died down so this became the main supplier of lavender plants and oil. In 1932, planting was done by three men and a boy in 18 days for a total cost of £15!

The lavender was subsequently harvested by hand and taken by horse and cart to the railway station at Heacham and transported to Long Melford in Suffolk for distillation (Figure 5.1).

Figure 5.1 'Old Major' helps to collect the harvest, around 1940.

Lavender perfume

The oil was used in a secret formula to produce a perfume, made originally for King George IV by Mr Avery, a chemist interested in French perfumery. The perfume was made in a garden shed and bottled by two of Linn Chilvers' sisters. Only hundreds were then made and sold. After the death of Mr Avery, the formula was purchased by Norfolk Lavender, and now many thousands are being made and sold.

In 1936, the company acquired three stills; one was new and held about an eighth of a ton of flowers; the other two were second hand, having been made in 1874, but held a quarter of a ton each. The stills were installed in Fring, which was nearer to the fields and were serviced by a boiler which made steam for a Great Yarmouth trawler.

The Queen's mother, old Queen Mary visited the company in 1936, and the following year some of the fields in Sandringham were leased out for lavender growing by the company.

The Second World War saw the beginning of exports of lavender to the United States, as a foreign exchange commodity, with the Home Guard defending the distillery. By the end of the war the company produced bunches of lavender, lavender oil and lavender water as well as the special lavender perfume. Next, talc was produced and bath salts.

Beginning of modernisation

In 1953, Linn died leaving the post of Managing Director to a former worker, Tom Collison and Adrian Head became chairmen; Adrian's son, Henry Head, is now in charge. The company began to change and in 1978, part of the old building became a tea-room and the stills were moved to Heacham. The fields of lavender and the stills can now be visited by the public for a good fee. There is now an enormous shop selling a good array of lavender plants, a modern warehouse from which mail orders are despatched, shops selling cosmetics, perfumes, tea-towels, various novelties and there are now eight full-time staff or equivalent, with forty-four more taken on during the season. The lavender fields now comprise 100 acres and one harvest of lavender is obtained. The residual plant material is burnt. The essential oil and products are exported all over the world.

Heel cuttings and newer methods of propagation

Until recently, heel cuttings were used during October for propagation (Mellish, 1996 – Head gardener, Norfolk Lavender). The stock plants were attacked with hedging shears and the debris sorted out into the good 13 cm cutting with a good heel and some eyes. The cuttings were bundled into groups of fifty and dipped into rooting powder, then inserted into cutting beds in slit trenches, with about 2 cm showing above the ground. There were sixteen rows of cuttings with only 2 cm between the rows in each cutting bed. The cuttings were left for 12 months, then lifted out and relined out. They were then either sold as bare-root plants, or potted up for later sale or planted out in the lavender fields.

Due to the increasing number of plants required, soft and semi-ripe cuttings are now used which are raised in heat beds. Micropropagation is also used.

Some varieties do not respond well to the new methods, so the old method is used with some slight changes: a strimmer is used for the cuttings and a tunnel used for propagation.

Field management

Fertilizers used are organically-based and low in nitrogen. Cultivation is used to keep out weeds, together with spot sprays for bindweed and brambles and pre-emergent sprays for weeds which could grow in the lavender rows. However, the company insists in their brochure that no pesticides are used!

Distillation

The amount of oil and the quality depends on the type of lavender and sunshine etc. The lavender is loaded into each still and treaded down by a worker to fill the still to capacity. Steam is passed through to obtain 500 ml of lavender oil from 250 kg of lavender (one large still) in 20 min.

Lavender products

About one-third of the lavender is dried for flowers, pot-pourri and sachets. The lavender is first loosely packed into sacks and laid on a special floor, with warm air circulating through them for 2–3 days. The stalks are separated from the flowers if needed.

Lavender cultivars developed

There are several cultivars of *L. angustifolia* developed by Norfolk Lavender for their own use for the production of lavender oil and *L. x intermedia* Grosso oil for lavandin oil.
There are numerous cultivars including:

L. angustifolia cv.	*L. stoechas* cv.
'Imperial Gem'	'Fathead'
'Little Lady'	'Helmsdale'
'Little Lottie'	'Kew Red'
'Miss Katherine'	'Pedunculata'
'Miss Muffet'	'Roxlea Park'
'Princes Blue'	'Snowman'
'Royal Purple'	'Willow Vale'
'Walburtons Silver Edge'	

L. lanata × *L. angustifolia*
L. 'Sawyers'
L. 'Goodwin Creek'

and other species and their cultivars, for example:

Section *Dentata*
L. dentata	*L. dentata* × *L. latifolia*
'Candicans'	*L. x* 'allardii'

Section *Pterostoechas*	Section *Subnuda*
L. x christiana	*L. subnuda*
L. canariensis	*L. aristobracteata*
L. buchii	*L. macra*
L. multifida	

References

Mellish, N. (1996) The Lavender Bag, 5, 20–1.
Norfolk Lavender Brochure (1999) Jarrold Publishing, Norwich.

6 The retail lavender nursery

Simon Charlesworth

Introduction

There may be romantic appeal to running a lavender nursery, but there is little room for this notion on the modern nursery. Downderry Nursery, however, offers a sensual experience which compensates for a great deal of time, worry and effort. Set in the peaceful beauty of a Victorian walled garden, in the heart of the Kent countryside, it is also the home of the Lavender and Rosemary National Plant Collection® in Kent (Figure 6.1). It affords both customer appeal due to its display gardens of lavender and is a successful commercial concern.

Business plan

Predominant factors involved in the operation of a retail lavender nursery, marketing and well organised nursery management, are prerequisites for the efficient production required to fulfil

Figure 6.1 Out-buildings and sales area fronted by the display of lavenders. (See Color Plate XII.)

sales projections, be competitive and make a profit. As with any business, there is no more important task than planning. Time spent at this stage will yield great dividends in the future, in both the short and long term. The tangible benefits of 'thinking time' may not flow for some time, but provides a sound infrastructure on which the execution of jobs, from propagation to selling, can operate smoothly and efficiently.

Once the initial business plan is in operation the rolling programme of production needs to be addressed annually in advance. Successful sales require that all aspects of the business are integrated to produce the right plant, at the right time, at the right price. To achieve this, production scheduling needs to work back from the sale date. For instance, a finished lavender in a 3 l pot for June requires striking a cutting in June the previous year. However, striking a cutting in February can achieve the same by July. By pruning a lavender four weeks before it is supposed to bloom, June flowering can be delayed until August. The options are numerous.

Marketing

Often given short shrift in horticulture, marketing is as important on the nursery as watering plants. There is no purpose in producing a plant for which there is no demand. Lavender is undergoing something of a renaissance. Demand is insatiable. This makes marketing considerably easier, but still requires attention to ensure that the market share is at least maintained and preferably, increased. In growing one genus there is a certain degree of vulnerability to fads and fashions. However, the extraordinary expertise of the specialist nurseryman provides a firm foundation to take advantage of niche market opportunities. The enduring popularity of lavender is a real benefit.

Initial market research aims to reveal gaps in the market for a product, after which honing methods to target the right market and marketing the right plant, at the right time and at the right price are all important. As a specialist nursery it is possible to respond swiftly to changes in tastes within the product range offered for sale, which the larger nursery may find comparatively more costly and complex.

The specialist is also able to command a higher price for a lavender, as the extra knowledge accumulated is available to the customer to make a more informed decision as to which lavender would best suit their situation. In this respect the dissemination of information from grower to customer increases customer aspirations and broadens their perception of what it is possible to grow. With the additional knowledge, the customer is prepared to experiment for themselves and try growing lavenders that previously would have seemed too great a challenge or of which they had little or no knowledge. The end result is a heightened profile for the specialist nursery and increased sales.

The creation of a brand image is of the utmost importance in a world where first impressions can make or break a business. Whether advertising, designing a catalogue, or web site, or displaying plants in a garden setting and plants for sale on the retail nursery, it is imperative that a theme runs through the entire business. User friendliness is a key aspect in presentation. The customer needs to be drawn into the experience, to feel relaxed and yet informed, without being confronted by a barrage of complex details.

A balancing act, of some dexterity, needs to be performed to cope with the logistics of moving the nursery forward, by taking advantage of all opportunities to increase awareness and hopefully, profitability. In order to achieve this it may be necessary to take the lavenders to potential customers. This can be achieved through mail order and by exhibiting at flower shows.

Advertising

The catalogue, web site and exhibiting, act as important advertisements in themselves, as to the quality and range of lavenders on offer. Advertising is often the interface between business and potential customers. Identifying the potential customer is a prerequisite of effective advertising. Returns on a broad untargeted approach may be considerable, but can be a costly exercise for the specialist nursery. It is often better to target the right consumer group. In the case of mail order, advertising in national gardening magazines is cost effective. High enquiry returns and subsequent orders can flow for moderate cost. This may also allow for sufficient funds to be available for one-off advertising opportunities specifically applicable to lavenders. The advertising base can be broadened to encompass bespoke plant labelling carrying the company logo, name and contact number and also carrier bags with similar information. Both have a dual function and keep the brand image in front of the consumer.

With business flowing from mail order, flower shows and from the retail nursery it is essential that all these strands can be interwoven effectively. This is one of the key functions of nursery management.

Nursery management

Efficient staff

The role of efficient staff in administration, production and sales cannot be underestimated. It is critical that each department of the business knows what the other is doing. Experience dictates that team work among a core staff of dedicated and loyal individuals, in a relaxed atmosphere, creates the healthiest and most productive working environment.

The office is the centre of all aspects of nursery operations. Mismanagement here, can be costly elsewhere. The computer is the most productive tool in efficient office management, but only if staff are fully conversant with software.

The first task of nursery management is to prepare a budget and cash flow forecast for the coming year. Preferably on a rolling monthly basis. This will highlight what lavenders need to be ready to supply income at given times of the year. The principle selling season is the beginning of May to the end of July with August and September providing additional income rather than substantial income.

The consumer has yet to be convinced that autumn planting is in many ways better than spring planting. The wholesale lavender nursery would undoubtedly find the selling season brought forward by a few months.

Production scheduling

Production scheduling requires close scrutiny by the retail nursery, but is less precise than for the wholesale nursery, for which orders for plants will generally have been placed well in advance. Flower shows are the most predictable element of scheduling. The lavenders in an exhibit are more in demand, than those that are not exhibited and the short, hardy lavenders are especially sought after. Beyond this, the mail order and retail sales at the nursery are more imprecise. It can never be anticipated with complete confidence that certain species or cultivars are going to sell in quantity. A surge in demand for less common lavenders may occur due to unsolicited publicity. There are, however, some garden stalwarts which seem to have endless appeal.

The skill of good management is to try to prepare for the surprise and ensure stocks of all lavenders are available throughout the season, in the right quantities, without overproducing. The previous year's sales are a reasonable guide to the forthcoming year, but should not solely be relied upon for forecasting. There is a need to approach the new season proactively. To feed consumer demand it seems always necessary to introduce a new lavender or look for a new angle on an existing lavender.

Once the lavenders to be marketed for the coming season are selected, it is necessary to put the production schedule into operation. Working back from the following year's sale date provides the starting point for scheduling.

Flower shows

For flower shows falling between mid-May and the end of July a ten-week lead-in time is sufficient to get a lavender from a cutting to a saleable liner. This allows for six weeks in propagation and weaning and a further four weeks as a liner. Stock plants need to be ready at the start of this ten-week period and potted accordingly.

Mail order and retail nursery

Mail order and retail nursery liners need to be ready from March to October, which makes scheduling a little more complex. To ensure a finished liner is ready in March it is necessary to prepare stock plants in early August for cutting material in September.

Bottlenecks are always experienced at some point during the season. Principally this occurs in May when the season is at its height and lavenders need to be available for flower shows, mail order and at the retail nursery. Area apportionment when planning the years stock levels becomes paramount, to avoid the need to move plants. Space is at a premium in spring. The need to produce pristine lavenders to flower exactly at the right time for large flower show exhibits still requires considerable moving and juggling of those lavenders reserved for shows. For this purpose it is often necessary to force lavenders, but lavenders do not respond well to being forced. Very often they become willowy, with pale, uncharacteristic blooms. It is certainly the most challenging period of the entire year, made more so by the inherent unpredictability of the weather.

By May much of the watering can be done using overhead spraylines, except for exhibition plants. Keeping flower spikes dry is crucial to their appearance. Watering these plants at the base with a hose pipe and hand lance also ensures they are regularly inspected to assess progress and quality.

Administration problems

While the jobs in production are burgeoning, administration is equally reaching critical mass. Handling scheduling, enquiries, orders, accounts and the endless stream of predominantly inane paperwork requires good organisation and a good software package.

Computerised tracking system for plant stock

There are some extremely comprehensive stock control packages available to cope with all of the above, but they often come at considerable cost and with many irrelevant features. Developing a bespoke system in-house using a widely available database is far more effective and efficient. Tracking stock is an essential part of scheduling to make sure targets are being met.

Rooted cuttings enter the computer system once they reach weaning, by which time most will survive. Any shortfall can be rectified within the required time to ensure finished plants are still available on time. With this information it is possible to track plants from a rooted cutting to liners and 3 l, knowing the availability at each stage. Lavenders can then be reserved in confidence, sold and despatched to mail order customers. Each day on the retail nursery and each flower show, can be treated as a separate customer for the purpose of stock control. Entering lavenders in the system 10 weeks prior to a show ensures that the manager can assess progress and take action to correct any problems.

Accounts information can be integrated with stock control to provide a snapshot of business at any time. Keeping accounts up to date ensures that any deviation, positive or negative, from the business plan can be identified and if necessary, rectified.

Sales

At the same time as all these backroom dramas, sales may be reaching fever pitch. The specialist nursery should provide more than a vending machine approach to sales, striving to give the consumer an experience beyond just buying a plant. This takes a considerable amount of time and patience, but is part of the service that makes the specialist nursery a real gem.

Maintenance of garden and plants

Not all jobs on the nursery are specifically directed toward sales. Maintenance is a time consuming part of the business, best reserved for the winter months. Some jobs can, and need, to be done during the slight sales plateau of August, taking advantage of the good weather. In this respect labour too needs to be flexible throughout the year, to take advantage of weather windows.

National Plant Collection®

Maintaining a garden display of lavender in addition to the nursery, especially if the nursery has a National Plant Collection®, is an extra responsibility.

It is possible to apply for status as a National Plant Collection® from the National Council for the Conservation of Plants and Gardens (NCCPG), a charity established to conserve garden plants for future generations to enjoy and to maintain botanical diversity.

Several aspects are considered in granting a National Plant Collection®. These include, the range of species and cultivars within a genus and evidence that the prospective collection holder is serious about correctly identifying plants and researching their history and provenance. The work of the collection holder in maintaining a collection is largely a labour of love, but it does give the nursery a certain kudos, which is difficult to measure economically.

Correctly identifying and correctly naming plants is a very important aspect of the work of a National Plant Collection® holder as there is a complex and confusing history in the naming (nomenclature) and taxonomy (classification) of lavenders.

Production

Propagation

There is no mystery to the art of propagation as long as it is remembered that the propagator provides the total life support for a cutting until it is rooted. Two efficient means of propagating lavender are to use a low tunnel in a greenhouse or to use a mist bench. The effect is much the same. Bottom heat at about 23°C can be supplied by either a soil warming cable in moist sand or a foil panel with a heating element, covered with polythene and capillary matting. The latter provides a more even heat, but rapidly cools if there is a power failure.

The humidity developed in a low tunnel provides similar conditions to a mist unit operated using an electronic wet leaf to control length of misting. With a mist system rooting is aided by keeping the base of the cutting warm and the top cool.

Cuttings

Cuttings are best taken in late August and early September for overwintering as a liner (9 cm pot). The plant material needs to be soft and about 5 cm in length preferably from a non-flowering shoot. The type of cutting whether nodal, internodal or heal, makes little difference at this level.

Cutting material should be taken from perfect specimens of the lavender required. Substandard plants either through pest, disease or with a particularly poor habit must be avoided. The need to maintain good quality stock plants is necessary for the quality and uniformity of the next generation of plants.

Method of taking cuttings

All cuttings first require their lower leaves to be removed to provide about 2 cm of clear stem, then dipped in a softwood hormone rooting powder and struck into a proprietary compost. A soilless rooting medium with about 20 per cent vermiculite has proven best. The use of plug trays greatly assists with the ease and speed of handling rooted cutting. The use of preformed plugs either using a gluing agent or an outer 'jacket' for the plug have advantages for handling, but tend to require greater vigilance in watering once removed from the propagation bench. It is also not possible to hold the cuttings in plug form for as long.

The speed of rooting varies according to the hardiness of the species. The tender and half-hardy lavenders root fastest and are usually ready to be removed from the propagation bench within two weeks. The frost hardy, hardy and very hardy lavenders take on average about two weeks. Successful rooting of between 95–100 per cent is typical.

Raising lavender from seed

Lavenders can be raised from seed, but cultivars are notorious for not coming true to type and are best raised from cuttings. Some species can be raised from seed with minimal isolation of the stock plants. It is easy to tell whether seed is set by merely tapping the seed head. If it rattles the seed can be collected, sown immediately or kept in a dark, dry place. When sown they can be placed on the mist bench and lightly covered with compost. Germination is usually rapid, approximately four days. Remove them to the weaning bench as soon as the majority have germinated to ensure the seedlings do not become leggy in the heat. Prick out to plugs when the first true leaves are visible. From then on they can be treated as for cuttings.

Once on the weaning bench the occasional mist will ensure that the plants are not unduly stressed. Light shading using fleece suspended 1–2 m above the plants also helps protect the plants from harsh sunlight. The air temperature at this stage should be maintained at about 15–20°C. The rooted cuttings should be kept to 5 cm so any top growth needs to be removed. Trimming will encourage side shooting and ensure plants branch immediately, giving good plant structure.

Nutrient availability is soon exhausted in any plug plant, so the addition of miniature pellets of slow release fertiliser with a 5–6-month release rate enables the life of a plug to be extended. An alternative is to use liquid feed. A balanced seaweed feed works well. Lavender species *Lavandula (L.) stoechas* and *L. viridis* and hybrids of the two, form very fibrous roots that require these to be potted on relatively quickly.

Two weeks in weaning should be sufficient to enable the rooted cuttings to survive a harsher environment of approximately 10°C and some may be ready for potting to liners.

Clearly labelling lavenders from propagation onwards is imperative, as many look very similar for much of the year. An error at this stage, or mislabelling stock plants from which cuttings are taken, can be extremely costly.

Potting

To avoid interruption to growth, lavenders should be potted as soon as a good rootball has formed. For lavenders a compost using young peat with a pH of 6–6.2 provides a good medium for healthy growth. The addition of slow release fertiliser at the rate of 2.5–3 kg per cubic metre of compost is sufficient nutrition. The required rate of release is determined by the potting season. Autumn potting for liners kept inside requires a 12–14-month release rate with reasonably balanced N:P:K ratios. This allows the plant to develop gradually and not become too lush and vulnerable to cold weather nor too susceptible to pest and disease before the spring.

Spring potting requires a different approach, particularly if liners are to be sold as soon as possible. By April it is perfectly feasible to use a release rate of 5–6 months with a balanced N:P:K. With this release rate a liner can be ready within 3–4 weeks of potting allowing for one trim on potting, or approximately 4–5 weeks allowing for two trims.

A liner is certainly the most cost effective plant size to sell. Once planted in the ground a liner will be a year ahead of a plug, although a 3 l plant will be a year ahead of the liner. However, it costs more in time, materials and space to pot and maintain a 3 l plant than a liner. Most customers prefer to purchase lavenders as liners because they establish better than a plug and cost about one-third the price of a 3 l lavender. This enables the customer to buy a selection, or sufficient for a lavender hedge, at reasonable cost.

There is however, a growing market for specimen plants of 5, 10 and 15 l. These are best prepared in late autumn with a proprietary coarse potting compost, using 12–14 month slow release fertiliser. The plants should be set out to allow for substantial growth during the following spring. If the existing rootball is well watered when the plants are potted to a specimen pot, there is often no need to water the plants until the following February. The dryer fresh compost surrounding the rootball will provide excellent insulation through winter and enable these plants to be overwintered with the minimum of protection. A polythene tunnel netted along the sides is sufficient. Fleece can always be used in the coldest periods, on the frost hardy and half hardy lavenders, which will keep the temperature under the fleece 5°C above the surrounding air temperature. The tender lavenders will always need to be protected to 5°C. Pruning specimen lavenders to 10–15 cm on potting will encourage rapid spring growth. Similar treatment can be given to stock plants for spring cutting material. At all stages of the production process there is a need to be vigilant for pest and disease.

Pest and disease control

Lavender is not prone to many pests and diseases. Good hygiene and regular stock inspections can save vast resources and reduce the need to apply expensive chemicals or biological controls on a regular basis. The first pest that may be encountered, in the life of a lavender on the nursery, is sciarid fly whose larvae attack the base of cuttings. This pest can be a problem for all lavenders on the mist propagation bench due to the moist atmosphere and can be summarily dealt with using nematodes. At the next stage vine weevil can be a particularly pernicious pest where prevention is definitely better than cure. Vine weevil larvae attack the roots of plants at the base of the stem. The addition of chemical additives when potting to a liner and larger sizes, will provide almost complete protection. Aphids can affect all lavenders, especially in the spring, but are easily controlled biologically with a soft soap spray, or chemically, if persistent.

Not all lavenders are affected by the same pests and diseases, nor to the same extent. *L. angustifolia* and cultivars can suffer with red, and two-spotted, spider mite if kept undercover in a dry atmosphere, in the summer months. *L. stoechas* and cultivars are particularly prone to leafhopper, another sap sucking pest and while this is often more cosmetic than harmful to the plant it will clearly affect the saleability of the plant. New spray applications on the market, specifically designed to kill sap sucking pests, are effective against aphids and leafhopper. The greatest pest problem of tender and half hardy lavenders, kept undercover in the summer, is whitefly and it is certainly the most difficult to eradicate so prompt action and regular treatment is necessary.

Good hygiene should avoid problems with the water borne disease phytophthora, then, the only fungal disease really affecting lavenders is botrytis. During winter months and warm wet summers botrytis can thrive. The first, preventative action, is to ensure that lavenders are not watered overhead in winter and that there is good air circulation. If botrytis does occur then alternating between two sprays with different active ingredients should be effective. This alternation is required because botrytis is very good at becoming resistant to the active ingredients.

Weeds, although not strictly a pest or disease, can harbour both. It is therefore, advisable and good nursery practice, to remove them from the growing area and from the surrounding area, particularly before they set seed.

Good garden lavenders

There are many lavender species and cultivars which make good garden and patio plants. Many new cultivars are coming to market to add to the 150+ already in existence. Most are the result of what could be described as passive breeding programmes, selections from a seed population with rather uncertain provenance. There are, however, some fine lavenders available as a consequence. More rigorous active breeding programmes are destined to result in greater improvements in commercial and garden cultivars.

The following selection is presented according to hardiness, one of the main criteria customers use when buying lavender. Experience has indicated that plants may survive temperatures about 5°C below those shown, provided it is not for long periods and the soil is relatively dry. Hence, the need for good drainage if lavenders are to thrive.

Very hardy lavenders

These traditional lavenders are the easiest to grow, most reliable and effective for the flower border or for hedging. Their unsurpassed scent and colour are evocative of summer. Hardy to at least −15°C, they will cope with most conditions. They fall into two groups, true lavender (*L. angustifolia*) and Lavandin (*L. x intermedia*).

L. angustifolia subspecies (subsp.) *angustifolia* (True Lavender) commonly and sensibly referred to as just *angustifolia* (although there is also an *L. angustifolia* subsp. *delphinensis* and an *L. angustifolia* subsp. *pyrenaica*), have a compact u-shaped habit and narrow grey-green leaves. Their short flower stalks and short round-topped flower spikes form about half the height of the plant. The flowers have a rich sweet scent in June and July. This is the most popular species grown in England for oil extraction, yielding high quality oils used for perfumes, aromatherapy, potpourri and for drying on the stalk. They are naturally distributed through Central and South West Europe.

'Ashdown Forest' is an extremely bushy lavender for a low hedge reaching 50 cm with pale purple flowers and a white eye to the corolla. It was introduced in the 1980s and although there are several similar cultivars, for instance 'Cedar Blue' and 'Little Lady', 'Ashdown Forest' was the first of the short, pale flowered cultivars.

'Bowles Early' which is also known by the synonyms 'Bowles Variety' and 'Miss Donnington', was introduced in 1913 and is a slightly taller bushy plant reaching 60 cm with pale purple flowers, although without the white eye to the corolla and therefore not as pale as 'Ashdown Forest'.

'Compacta', also referred to as 'Nana Compacta' was introduced from the United States in 1901. It is similar to 'Munstead', but with larger steely-purple flowers and greyer foliage. It grows to 55 cm.

'Folgate', introduced in 1933, is a little known, but good all-rounder with a colour midway between 'Hidcote' and 'Munstead'. It reaches 60 cm with mid-purple-blue flowers. It is the blue tinge to the flowers that make this cultivar distinctive.

'Hidcote' is probably the most well-known lavender in cultivation, has deliciously dark purple flowers with grey foliage and reaches 50 cm (refer to Figure 2.6). It was introduced in 1950.

'Imperial Gem', introduced in the 1980s is very similar, growing slightly taller and with more silvery foliage.

'Hidcote Pink', introduced in 1958 is a very pale pink lavender which looks fantastic with 'Hidcote', growing to the same height. The leaves are more grey and narrow and flowers lighter, than 'Rosea'.

'Little Lottie' is a lovely little lavender with masses of pale pink flowers growing to just 40 cm. It was introduced in 1998.

'Loddon Blue', introduced in 1963, is an excellent short hedging lavender growing to just 45 cm with mid-purple-blue flowers.

'Miss Katherine', the darkest pink hardy lavender available, has star-shaped markings on the flowers when they first open. It grows to 60 cm and was introduced in 1992.

'Munstead' has many synonyms, among them 'Munstead Variety', 'Munstead Dwarf', 'Munstead Blue'. It was introduced in 1916 and has become a garden stalwart. Very bushy it reaches 60 cm with mid-purple flowers and green-grey foliage.

'Nana Alba', also known as 'Dwarf White', is a wonderful little plant slowly reaching just 40 cm with white flowers and green-grey foliage. It was introduced in 1938.

'Princess Blue', introduced in the 1980s, is a fine, upright plant growing to 60 cm with long, pale blue flowers some of which seem to linger longer than other *L. angustifolia* cultivars.

'Rosea' has brilliant green foliage in the spring making this pale pink lavender distinctive. It grows to 60 cm and was introduced in 1937. It is very similar to, if not the same as both 'Jean Davis' and 'Loddon Pink'.

'Royal Purple', introduced in the 1940s, is a gorgeous plant and tall for an *L. angustifolia* growing to 75 cm with long, purple flowers.

L. x intermedia (Lavandin) are sterile hybrids of *L. angustifolia* and *L. latifolia* (Spike Lavender). They have a narrow base and an upright v-shaped habit. The grey-green leaves are broader, flower stalks and flower spikes longer and more pointed than *L. angustifolia*. They also have lateral flowering shoots. The flowers are strongly scented, mildly camphoraceous and appear during

July and August, but often continue into autumn. This is the most popular for oil extraction, higher yielding than *L. angustifolia*, but producing lower quality camphoraceous oils, used in soaps, cosmetics and detergents. Widely used for drying off the stalk and for pot-pourri. They are naturally distributed in France, Italy and Spain.

'Alba' has been known in Europe since 1880. A marvellous plant with a fan-shaped habit forming a perfect dome of white flowers. It reaches 75 cm.

'Dutch Group' is the most widely grown of the lavandins. A real old favourite, introduced in 1920 and popularly called 'Vera', it is superb as a tall hedge, growing to 90 cm with pale purple flowers and distinctive grey foliage which looks good all winter.

'Grappenhall', one of the earliest cultivars was introduced in 1902. It has unusual lilac-purple coloured flowers and very broad green-grey foliage which make this lavender quite different. It grows to 75 cm.

'Grosso', a French introduction in 1972, is a profuse flowering lavender, the most widely grown lavender for oil in the world. It reaches 75 cm with blue-purple flowers.

'Hidcote Giant' has very distinctive bushy purple flowers on stout stems which make it great as cut flower. It grows to 90 cm and has grey-green foliage. Introduced in 1958.

'Old English', introduced in the 1930s, is a lavender with typical cottage garden appeal, growing to 100 cm with clear pale-purple flowers.

'Sussex', has the longest flowers of any hardy lavender and makes this a remarkable plant. It grows to 90 cm with pale purple flowers.

Hardy lavenders

These lavenders are sufficiently tough to withstand all but the most severe weather, hardy as they are to −10°C. There are several cultivars that fall into this category which are a hybrid of the species *L. lanata* and the species *L. angustifolia*. The two described below are particularly good. Both are only known from cultivation.

The *L. lanata × angustifolia* are similar in habit too, but more robust than, *L. lanata*. The soft silvery-grey foliage provides a marvellous contrast with the strongly scented purple flowers which appear from late June to late July. These lavenders are sterile.

'Richard Gray' is a very tidy plant and keeps its shape well. It has round-topped purple flowers and grows to 50 cm. Introduced in the 1980s.

'Sawyers', also introduced in the 1980s, is a taller form, reaching 70 cm, with large conical bushy purple flowers.

Frost-hardy lavenders

These are some of the most spectacular and increasingly popular lavenders. The flowers all have 'ears' on top. These 'ears' are sterile bracts (coma) and so have no flowers at their base, unlike the rest of the flower spike. With milder winters these lavenders are now more widely grown. They will survive to −5°C and often to temperatures several degrees lower. They have a very long flowering season if dead-headed, from early May through to September and beyond, if there are no severe frosts. Most have a camphoraceously scented pale green foliage and a bushy habit, but no appreciable scent to the flowers. Those lavenders of interest in this range include some of the subspecies of *L. stoechas* and the species *L. viridis*. They also include a few of the astonishing number of hybrids of *L. stoechas* with *L. viridis*.

L. stoechas subsp. *pedunculata* often referred to as Spanish Lavender, but also known as 'Papillon'. It is a graceful upright lavender. The long flower stalks (peduncles) are topped with purple flowers and beautiful long pale purple 'ears' that look magical fluttering in a summer's breeze. They have very narrow pale green-grey leaves and will grow to 75 cm. It has a natural distribution across Central Spain and Portugal.

'Willow Vale', introduced from Australia in 1992, has a similar habit to *L. pedunculata*, although not quite so upright and with crinkly 'ears' that tend to lay more horizontal. The flowers are noticeably more purple. It grows to 60 cm.

L. stoechas subsp. *stoechas* commonly known as French Lavender is a compact plant with purple flowers and short 'ears' on very short flower stalks. It grows to just 45 cm. This subspecies has a wide natural distribution including Madeira, the Middle East, North Africa, North East Spain to Turkey and Tenerife.

'Kew Red' is a most remarkable lavender with uniquely coloured cerise-crimson flowers and pale pink 'ears' (Figure 6.2). If grown as a perennial it is best grown in a pot as it is not too tolerant of wet winters. It grows to 40 cm.

forma leucantha the white French Lavender also known as 'Snowman' has a tidy rotund habit and grows to 45 cm.

Typically the hybrids of *L. stoechas* with *L. viridis* have masses of flowers and vigour.

'Avonview' is a splendid lavender with the longest and darkest purple flowers of all the crosses and broad pale purple 'ears' (Figure 6.3). It grows to 60 cm and was introduced from New Zealand in 1992.

'Fathead', introduced by Downderry Nursery in 1997, has plump round flowers which distinguish this cross. The masses of long lasting dark purple flowers fade to pink with age. It reaches just 45 cm.

'Helmsdale' is a sensational robust lavender with unique, burgundy-purple flowers (Figure 6.4). It grows to 70 cm and was introduced from New Zealand in the 1990s.

'Marshwood' is another New Zealand introduction in the 1990s. A striking, vigorous lavender with absolutely enormous purple flower spikes and pink 'ears'. It reaches 90 cm.

Figure 6.2 *L. stoechas* subsp. *Stoechas* 'Kew Red'. (See Color Plate XIII.)

Figure 6.3 L. stoechas × viridis 'Willow Vale' *L. angustifolia* Hidcote. (See Color Plate XIV.)

'St. Brelade' a sweetly scented green-leaved lavender smothered with bright, pale purple flowers with dusky pink 'ears'. It grows to 60 cm. Introduced from Jersey in 1995.

L. viridis, commonly known as green lavender, has amazing yellow-green flowers which makes this lavender distinctly different. The very green foliage has a strong and unusual lemon scent, particularly when bruised. It reaches 60 cm (24 in). It is naturally distributed in Madeira, South Portugal and South West Spain.

Half-hardy lavenders

These lavenders will thrive if given just a little protection to keep them above 0°C. A dry bed at the base of a sheltered south facing wall is often sufficient.

Well worth considering are two species *L. dentata* and *L. lanata* and one hybrid *L. lanata* with *L. dentata*.

Figure 6.4 L. stoechas × viridis 'Helmsdale'. (See Color Plate XV.)

L. dentata are known as fringed lavender (refer to Figure 2.9). They derive their name from the toothed (dentate) leaves which have a richly aromatic lavender-rosemary scent. They will flower almost continuously if dead-headed, the pale purple flowers, with their short tuft of 'ears' atop, rise above a spreading, but upright bush. It is native to Algeria, Arabia, Balearic Islands, Morocco, South and East Spain, Tenerife and Yemen.

'Linda Ligon' is a delightful and unusual variegated lavender, growing to 60 cm with pale purple flowers and green splashed cream toothed foliage. It was introduced from the United States in 1996.

'Ploughman's Blue', introduced from New Zealand in 1996, has luxuriant green foliage and grows to 75 cm with pale purple flowers. It is a particularly hungry cultivar in a pot, but grows splendidly in the garden.

Varietas candicans is a stunningly attractive soft, silver leaved *L. dentata* growing to 75 cm with pale purple flowers. This is the toughest of all the *L. dentata* species and will survive outside in a pot in a sheltered position against a south-facing wall, if the compost is kept dry.

L. lanata, known as Woolly Lavender due to its broad woolly leaves, has beautiful strongly scented slender violet flowers rising above silver foliage from late June to late July. It reaches 50 cm and is naturally distributed in Southern Spain.

L. lanata × dentata 'Goodwin Creek Gray' is a remarkable bushy lavender that deserves to be more widely grown. The broad toothed foliage is velvet to the touch and long, slender, blue flowers appear from June to September if dead-headed. It grows to 75 cm. Introduced from the United States in the 1990s.

Figure 6.5 L. minutolii. (See Color Plate XVI.)

Tender lavenders

These include some of the most beautiful and delicate lavenders. Their spectacular flowers can be enjoyed all year round if grown in pots. They make a delightful display outside from May to October, but need to be brought in before the first frosts and kept warm at around 5°C. All have spiralling triple flower spikes in a trident formation on each long flower stalk, but have no scent. Most have unusually scented lacy (pinnate) foliage and a spreading, but upright habit. Regularly dead-heading keeps the blooms coming. Five species are worth considering.

L. buchii varietas *buchii* is a gorgeous silver leaved lavender with blue-purple flowers reaching 60 cm and is native to Tenerife.

L. canariensis is a breathtakingly beautiful lavender. The intense blue flowers above fresh green foliage are an astonishing sight all summer. It reaches 60 cm and is native to the Canary Islands.

L. x christiana is a sterile hybrid of *L. canariensis* and *L. pinnata*, it is often mistakenly sold as *L. pinnata*. This exceptionally vigorous lavender has the most enormous blue flower spikes above grey foliage. It can grow to 100 cm, but is more typically 75 cm native to Tenerife.

L. minutolii is a very attractive lavender with deliciously sweet-scented foliage (Figure 6.5). The green-grey felty and deeply-cut arrow-head shaped leaves are topped with blue-purple flowers. It grows to 60 cm and is naturally distributed on the Canary Islands.

L. pinnata is a delightful compact pale purple-pink flowered lavender reaching just 40 cm with flat, lobed foliage. Naturally distributed on La Palma, Lanzarote, Madeira.

Caring for lavenders in the garden

Providing care advice to the gardener is always welcome and good policy. Once established lavenders thrive on neglect except for their annual prune, which is the subject of great confusion and is explained in some detail.

Soil

Lavenders require well-drained neutral to alkaline soil, although *L. stoechas* subsp. *stoechas* (which always grows in acid soil in the wild) and to a lesser extent *L. x intermedia*, can thrive in a slightly acid soil. In heavy soil adding grit at the rate of 25 kg/m^2 when planting will improve drainage as will planting on a slight mound. Wet soil in winter can have a particularly deleterious effect on half-hardy and frost-hardy lavenders and it is this additional wet soil, rather than just a frost, that is more likely to kill these plants.

Site

Plant lavenders in a sunny position or at least where they are in the sun for most of the day. Do not grow them under a leaf canopy. Tender, half hardy and dwarf lavenders are ideal for 30–40 cm terracotta pots and look particularly impressive as patio plants.

Spacing

Space lavenders 45–90 cm between plants for informal plantings, depending on their eventual size. Planting in groups of three is very effective. For hedging, dwarf lavenders to 60 cm are best planted 30–40 cm apart. Tall lavenders over 60 cm may be planted 40–45 cm apart. For a formal hedge use the same lavender.

Planting in the garden

Ensure the soil and site are as described above. Moisten the plant compost, but do not waterlog. Dig a hole and add a dusting of bone meal to the hole and the soil removed from it and mix together. Fill the hole with water and allow to drain away. Place the plant in the hole and fill to the level of compost around the plant stem. In dry conditions water the soil around the plant, but do not over water. Be attentive to lavenders in the first few weeks after planting, especially if the weather is dry as the compost in which the plant was originally potted will dry out very quickly.

Planting in pots

Use a mix of one-third each of a soilless compost, John Innes No. 2 or 3 and coarse grit. For feeding, add slow release fertiliser at the recommended rate. One application should last all season.

Watering

This should be unnecessary after establishment, except plants in pots!

Feeding

Little feeding is required, although a sprinkling of potash 35 g/m^2 or rose fertiliser 60 g/m^2 around the base of plants in spring will encourage more prolific flowering and improved flower colour. Adding bulky manures may well lead to sappy growth and few flowers.

Harvesting

To use lavender for drying and pot-pourri, harvest just as the flowers are opening and hang upside down in bunches in a dry dark room.

Overwintering

Tender and half-hardy lavenders (and frost-hardy lavenders grown in pots) should be kept under glass in light, airy conditions. These plants need very little water from November to February. Wait until the pot is noticeably lighter or even until plants start to droop and then water only on top of the compost. Never water over the foliage in winter. These plants suffer in still, moist air.

Pruning

This is often the most misunderstood aspect of growing lavender and needs some clarification. It is a very important task that demands a strong constitution, because generally the harder lavenders are pruned, the longer they will last. They require different treatment according to hardiness.

Hardy lavenders which normally flower just once, may have a weak second bloom after pruning. To keep them in shape they should be pruned to just 22 cm immediately after flowering. It is particularly important to be severe with the tall growing lavenders, even if you have to sacrifice some late flowers. If there is a reasonable number of small shoots visible below where you cut they will grow strongly even from old wood, but if there are no shoots below the cut lavender will die. If pruned at the correct time new growth should leave lavenders overwintering as leafy hummocks.

It is possible to save old gnarled lavender, which has much bare wood topped with a mass of growth. Prune to within 10 cm of the bare wood to see if this encourages shoots to sprout further down the plant. If it does, then when next pruned do the same again, until the new growth starts at ground level.

Frost hardy lavenders, typically have 'ears' and because they flower from spring to autumn it is difficult to know when to prune. A general guide is to prune hard to 22 cm immediately after the first flowering. Dead-head for the rest of the flowering period, with possibly just a light trim in early September.

Half-hardy and tender lavenders are the toothed and trident-headed lavenders that flower almost continuously. Generally, dead-head throughout the year with the occasional severe prune, as outlined above, to keep especially the more vigorous forms in shape. After a severe pruning keep the compost quite dry until a moderate flush of new growth appears.

7 Lavender growing in Australia

Rosemary Holmes

Growing requirements

General conditions

Lavender is a very old herb which was used as a disinfectant, antiseptic and relaxant, and for culinary, medicinal, and therapeutic purposes as far back as Roman times.

Lavender is a hardy, herbaceous, evergreen plant that can thrive under a wide range of soil and climatic conditions, but it prefers a neutral to alkaline soil of pH 7–7.5. It will tolerate drought once the roots are established in the ground, as has been proven here during two very severe droughts. The first planting during 1982 saw weeks of drought with only bare soil to be seen, and despite some efforts at watering (not very successfully) we only lost twenty-five plants out of the 1000 planted out proving that lavender was indeed a xerophyte.

Water

With the great variation in weather patterns these days the normal seasonal conditions do not always apply and in the drier regions it may be necessary to drip feed young plants if there is a prolonged dry spell, during their first year of growth, and while the roots are becoming established.

Four basic requirements for growth

There are four basic requirements for growing good lavender. Being of Mediterranean origin it likes to grow out in full sun and part shade will not produce the best quality plants.

Lavenders must have good drainage. The quickest way to kill a lavender plant is to have poor root drainage and if there is any doubt about free drainage, raise up the garden bed or if out in field conditions, bench up the rows so the water can drain away quickly.

Lavenders must also be pruned hard each year or they will become very woody at the base after a couple of years. When pruning, do not cut back hard into the old wood as they may not shoot again. Leave two or three pairs of green leaves on the bushes and the rest of the plant material is ideal for taking cuttings.

When pruning in the autumn add a little lime or dolomite to the soil, avoiding the main stem and sprinkling around the base of the plant. In spring add only a small amount of fertiliser to the soil and this will depend on the results of soil analysis which should be done annually or when stable every couple of years.

So it is full sun, good drainage, pruning and lime application in the autumn and not too much fertiliser.

Soil conditions

Lavender will tolerate poor soils but before any commercial growing is planned it is advisable to have a total soil analysis done before any crop is planted out, if it is to be grown professionally. Too much nitrogen will cause excess growth of foliage at the expense of the flowers, but some potash is needed to produce good quality flowers. The amount of fertiliser required will depend totally on the initial soil analysis and for ensuing years.

If lavender is to be grown for commercial purposes then it is a wise idea to trial several varieties in your own ground prior to the final decision on which lavenders to grow and for what purpose. Microclimates can vary over a short distance and while lavenders are hardy plants, can tolerate some frost, heavy snows, winds and intense heat, some varieties will do better near the coast, others better in humidity and others better at higher altitudes.

Common lavenders to grow

The three groups most commonly grown for commercial use are, the *L. angustifolia* group, the *L. x intermedia* group and the *L. stoechas* group which can have weedy tendencies and are mainly grown for fresh flowers. Another group, the *L. pterostoechas* lavenders are what are known as the 'Fern-Leaf' lavenders and they are also mainly used in fresh floral work.

There are lavenders available in the many diverse shades of blue/mauve/purple as well as pink, white and green colours. During spring and summer when most of the lavenders are in bloom, it is wonderful to create a massed arrangement of lavender flowers in a vase and people are still surprised to see the entire arrangement is made up of lavender flowers.

Different markets for lavenders

Lavenders can be grown for the fresh flower market, the dried flower market, kilos of loose stripped lavender, or for oil production. Those interested in plants can also propagate from their parent stock and sell plants from a retail nursery. There is also the opportunity to add value by making craft products using lavender, such as the traditional lavender bags, and anything can be created by those clever at sewing.

There is also the culinary market, which is increasing, and a clever chef can create gourmet lavender produce as long as they realise you must use the *L. angustifolia* varieties that have a light sweet perfume which enhances food.

The *L. x intermedia* group of lavenders have a camphoraceous aroma (in the essential oil) and these varieties should only be used for craft purposes. Lavender skin care products and the use of lavender in aromatherapy is becoming more popular and many more people are looking to grow lavender for its oil production.

The *L. angustifolia* lavenders produce lavender oil which is the best quality oil for aromatherapy or medicinal uses. The *L. x intermedia* lavenders produce an essential oil known as lavandin oil which has a high camphor content and is mainly used in fragrance and soap production. It is important when deciding which lavender variety to grow for oil to realise that the lavender oil type may produce better quality but fewer kilograms of oil, while the lavandin plants will produce more kilograms per acre, but the oil is of an inferior quality compared to the lavender essential oil.

Harvesting lavenders

In small holdings it is financially practical to harvest the lavender with a sickle, whereas in large farms it is more economical to use machinery. Where machinery is not practical to use, such as on steep rocky slopes, then the sickle comes to the forefront.

When to harvest the lavender is always a pertinent question. If the flowers are to be used for a fresh market, then they should be picked in the early morning after any dew had dried, but before it becomes too hot during the middle of the day. Pick the blooms when the first couple of flowers have opened at the top of the inflorescence.

If it is for the dried flower market, the blooms should be harvested just prior to opening, when the flower head is still tight and showing full colour. If it is harvested later, when the flowers have opened and the bunches are to be dried, they will drop as they hang to dry.

If you are growing for oils then the flowers can be harvested much later and can even be browning off a little at the edges.

Bunching lavender for retail

For retail sales bunches should be about the width of the thumb and first finger around the middle when joined in a circle. The size will shrink as the bunches dry. Use an elastic rubber band size thirty-two (seen on bunches in England as well as in Australia). If you are not stripping on a large commercial basis, then put a dust mask over your face and wear rubber gloves when stripping the dried flowers off the stem.

Operate in an open-air environment where you have plenty of fresh air around you. While one of the benefits of lavender is soothing and relaxing and it can help relieve headaches, it will give you a big headache if you breathe in the concentrated oil in an enclosed room. Store loose-stripped lavender in an airtight container in a cool dark place in the house.

If lavender is to be left hanging for some months, make sure it is in a cool, airy place out of direct sunlight. The sunlight will bleach the stems and flowers if it is left out in the open where the sun can reach it every day.

Cuttings

If the interest in lavender is to operate a nursery, then it is important to only propagate by heel or tip cuttings, especially for the *L. angustifolia* and *L. x intermedia* varieties. DO NOT grow these from seed as they will not grow true to form and will cause great confusion if a botanically-accurate retail nursery is to be established.

Use a very open drainage propagating mix such as with peat moss and coarse river sand or whatever combination you require as long as it is only kept damp: do not over water cuttings or they will quickly rot. When potting up, after the cuttings have rooted, use the best quality potting mix you can find. It is uneconomic to not use good quality mixtures due to heavy losses of cuttings. It is also essential to have a very good pair of propagating secateurs, and to keep them clean.

Business plan

If you want to grow lavender, a business plan to estimate your expenses and income is required.

The average guide for bunches from an *L. angustifolia* bush is about 3–4 bunches per bush. The average guide for bunches from an *L. x intermedia* bush, which usually flowers up to two weeks later, and is a larger plant with longer stems or peduncles, is about 6–7 bunches per bush.

It takes about seventy-two bunches, depending on the variety, to produce one kilo of stripped lavender.

It takes one kilo of lavender to produce about 8–12 mls lavender oil (again depending on the variety).

Pests and diseases

There are very few pests and diseases that affect lavender plants. The spittle bug (Philaenus spumarius) may be detected, usually in the spring, as small areas of spittle on the stem of the plant. A small green insect can be found in the spittle but these do little damage and can be ignored. It disappears when the bushes are harvested. The alfa mosaic virus can cause yellow patches on leaves, which is an infection that can spread and should be removed when seen: take out the branch or the entire bush and disinfect the secateurs or sickle with a viscous soap brew. It is wise not to have Lucerne growing in the immediate area as the Aphids can breed there. The disease called Shab does not occur in Australia but has been found in England.

Organic farming

There is an increasing interest in growing lavender organically, but whichever way it is farmed, it is important to keep the weeds down, and there is always some hand weeding to do in the centre of the bushes, because it is not practical to use chemical sprays on the actual bushes, as so much is now used for culinary purposes. If in a farming situation, grassed areas between the rows can be mown, and look attractive when the flowers are in bloom.

You need to have some skill, some faith, but above all the ambition to create a profitable enterprise: to be an entrepreneur, to create an innovative new era of lavender farming. Lavender is currently being grown in all states in Australia, excluding the Northern Territory.

References

Hyde, K. W. (ed.) (1997) *The New Rural Industries. A Handbook for Farmers and Investors*, p. 258. Published by Rural Industries Research & Development Corporation.

McGimsey, J. A. and Rosanowski, N. J. (1996) *Lavender: A Growers Guide for Commercial Production*. New Zealand Institute for Crop & Food Research Ltd.

McNaughton, V. (1999) *The Essential Lavender. Growing Lavender in New Zealand*. Timber Press, New Zealand.

8 Naming and misnaming of lavender cultivars

Maria Lis-Balchin

Introduction

Lavender cultivars have been increasing since the early 1600s. Most of the earlier hybridisations involved *Lavandula angustifolia, L. latifolia* and later *L. stoechas*, giving rise to the numerous cultivars of colours ranging from white, yellow, blue to mauve and purple. Many different cultivars have been produced in different parts of the world, often giving rise to the same cultivar being produced with a different appearance and name. There is at present no International Register of Lavenders, but there is frequent talk among growers and enthusiasts about setting up such a register (Head, 1999). The International Registration Authorities are appointed through the International Society for Horticultural Science (ISHS) in London and operate within the provisions of the current edition of the International Code of Nomenclature for Cultivated Plants. This results in the publication of all the registered cultivars and all the relevant data on the characteristics, history and origin of each cultivar is kept.

Some cultivars of *L. angustifolia* in cultivation

Andrews (1994) listed a large number of *L. angustifolia* cultivars, some of which are presented, with their originators and probable dates of origin.

'Alba'	Europe, since 1600s
'Ashdown Forest'	UK, 1985 (Nutlin Nursery)
'Blue Cushion'	UK, 1992 (Blooms)
'Miss Donnington'	UK, *c.* 1913 (A. Perry)
'Dwarf Blue',	
'Blue Dwarf'	*c.* 1911
'Baby Blue'	
'Fring'	UK (Norfolk Lavender)
'Heacham Blue'	UK (Norfolk Lavender)
'Gray Lady'	USA, before 1967 (Grullman)
'Hidcote'	UK, before 1950 (L. Johnson)
'Hidcote Blue'	
'Hidcote Pink'	
'Imperial Gem'	UK, 1980s (Norfolk Lavender)
'Irene Doyle'	USA, 1983 (de Baggio)
'Lady'	USA, 1993 (Burpee & Co.)
'Loddon Pink'	UK, 1950 (T. Carlile)

'Loddon Blue'	UK, before 1963 (T. Carlile)
'Maillette'	France (P. Grosso)
'Munstead'	UK, 1916 (Barr)
'Munstead Blue'	
'Munstead Variety'	
'Dwarf Munstead'	
'Nana Alba'	UK, before 1938 (Musgrave)
'Dwarf White'	
'Baby White'	
'No. 6', 'No. 9'	UK (Norfolk Lavender)
'Princess Blue'	UK, late 1980s (Norfolk Lavender)
'Provencal'	France (D. Christie)
'Rosea'	UK, before 1940s (Norfolk Lavender)
'Nana Rosea'	
'Munstead Pink'	
'Royal Purple'	UK, 1940s (Norfolk Lavender)

Cultivars of *L. x intermedia*

Andrews (1994) provided this list of hybrid lavender cultivars:

'Abrialii'	France, before 1935
'Abrial'	
'Abrialis'	
'Alba'	Before 1880
Dutch group	
'*L. vera*'	Europe, before 1920s
'*L. hortensis*'	
'Dutch'	
'Early Dutch'	
'Giant Blue'	Europe, before 1935
'Glasnevin'	Ireland, before 1935
'Grappenhall'	UK, *c.* 1902 (Clibran)
'Grosso'	France, *c.* 1972
'Hidcote Giant'	UK, before 1958 (L. Johnson)
'Old English'	UK before 1930s (Herb Farm, Seal)
'Provence'	France
'Seal'	UK before 1935 (Herb Farm, Seal)
'Super'	France, *c.* 1956
'Twickle Gem'	Before 1934

Descriptions of some of the cultivars

McNaughton (1999) gave descriptions of large numbers of cultivars, from both UK and New Zealand origins, which included the *L. angustifolia* cultivars, which all resemble the species of origin and are the hardiest and most fragrant of all the cultivars:

	Corolla colour
L. angustifolia 'Ashdown Forest'	bright violet-blue 93A
L. angustifolia 'Backhouse Purple'	violet-blue 90C

L. angustifolia 'Blue Bun'	bright violet-blue 90A
L. angustifolia 'Blue Cushion'	bright violet-blue 90B
L. angustifolia 'Bosisto'	dark lavender-violet 92A
L. angustifolia 'Budakalaszi'	vibrant violet-blue 90A
L. angustifolia 'Celestial Star'	white
L. angustifolia 'Coconut Ice'	dark mauve-pink 75A or white
L. angustifolia 'Fiona English'	dark lavender 90A
L. angustifolia 'Fring'	bright violet-blue 90A
L. angustifolia 'Granny's Bouquet'	vibrant violet 88A
L. angustifolia 'Hidcote' (several types)	vibrant violet 88A/ violet-blue 90C
L. angustifolia 'Lady'	bright violet-blue 90A
L. angustifolia 'Martha Roderick'	bright violet-blue 90B
L. angustifolia 'Melissa'	soft lavender-pink 69C
L. angustifolia 'Mystique' (syn. 'Bedazzled')	vibrant violet 88A
L. angustifolia 'Okamurasaki'	bright violet-blue 90A
L. angustifolia 'South Pole'	softest lavender-lilac 85D
L. angustifolia 'Trolla'	vibrant violet-blue 90A
L. angustifolia 'Violet Intrigue'	vibrant violet 88A

Numerous cultivars of *L. x intermedia* are also given including:

	Corolla colour
L. x intermedia 'Grappenhall'	lavender-violet 91A
L. x intermedia 'Jaubert'	violet-blue 90C
L. x intermedia ' Margaret'	violet-blue 94C
L. x intermedia 'Sumian'	vibrant violet' 88A
L. x intermedia 'Abrialii'	violet-blue 90D
L. x intermedia 'Dutch'	blue-violet 94B
L. x intermedia 'Grosso'	violet-blue 90C
L. x intermedia 'Super' (several)	violet-blue 90C/94C/92A + 94C

Cultivars of *L. stoechas* include:

	Sterile bract
L. 'Atlas'	vibrant purple-violet 87A, 88C
L. 'Blueberry Ruffles'	vibrant light blueberry 89D to 88D
L. 'Helmsdale'	rich burgundy 79A/B
L. 'James Compton'	vibrant red-violet 88C
L. 'Marshwood'	vibrant red-violet 88C
L. 'Plum'	vibrant purple-violet 87A/B
L. 'Pippa White'	white with green veining

Special cultivars round the world

Australia

Spencer (1995) suggested the following:

- *L. angustifolia* 'Yuulong' (bred in Australia by Rosemary Holmes and Edythe Anderson).

- Mitcham Lavender could have come from UK, but could also be *L. x allardii* and was also referred to as *L. spica* var. gigantea in 1956 Herbarium.
- *L. angustifolia* 'Lavender Lady'
  Introduced from United States in 1993, grown from seed.
- *L. stoechas* subsp. *pedunculata* 'Purple Crown'
  Plum flower with pink bracts, with longer bracts and taller generally than both *L. stoechas* and *L. pedunculata*.
- 'Lavender Sidonie' named after founder Sidonie Barton in NSW, who discovered this chance hybrid between *L. pinnata* and an unknown species in her garden. It grows very quickly and has rich violet flowers surrounded by soft feathery foliage.

France

France was one of the original producers and exporters of lavender oil to the cosmetics industry round the world, but its plantations of lavender have now decreased. Lavandins were first planted in the 1920s and originally there was the *L. x intermedia* 'Abrialii named after Prof. Abrial who developed it in the 1930s. There was however the incidence of 'Yellow decline' which devastated the crops. So, in the 1950s came 'Super' which was resistant to the 'Yellow' and in the 1960s came 'Grosso'. The main areas of production are in the Drome, Vaucluse and Provence, but the Haute Alps are almost abandoned.

Japan

McNaughton (1999a,b) recorded that the Japanese lavender history began in 1937 when the late Mr Seiji Soda, of the Soda Perfumery brought back 5 kg of lavender seeds from Marseilles (France). Full cultivation began in 1948 in the Furrano district of Hokkaido. The Tomita Farm started to grow lavender in 1958 and later in 1970. The lavender oil production reached 5 tons, after help from the government, but soon declined due to the cheaper alternatives provided by synthetic components. Most production of lavender ceased except fot the Tomita farm which became a place for visitors to tour in 1998.

Among the cultivars raised in Japan are:

- Four *L. angustifolia* cultivars named 'Hayasaki', 'Youtei', 'Hanamoiwa' and Okamurasaki'.
- In northern Japan, in the alps, covered with snow during winter, *L. angustifolia* and *L. x intermedia* cultivars are raised.
- In southern Japan, high temperatures occur and *L. angustifolia* is not suitable, only some *L. x intermedia* cultivars. These are also grown at the base of Mount Fuji.

New Zealand

McNaughton (1995) suggested the following cultivars for the enthusiast:

Section *Stoechas*
- *Lavandula* 'Evelyn Cadzow'
  Reddish purple sterile bracts, fertile bracts are reddish-purple and green and corollas a dark purple, bright green foliage, short (named after the breeder).
- *Lavandula* 'Marshwood'
  Dark purple sterile bracts, corollas being dark purple, fertile bracts are green tinged with burgundy-purple, greyish-green foliage and tall height (bred in Marshwood Gardens, Southland).

- *Lavandula* 'Helmsdale'
  Burgundy flower spikes, but fertile bracts of green edged with purple tips, greyish-green leaves, medium, bushy (bred in Marshwood Gardens).
- *Lavandula* 'Plum'
  Dark purple corollas with reddish-purple sterile bracts, bushy, medium height, lovely and robust (bred in Auckland).
- *Lavandula* 'Pippa'
  Sprawling, spectacular plant, long hairy peduncles and spikes with purple corollas and green fertile bracts, sterile bracts are creamy white and yellow-veined (bred in S. Auckland by P. Carter).

Section *Spica*
- *L. angustifolia* 'Blue Mountain'
  Corollas deep violet with dark calyces, grey foliage, used as hedging (bred in Central Otago).
- *L. angustifolia* 'Avice Hill'
  Sweet-scented, large violet corollas with dark violet calyces and broad green fertile bracts, grey foliage (bred in Christchurch).

United States

The popularity of lavender bagan in 1933 with the formation of the Herb Society of America. In 1981, Nancy Howard imported a collection of lavenders from England and others imported the lavandins from France. All cultivars were identified by Arthur O. Tucker by analysis of their esssential oil content.

McNaughton (1999) gives several different cultivars including:

- Lavender 'Betty's Blue' (*L. angustifolia* 'Betty's Blue')
  Dark purple flowers, taller and wider than most in this species.
- Lavender 'Buena Vista' (*L. angustifolia* Buena Vista)
  For pot-pourri, fragrant and deep purple flowers. Reblooms.
- Lavender 'Lisa Marie (*L. angustifolia* 'Martha Roderick × *L. lanata*)
  Silver downy foliage, frosted grey flower heads open to a blue-violet colour on spikes up to 8 inches long.
- Lavender 'Royal Velvet' (*L. angustifolia* 'Royal Velvet')
  From seed collected in UK. Intensely purple flower heads.
- Lavender 'Silver Frost' (*L. angustifolia* × 'Kathleen Elizabeth')
  A natural cross between *L. angustifolia* and *L. lanata*. White wooly leaves, covered with sprays of 15-inch spikes topped with dark violet flowers clothed in whitish woolly calyces.

Misnaming of cultivars

There is often confusion in naming of plants for a number of reasons:

- Mislabelling of stock
- Intentional re-naming of plants for better sales
- Grower believes that they have discovered new plant and gives it a new name.

It is often difficult to prove that species, hybrids and cultivars are misnamed, unless DNA tests are used and even those have not been perfected.

The reasons for difficulties are that the same plant grown in a different area under different conditions of climate, for example, rainfall, sunlight time and intensity, geographic location, fertilizers etc. can look totally different.

Lavenders and lavandins in New Zealand

Lavenders and lavandins in New Zealand were studied by McNaughton (1998; 1999) and many found to be exactly similar to well-known hybrids imported from France, for example, *L. x intermedia* 'Grosso' was found named as *L. x intermedia* 'Dilly Dilly' and *L. x intermedia* 'Wilson's Giant'.

There is a possibility that another French hybrid 'Super' was at some stage mixed up with the 'Grosso', so giving a mixture of taller and more blue flowered plants (the former) and the shorter plants of the latter. *L. x intermedia* 'Super' was found named as *L. x intermedia* 'Arabian Night' and *L. x intermedia* 'Sussex'.

The 'Super' imported from New Zealand has produced an essential oil similar more in composition to that of *L. x intermedia* 'Abrialis'.

Examples of name changes

Other name changes abound (McNaughton, 1998) like *L. x intermedia* 'Yuulong' which in some nurseries becomes *L. angustifolia* 'Yuulong'. Sometimes a single word is added, for example,

– *L. angustifolia* 'Lady' becomes *L. angustifolia* 'Lavender Lady'.
– *L. angustifolia* 'Mausen Dwarf' arose from mistaken reading of the label 'Munstead'.
– *L. angustifolia* 'Hidcote Blue' for *L. angustifolia* 'Hidcote.
– *L. x intermedia* 'Hidcote White' for *L. x intermedia* 'Alba'.

The name '*vera*' originally used for *L. angustifolia* has caused much confusion and one plant was even named *L. angustifolia* 'Vera' and has recently being renamed as *L. angustifolia* 'Egerton Blue'.

Some cultivars sometimes have different shades of the same colour and therefore get a new name, for example, *L. dentata* (dark-flower) becomes *L. dentata* 'Monet'.

L. angustifolia 'Alba' (McNaughton, 1999) is a typical example of a cultivar raised in New Zealand (Blue Mountain Nursery, 1950s) which differs from the one described by Chaytor (1937). It is a medium-sized bushy plant with mid-green foliage, with medium peduncles which are mid-green and semi-upright; spikes are medium, 3–5 cm, cylindric and interrupted, with the base-whorl 1–2 cm below the main spike; fertile bracts are broad and bracteoles are narrow; the corollas are white with buds and calyces densely pubescent and green.

References

Head, J. (1999) International Registration of Cultivars. *Lavender Bag*, 11, 10–11.
McNaughton, V. (1995) Some New Zealand Cultivars for the Enthusiast. *Lavender Bag*, 3, 6–10.
McNaughton, V. (1995a) *Lavender: The Growers Guide*. Garden Art Press, Suffolk, UK.
McNaughton, V. (1995b) Lavender in Japan. *Lavender Bag*, 11, 1–5.
Spencer, R. (1995) Notes on Lavender in Australia. *Lavender Bag*, 3, 11–14.

9 Phytochemistry of the genus *Lavandula*

Jeffrey B. Harborne and Christine A. Williams

Introduction

The annual production of lavender oil has been estimated at 462 tonnes (Lawrence, 1992). It is not surprising therefore that our knowledge of the phytochemistry of *Lavandula* is centred on the essential oils (mono- and sesquiterpenoids) present in the genus. Studies of the volatile constituents of the commercially important species, *L. angustifolia* and *L. latifolia*, and their hybrid has revealed the presence of well over 150 chemicals, many of them as trace components. The substances which contribute most to the characteristic lavender scent are described here under the terpenoid heading. One intriguing structure reported from the cultivated lavender plant could, however, be considered as an alkaloid. This is the substance 2-N-phenylaminonaphthalene, the structure of which was confirmed by synthesis (Papanov *et al.*, 1985). This would seem to be the only account of an alkaloid-like substance in *Lavandula*. However, simple organic bases such as stachydrine and betonicine have been described as occasional constituents of the family, the Labiatae, to which *Lavandula* belongs (Hegnauer, 1966).

Monoterpene lactones, such as nepetalactone from catmint, and diterpenoids, which are richly present in some genera of the Labiatae such as *Salvia*, are apparently not represented in the terpenoid fractions of *Lavandula* species. However, triterpenoids which occur widely in the family, are reported in *Lavandula* and some fifteen structures are listed in this review.

This genus *Lavandula* is relatively rich in phenolic constituents. Some nineteen flavones and eight anthocyanins have so far been found in these plants. None of them is unique to *Lavandula*, since they have been reported to occur elsewhere in the Labiatae. Some are rather characteristic of the family, such as the various glycosides of hypolaetin and scutellarein. One phenolic, which at first appeared to be unique to *Lavandula* grown in cell culture, is a blue pigment, the iron complex of a caffeic acid ester. Subsequent studies showed that although it did not occur in the intact plant of *L. angustifolia*, it could be found in low amount in several related genera (Banthorpe *et al.*, 1989).

The medicinal properties of lavender, for example, as an antiseptic, rest chiefly on the essential oils present and therefore this account of the phytochemistry of *Lavandula* opens with a brief record of the mono- and sesquiterpenoids present, a subject which is continued in other chapters in this book. The remainder of this review is devoted to the phenolic constituents, and especially the flavonoids, since it is possible that they also contribute to the use made of these plants by mankind.

Terpenoids

Essential oils

All *Lavandula* species and hybrids are highly aromatic plants, which produce complex mixtures of essential oils from glands on the surface of the flowers and leaves. However, only three taxa are important in the commercial production of essential oils for use in the perfume and cosmetic industries. These are L. angustifolia, L. latifolia and L. hybrida (L. latifolia × L. angustifolia), which produce lavender oil, spike lavender oil and lavandin oil, respectively. The amount of oil produced by these commercially grown taxa has been increased by plant breeding and the composition varies from oil to oil, from one country to another and with the age of the plant. For example, the major components of lavender oil (Figure 9.1) are linalyl acetate (1), linalol (2), *cis*-ocimene, and lavandulyl acetate, those of spike lavender oil are linalol, 1,8-cineole (3), camphor (4), α- and β-pinene and borneol and of lavandin oil are linalol, linalyl acetate, camphor, 1,8-cineole and borneol (Table 9.1). Lavandin oil from the hybrid plant can be seen to have inherited the major essential oil characteristics of both parents. However, there are many minor components giving totals of over 100 constituents in lavender oil, more than eighty in lavandin oil and some sixty in spike lavender oil. Lavender oil is the most variable in qualitative composition; for example, significant amounts of citronellol have been reported from a lavender oil from India, borneol from Russian oils, 1,8-terpineol from Italian oils and camphor from some Chinese and French oils (Table 9.2). There is also variation in the concentration of the major components, linalyl acetate and the ocimenes. Thus, if linalyl acetate is relatively high the ocimenes will be lower and *vice versa*. Naef and Morris (1992) in their evaluation of lavandin oil suggested that the minor components, fenchone, *iso*-fenchone and 5,5,6-trimethyl-bicyclo[2.2.1]heptan-2-one also make a contribution to the odour characteristics of this oil. According to Prager and Miskiewicz (1979) of the US Customs Laboratories, imported spike lavender oils can be distinguished from lavender and lavandin oils by the larger percentage of α- and β-pinenes, camphene, limonene, 1,8-cineole and camphor. There are numerous other references giving the detailed composition of these three oils from different countries, which have been summarised by Boelens (1995; see also chapter 6).

Data are also available for the composition of the essential oils of eight further *Lavandula* species and subspecies from Spain and Portugal, together with another Spanish sample of L. latifolia (Garcia et al., 1989). These results are summarised in Table 9.3. Each taxon has a distinct essential oil profile based either on qualitative or quantitative composition. The oil of L. angustifolia spp. *pyranaica* has a much higher percentage of borneol and camphor than the commercially produced lavender oils, whereas the oil of the Spanish sample of L. latifolia contains average amounts of the major constituents of spike lavender oil: linalol, borneol and camphor. However,

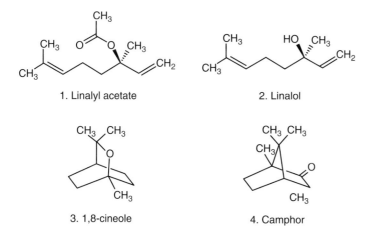

Figure 9.1 *Lavandula* monoterpenes.

Table 9.1 The major constituents* of lavender, lavandin and spike lavender oils

Essential oil	Percentage of essential oil in		
	Lavender oil	Lavandin oil	Spike lavender oil
Linalyl acetate	12–54	19–26	0–1.5
Linalol	10–50	20–23	26–44
cis- and/or trans-Ocimene	1.0–17	1.0–3.0	0–0.3
Lavandulol and acetate	0.1–14	0.5–0.8	0.2–1.5
1,8-Cineole	2.1–3.0[a]	10	25–36
Camphor	0–0.2[b]	12	5.3–14.3
α- and β-Pinene	0.02–0.3	0.6–0.9	1.6–3.6
Borneol	1.0–4.0[c]	2.9–3.7	0.8–4.9
Caryophyllene and/or its oxide	3.0–8.0	2.7–6.0	0.1–0.3
Myrcene	0.4–1.3	1.2–1.5	0.2–0.4
Farnesene	Trace	1.1	0.2–0.3
Germacrene D	0.2–0.9	1.0–1.2	—
Camphene	0.1–0.2	0.3–0.6	0.2–1.8
Limonene	0.2–0.4	0.9–1.5	1.0–2.2

Source: Data taken from Boelens (1995).

Notes
* Major constituents are those which make up >1% of the total oil of one or more of three commercial oils.
a a major constituent of some Bulgarian samples (Ognyanov, 1984).
b reported from a Chinese sample (Cu, 1988).
c a major constituent of some Russian samples (Karetnikova et al., 1969).

Table 9.2 A comparison of the main constituents of lavender oil from different countries

Country of origin	Percentage of major essential oil constituents					Reference
	Ocimenes	Linalol (2)	Linalyl acetate (1)	Lavandulol and acetate	Others	
Bulgaria	6.8–7.7	30–34	35–38	nd	1,8-cineole; 2.0–3.0	a
China	nd	24–36	29–36	1.6–1.7	camphor; 1.0–4.0	b
France	1.2–17	18–50	27–54	0.3–13	camphor; 1.0–2.0	c–h
India	nd	10	45	0.1	citronellol; 10	i
Italy	nd	30–50	12–42	0.1–1.8	α-terpineol; 1.0–8.0	i
Russia	5.0	10–37	31	14	borneol; 1.4–5.0 iso-borneol	j, k
Spain	3.5	31	29	7.0	terpineols; 3.0	l

Source: Data taken from Boelens (1995).

Notes
a Ognyanov (1984).
b Cu (1988).
c Hoffman (1979).
d Touche et al. (1981).
e Lalande (1984).
f Jean et al. (1991).
g Bernard et al. (1989).
h Naef and Morris (1992).
i Tajuddin et al. (1983).
j Karetnikova et al. (1969).
k Rabotyagov and Akimov (1990) and l = Boelens (1995).
nd Not determined.

Table 9.3 The major essential oils of some *L. latifolia* and some wild *Lavandula* taxa from Spain and Portugal

Lavandula taxon	Essential oil as a percentage of the total oil							
	Linalol (2)	Borneol	Camphor (4)	1,8-Cineole (3)	β-Pinene	Carvacrol	Bisabolene	Fenchone
Section *Lavandula*								
L. angustifolia Mill. spp. pyranaica (DC) Masclens	38	20	8.0	—	—	—	—	—
L. latifolia Medic.	34	—	11	31	—	—	—	—
Section *Dentata*								
L. dentata L.	—	—	—	55	12	—	—	—
Section *Stoechas*								
L. multifida L.	—	—	—	—	—	25	23	—
L. stoechas L. spp. *stoechas*	—	—	23	—	—	—	—	42
L. stoechas L. spp. *pedunculata* (Mill.) Samp. ex. Rozeira	—	—	24	—	—	—	—	20
L. stoechas L. spp. *sampaioana* Rozeira	—	—	38	—	—	—	—	20
L. stoechas L. spp. *luisieri* (Rozeira) Rozeira[a]	—	—	—	22	—	—	—	—
L. viridis L'Hér	—	—	13	42	—	—	—	—

Source: Data taken from Garcia Vallejo et al. (1989).

Notes
a A Portuguese endemic with 25% unknown esters present in the oil.
— Not detected as a major constituent.

camphor is a major component of the oil of three of the subspecies of *L. stoechas* but these taxa are distinguished by the presence of a high concentration of fenchone. The oil of the other subspecies of *L. stoechas*, spp. *luisieri* from Portugal, is chemically distinct in producing only 1,8-cineole and some unidentified esters as the main constituents. Similarly, *L. multifida* is unique in producing large amounts of carvacrol and bisabolene, while *L. dentata* is distinguished by the high percentage of 1,8-cineole and β-pinene in its oil. 1,8-cineole together with camphor is also a main component of the *L. viridis* oil. Linalol makes an important contribution to the odour of lavender, lavandin and spike lavender oils, but is not a major constituent of taxa belonging to sections other than section *Lavandula*. In a later study of four wild populations of *L. stoechas* from Crete, Skoula et al. (1996) found the main essential oil constituents to be 1,8-cineole, fenchone, camphor, α-pinene and myrtenyl acetate. However, they found significant variation in the composition of the essential oils of the different populations and between leaf and inflorescence. Thus, while fenchone was a major oil component of all four populations, three were also rich in camphor and the fourth in 1,8-cineole. In all these accessions, inflorescences of *L. stoechas* contained mostly fenchone, α-pinene and myrtenyl acetate while the leaves synthesised mostly 1,8-cineole and camphor. The inflorescences also produced larger amounts of essential oil than the leaves.

As with many other essential oil containing plants, investigations have centred on the attempted production of these constituents in plant cell culture. Tissue culture of *L. angustifolia* led to the establishment of four callus lines, which contained up to 20 per cent of the mono- and sesquiterpenoids found in the leaves or inflorescences of the original plant (Banthorpe et al., 1995).

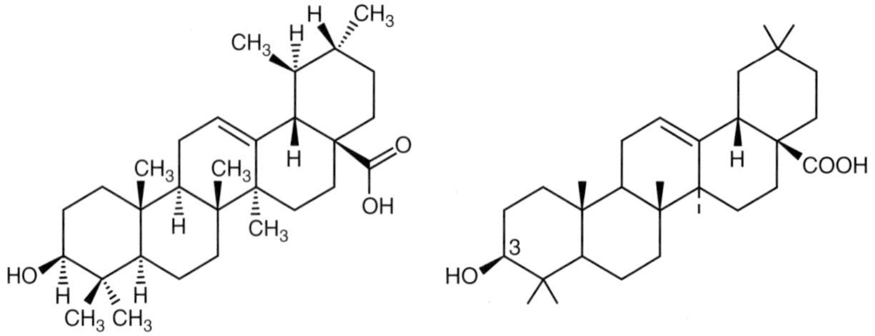

Figure 9.2 Lavandula triterpenoids.

Pretreatment of the callus by pulse feeding (flooding) with mevalonate to stimulate terpenoid accumulation had little effect but the derived cell suspensions stored monoterpenoids at concentrations 10^3-fold those of the controls and this could be increased by a further 10^2-fold by using a two-phase medium. Cell-free extracts of both productive and non-productive callus lines converted [^{14}C]isopentenyl pyrophosphate into predominantly 2E,6E- and 2Z,6E-farnesols, humulene and caryophyllene, all sesquiterpenoids, rather than the expected monoterpenoids accumulated by the parent callus. Pre-incubation of the extracts with NADPH or NADP$^+$ increased the accumulation of sesquiterpenoids by up to 20-fold possibly as a consequence of the redox conversion of 2E,6E into its 2Z,6E-isomer,which is needed for cyclisation but this has not been confirmed.

Triterpenoids

Ursolic acid (3-β-hydroxy-12-ursen-28-oic acid (5)) has been reported as 0.7 per cent of the dried leaf weight of *L. angustifolia* (named as *L. spica* L. in the reference: Le Men and Pourrrat, 1953 and as *L. vera* DC by Hegnauer, 1966) and of 1.0–1.9 per cent of the dried leaf weight of *L. latifolia* (Le Men and Pourrat, 1953; Brieskorn *et al.*, 1953) together with 0.5 per cent of oleanolic acid (6) in one sample (Brieskorn *et al.*, 1953). Ursolic and oleanolic acids (Figure 9.2) have also been reported from a Bulgarian sample of *L. angustifolia* (Papanov *et al.*, 1984), together with 3-epiursolic acid (the 3α-isomer). Papanov *et al.* (1992) later identified six minor triterpenoid constituents from the same plant including the new compound, formylursolic acid and the five known compounds: ursolic acid lactone, 3-oxo-12-ursene-28-oic acid, pomilic acid (3,19-dihydroxy-12-ursen-28-oic acid), betulin (lup-20(29)-ene-3β,28-diol) and betulinic acid (3β-hydroxylup-20(29)-ene-28-oic acid). In *L. pedunculata*, a further member of section *Stoechas*, the triterpene alcohols, α- and β-amyrin, uvaol and the triterpenes, ursolic and micromeric acids have been characterised (de Pascual Teresa *et al.*, 1978). From *L. canariensis* (section *Pterostoechas*), Breton Funes amd Jaudenes Ruiz de Atauri (1986) reported 2α-hydroxyursolic acid (2α,3β-dihydroxy-12-ursen-28-oic acid) and 23-hydroxytormentic acid (2α,3α,19α, 23β-tetrahydroxy-12-ursen-28-oic acid) for the first time in the Labiatae together with α-and β-amyrin, ursolic and oleanolic acids. In a sample of *L. dentata* (section *Dentata*) from Saudi Arabia, Khalil *et al.* (1979) isolated ursolic and micromeric acids. Ursolic and oleanolic acids are characteristic triterpenoid components of the Labiatae (see data in Hegnauer, 1966) and like α- and β-amyrin, betulin and betulinic acid are of widespread occurence in the angiosperms.

Flavonoids

Leaf flavonoids

The major leaf flavonoid constituents of *Lavandula* species are flavone glycosides. Five different subclasses are represented, that is, simple flavone glycosides, flavone C-glycosides, 6-hydroxyflavone 7-glycosides, and 8-hydroxyflavone 7- and 8-glycosides, and their distribution among the taxa shows some taxonomic significance. This complex flavonoid profile is typical of members of the Labiatae and especially the frequent occurrence of flavones with extra hydroxylation at the 6- and 8-positions. Most of the earliest flavonoid studies of *Lavandula* were restricted to detailed analyses of one or two species. Thus, in a study of *L. dentata*, Ferreres *et al.* (1986) reported apigenin, genkwanin (apigenin 7-methyl ether), luteolin, apigenin 7-glucoside, luteolin 7-glucoside and 7-rutinoside, and the flavone C-glycosides, vitexin and vicenin-2. In a general survey of the Labiatae for flavones with extra substitution in the A-ring, Tomás-Barberan *et al.* (1988), examined seven *Lavandula* species from Spain. They recorded 6-hydroxy- and 6-methoxy-substituted flavones only from *L. multifida*. In the remaining species, *L. stoechas*, *L. viridis*, *L. dentata*, *L. angustifolia*, *L. latifolia* and *L.lanata*, no flavonoids with extra 6- or 8-hydroxylation or 6-methoxylation were detected. The major leaf flavonoids of *L. stoechas* were identified as

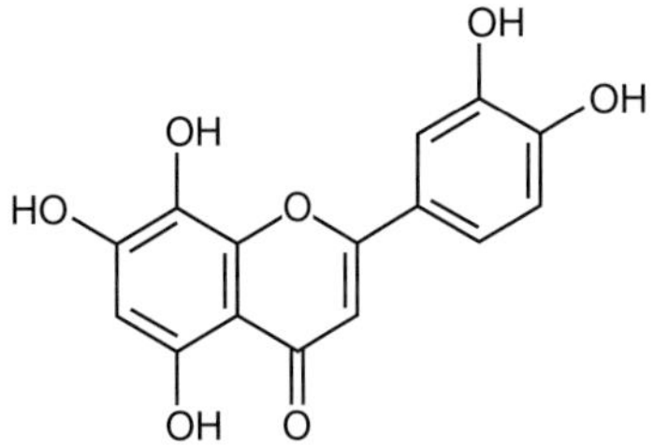

7. Hypolaetin

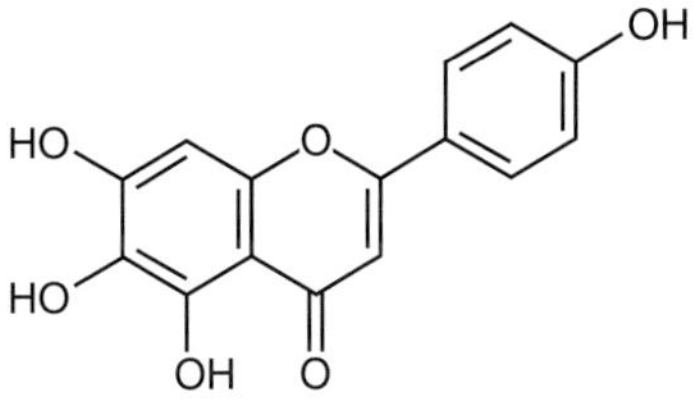

8. Scutellarein

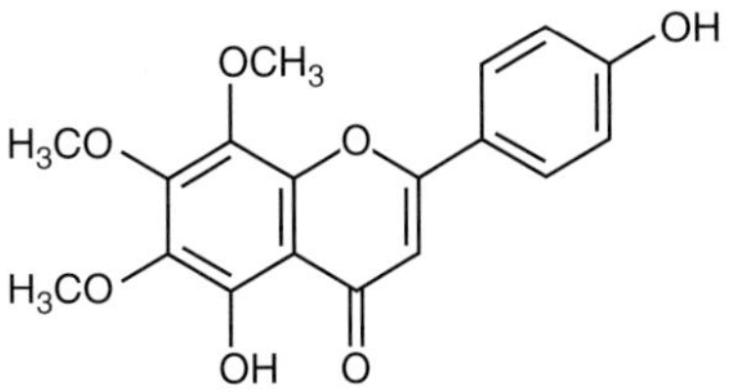

26. Xanthomicrol

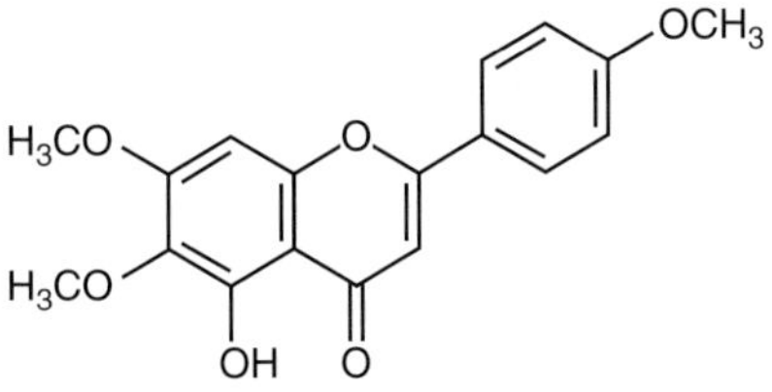

27. Salvigenin

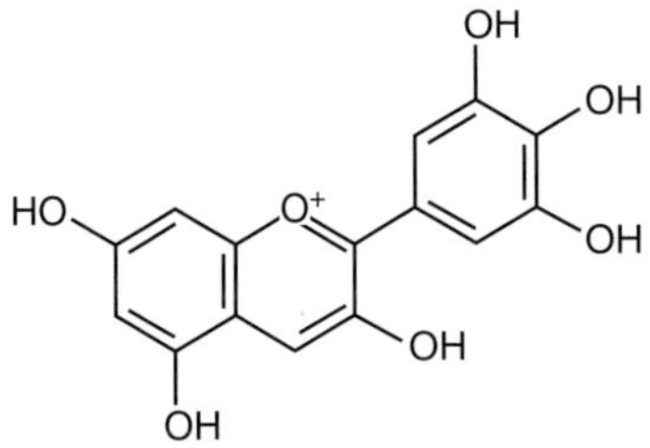

28. Delphinidin

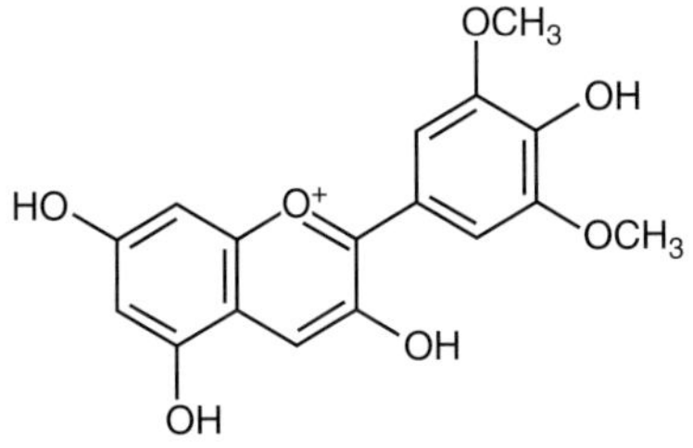

29. Malvidin

Figure 9.3 *Lavandula* flavonoids.

apigenin 7-glucoside, luteolin, luteolin 7-glucoside and 7-rutinoside (Xaver and Andary, 1988). More recently, the 7-glucoside and 7-glucuronide of hypolaetin (8-hydroxyluteolin (7)) (Figure 9.3) and the 8-glucuronides of hypolaetin 4′-methyl ether and scutellarein (6-hydroxyapigenin (8)) have been reported from *L. pubescens* and *L. coronipifolia* both members of section *Pterostoechas* (El Garf *et al.*, 1999). Flavone glycosides with 8-hydroxylation are rare plant constituents, which have proved useful as taxonomic markers at sectional level in other Labiate genera. In *Lavandula* they were found to be characteristic components of sections *Pterostoechas, Subnuda* and *Chaetostachys* and to be absent from sections *Stoechas, Dentata* and *Lavandula* (Upson *et al.*, 2000) (Table 9.4). In this survey of twenty *Lavandula* species, subspecies and varieties, fifteen flavonoid glycosides from five different flavone subclasses were detected and are listed in Table 9.5. It includes new reports of luteolin 7,4′-diglucuronide and luteolin 7-glucoside-4′-glucuronide as trace constituents in all the taxa analysed from section *Lavandula*: three subspecies of *L. angustifolium, L. lanata* and *L. latifolia*. However, the ordinary flavone glycosides, apigenin and luteolin 7-glucosides and 7-glucuronides are the main flavonoid components of section *Lavandula* and also of sections *Stoechas* and *Dentata*. A chrysoeriol 7-glycoside (with 3′-O-methylation) was detected in two taxa of section *Stoechas* and as a trace constituent in all the taxa examined in section *Lavandula*. The flavone C-glycoside, vitexin was found in trace amount in the three taxa of section *Stoechas* and as a major constituent of *L. dentata* (section *Dentata*). A preliminary survey of fourteen further *Lavandula* species from all the sections except *Lavandula* (Table 9.6) confirmed the sectional differences, which are summarised in Table 9.7. External flavonoids were detected in trace amount on the leaf surface of all the taxa of sections *Stoechas, Dentata* and *Lavandula* and in four taxa from the remaining sections (Upson *et al.*, 2000). The compounds identified were apigenin, genkwanin (apigenin 7-methyl ether), xanthomicrol (5,4′-dihydroxy-6,7,8-trimethoxyflavone (26)) and salvigenin (scutellarein 6,7,4′-trimethyl ether (27)) (Table 9.4).

Floral flavonoids

As part of a survey of anthocyanin pigments in forty-nine species of the family Labiatae, the floral constituents of *L. dentata* and *L. stoechas* were characterised, with the results shown in Table 9.8 (Saito and Harborne, 1992). The eight pigments identified all occur variously in other species of the Labiatae. So it is clear that the pattern in *Lavandula* conforms to that found in the family as a whole. As expected in petals where the colour is purple-violet or violet, the main pigments are based on delphinidin (28) and malvidin (29). All pigments without exception have a *p*-coumaric acid residue attached to the 3-glucose and most of them have additional acylation through malonic acid, usually attached to the 5-glucose. Both these acylations contribute to the stabilisation of the anthocyanin pigments and may also contribute to the colour properties of the petals and bracts.

Hydroxycinnamic acids and coumarins

Two hydroxycinnamic acid esters, rosmarinic acid (30) and chlorogenic acid (31), are regularly present in the leaves of *Lavandula* species (Table 9.9). Rosmarinic acid was detected in leaves of five out of six species, while chlorogenic acid occurs in four out of the six (Jensen, 2000). Both these acids are widespread throughout the Labiatae family.

A third acid, the 2-(3,4-dihydroxyphenyl)ethenyl ester (32) of caffeic acid has a more striking occurrence in callus tissue of *L. angustifolia* (Banthorpe *et al.*, 1985) (Figure 9.4). It is remarkable in not being detectable in the plant from which the callus is derived, although it has

Table 9.4 Flavones* detected in *Lavandula* species, subspecies and varieties

Taxon	Simple flavone glycosides			6-Hydroxyflavone glycosides		Flavone C-glycosides	8-Hydroxyflavone 8-glycosides			8-Hydroxyflavone 7-glycosides	
	9–12	13	14 and 15	16	17	18	19	20	21	22	23
Section *stoechas*											
L. stoechas L. spp. *stoechas*[ab]	+	+	–	–	–	(+)	–	–	–	–	–
L. stoechas L. spp. *luiseiri* (Rozeira) Rozeira[abd]	+	–	–	–	–	(+)	–	–	–	–	–
L. viridis L'Hér.[abd]	+	+	–	–	–	(+)	–	–	–	–	–
Section *dentata*											
L. dentata L. var. *dentata*[ab]	+	–	–	(+)	+	+	–	–	–	–	–
Section *Lavandula*											
L. angustifolia Mill. spp. *angustifolia*[ab]	+	(+)	(+)	–	–	–	–	–	–	–	–
L. angustifolia Mill. spp. *pyrenaica* (DC.) Masclans[ab]	+	(+)	(+)	–	–	–	–	–	–	–	–
L. angustifolia Mill. spp. *delphinensis* (Jord) O. Bolòs and Vigo[ab]	+	(+)	(+)	–	–	–	–	–	–	–	–
L. lanata Boiss.[ab]	+	(+)	(+)	–	–	–	–	–	–	–	–
L. latifolia Medic.[ab]	+	(+)	(+)	–	–	–	–	–	–	–	–
Section *Pterostoechas*											
L. canariensis Mill.	(+)	–	–	–	–	–	+	–	+	(+)	+
L. coronopifolia Poir	(+)	–	–	–	–	–	+	–	+	+	+
L. maroccana Murb.	(+)	–	–	–	–	–	+	(+)	+	+	+

Table 9.4 (Continued)

Taxon	Simple flavone glycosides			6-Hydroxyflavone glycosides		Flavone C-glycosides	8-Hydroxyflavone 8-glycosides				8-Hydroxyflavone 7-glycosides	
	9–12	13	14 and 15	16	17	18	19	20	21	22	23
L. mairei Humbert var. *mairei*[c]	(+)	(+)	–	–	–	–	–	–	–	+	+
L. minutifolia Bolle var. *minutifolia*[acd]	–	–	–	–	–	–	+	–	+	(+)	+
L. multifida L.	(+)	–	–	–	–	–	+	–	+	+	(+)
L. rotundifolia Benth.	(+)	–	–	–	–	–	+	+	+	+	+
Section *Subnuda*											
L. aristibracteata A. G. Mill.[d]	–	–	–	–	–	–	+	+	(+)	(+)	(+)
L. dhofarensis A. G. Mill. spp. *dhofarensis*	–	–	–	–	–	–	+	+	–	(+)	–
L. subnuda Benth.	–	–	–	–	–	–	+	+	+	(+)	(+)
Section *Chaetostachys*											
L. bipinnata Kuntze[d]	(+)	–	–	–	–	–	+	+	(+)	(+)	(+)

Sources: Data taken from Upson *et al.* (2000).

Notes
* See Table 9.5 for key to structure numbers.
a apigenin.
b genkwanin.
c xanthomicrol.
d salvigenin detected on the leaf surface in these plants.

Table 9.5 Some flavones detected in *Lavandula* species

Subclass	Structures
Simple flavone glycosides	Luteolin 7-glucoside (9) and 7-glucuronide (10)
	Apigenin 7-glucoside (11) and 7-glucuronide (12)
	Chrysoeriol 7-glycoside (13)
	Luteolin 7,4'-diglucuronide (14)
	Luteolin 7-glucoside-4'glucuronide (15)
6-Hydroxyflavone glycosides	6-Hydroxyluteolin 7-glycoside (16)
	Scutellarein 7-glycoside (17)
Flavone C-glycosides	Apigenin 8-C-glucoside (vitexin) (18)
8-Hydroxyflavone 8-glycosides	Hypolaetin 8-glucuronide (19)
	Hypolaetin 4'-methyl ether 8-glucuronide (20)
	Isoscutellarein 8-glucuronide (21)
8-Hydroxyflavone 7-glycosides	Hypolaetin 7-glucoside (22)
	Scutellarein 7-glycoside (23)
External flavones[a]	Apigenin (24)
	Genkwanin (25)
	Xanthomicrol (26)
	Salvigenin (27)

Source: Data taken from Upson et al. (2000).

Note
a Isolated from the leaf surface by washing with acetone; other compounds occur internally and are extracted into 80% methanol.

Table 9.6 The distribution of the major flavone glycosides* in *Lavandula* sections *Stoechas*, *Dentata*, *Pterostoechas* and *Subnuda*

Taxon	Simple flavone glycosides		8-Hydroxyflavone glycosides	
	9	10–12	19	22
Section *Stoechas*				
L. stoechas L. spp. atlantica Braun-Blanq.	+	+	−	−
L. stoechas L. spp. cariensis (Boiss.) Rozeira	+	+	−	−
L. stoechas L. spp. lusitanica (Chaytor) Rozeira	+	+	−	−
L. stoechas L. spp. pedunculata (Mill.) Samp. ex. Rozeira	+	+	−	−
L. stoechas L. spp. sampaiana Rozeira	+	+	−	−
Section *Dentata*				
L. dentata L. var. candicans Batt.	+	+	−	−
Section *Pterostoechas*				
L. antinae Maire	(+)	−	+	+
L. buchii Webb & Berthel var. buchii	(+)	−	+	+
L. buchii Webb & Berthel var. gracile León	(+)	−	+	+
L. mairei Humbert var. antiatlantica Maire	(+)	−	−	+
L. minutoli Bolle var. tempupina Svent.	−	−	+	+
L. pinnata L. f.	(+)	−	+	+
L. tenuisecta Coss. ex Ball	(+)	−	+	+
Section *Subnuda*				
L. nimmoi Benth.	(+)	−	+	+

Sources: Data taken from Upson et al. (2000).

Note
* See Table 9.5 for key to structures.

Table 9.7 The distribution of different flavone subclasses* in the six sections of the genus *Lavandula*

Lavandula section	Flavone 7-O-monoglycosides	Flavone di-O-glycosides	6-hydroxy-flavone 7-O-glycosides	Flavone C-glycosides	8-hydroxy-flavone 8-O-glycosides	8-hydroxy-flavone 7-O-glycosides	External flavones
Stoechas	+	−	−	+	−	−	+
Dentata	+	−	+	+	−	−	+
Lavandula	+	+	−	−	−	−	+
Pterostoechas	−	−	−	−	+	+	+
Subnudae	−	−	−	−	+	+	+
Chaetostachys	−	−	−	−	+	+	+

Source: Data taken from Upson et al. (2000).

Note
* For key to structures see Table 9.5.

Table 9.8 Anthocyanins of *Lavandula* flowers: pigments present

L. dentata L.
Petals
Malvidin 3-(6″-*p*-coumaryl glucoside)-5-glucoside
Malvidin 3-(6″-*p*-coumarylglucoside)-5-(6‴-malonylglucoside)
Delphinidin 3-(6″-*p*-coumarylglucoside)-5-(4‴,6‴-dimalonylglucoside)

L. stoechas L. spp. *pedunculata*
Petals
Delphinidin 3-(6″-*p*-coumarylglucoside)-5-(6‴-malonylglucoside)
Delphinidin 3-(6″-*p*-coumarylglucoside)-5-glucoside
Cyanidin 3′-(6″-*p*-coumarylglucoside)-5-(6‴-malonylglucoside)

Bracts
Malvidin 3-(6″-*p*-coumarylglucoside)-5-(6‴-malonylglucoside)
Malvidin 3-(6″-*p*-coumarylglucoside)-5-(dimalonylglucoside)

Table 9.9 Distribution of hydroxycinnamic acid esters in the genus *Lavandula*

Lavandula species	Presence of	
	Rosmarinic acid	Chlorogenic acid
L. lanata	+	Trace
L. minutolii	−	+
L. multifida	−	+
L. canariensis	+	−
L. pedunculata	+	−
L. pinnata	+	+
L. stoechas	+	−

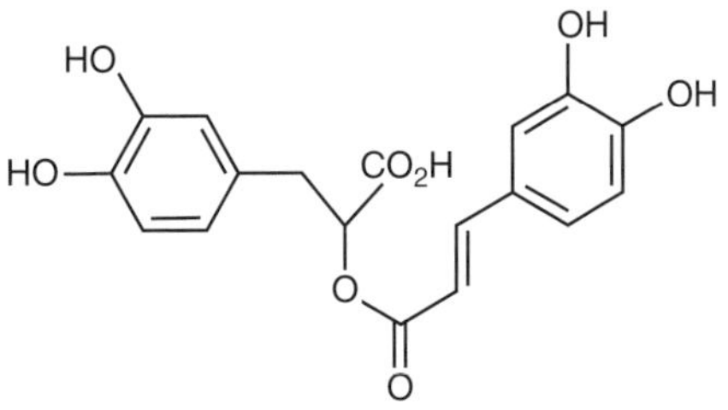

30. Rosmarinic acid

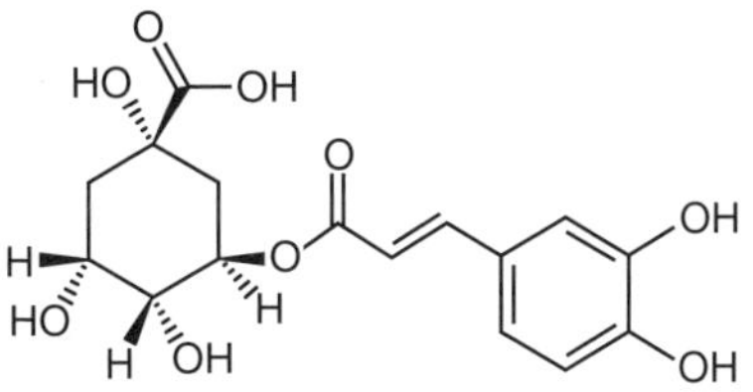

31. Chlorogenic acid

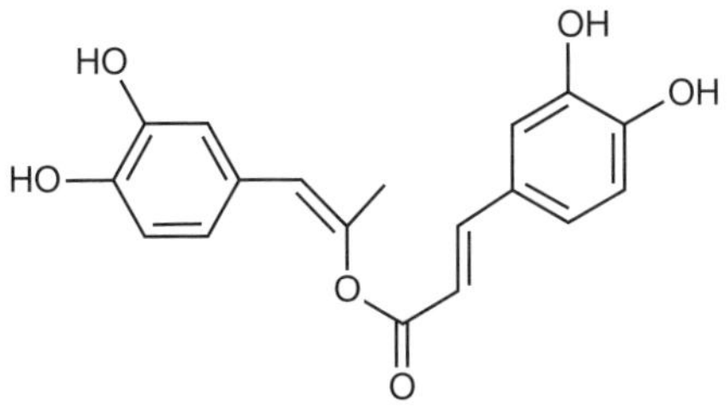

32. Caffeic acid 2-(3,4-dihydroxyphenyl)ethenyl ester

Figure 9.4 *Lavandula* hydroxycinnamic acid esters.

been recorded in several other genera of Labiatae in low amount (Banthorpe et al., 1989). This enol ester (32) exists in *cis*- and *trans*-forms. It forms a deep blue pigment with Fe^{2+} and this frequently happens in the callus culture, which then appears blue as a result. The enol ester is fungitoxic and at the 1 μg level is toxic to the fungal pathogen *Cladosporium herbarum*.

Two coumarins, coumarin itself and 7-methoxycoumarin (herniarin), have been detected in the volatile oil fraction of several *Lavandula* species. They appear to occur also in bound form as glucosides, since they have been detected after emulsin hydrolysis of *L. angustifolia* extracts. The free coumarin content of *L. latifolia* leaf oil is 1.07 per cent for herniarin and 9.04 per cent for coumarin. Analysis of *L. dentata* extracts has yielded some umbelliferone (7-hydroxycoumarin), together with herniarin and coumarin (Khalil et al., 1979).

Conclusion

The present review of the phytochemistry of *Lavandula* is based chiefly on the chemical analysis of the commercially important species and their cultivated forms, as grown in Europe. However, there are over thirty species in the genus, some of which are native to Somalia and India. Our present knowledge of the phytochemical potential of these plants is therefore somewhat limited. Undoubtedly, wider chemical studies in the genus, particularly of the essential oil components, could be rewarding.

References

Banthorpe, D. V., Bates, M. J. and Ireland, M. J. (1995). Stimulation of accumulation of terpenoids by cell suspensions of *Lavandula angustifolia* following pre-treatment of parent cells. *Phytochemistry*, 40, 83–7.

Banthorpe, D. V., Bilyard, H. J. and Brown, G. D. (1989). Enol esters of caffeic acid in several genera of Labiatae. *Phytochemistry*, 28, 2109–13.

Banthorpe, D. V., Bilyard, H. J. and Watson, D. G. (1985). Pigment formation by callus of *Lavandula angustifolia*. *Phytochemistry*, 24, 2677–80.

Bernard, T., Perineau, F., Delmas, M. and Gaset, A. (1989). Extraction of essential oils. Part III. Two-stage production of the oil of *Lavandula angustifolia* Mill. *J. Essent. Oil Res.*, 1, 261–7.

Boelens, M. H. (1995). Chemical and sensory evaluation of *Lavandula* oils. *Perf. Flav.*, 20, 23–51.

Breton Funes, J. L. and Jaudenes Ruiz de Atauri, I. (1986). Triterpenes from *Lavandula canariensis* Mill. *J. Nat. Prod.*, 49, 937.

Brieskorn, C. H., Eberhardt, K. H. and Briner, M. (1953). Biogenetische Zusammenhänge Zwischen Oxytriterpensäuren und ätherischem Öl bei einigen pharmazeutisch wichtigen Labiaten. *Archiv. Der Pharmazie*, 286, 501–6.

Cu, J.-Q. (1988). Yunnan – the kingdom of essential oil plants. In: B. M. Lawrence, B. D. Mookerjee and B.J. Willis (eds), *Flavors and Fragrances*, Proceedings of the 10th International Congress of Essential Oils, Fragrances and Flavors, Washington DC, USA, 16–20 Nov., 1986, Elsevier Science Publishers BV, Amsterdam, pp. 231–41.

El-Garf, I., Grayer, R. J., Kite, G. C. and Veitch, N. C. (1999). Hypolaetin 8-O-glucuronide and related flavonoids from *Lavandula coronopifolia* and *L. pubescens*. *Biochemical Systematics and Ecology*, 27, 843–6.

Ferreres, F., Barberán, F. A. T. and Tomás, F. (1986). Flavonoids from *Lavandula dentata*. *Fitoterapia*, 57, 199–200.

Garcia Vallejo, M. C., Garcia Vallejo, I. and Negueruela, A. (1989). Essential oils of genus *Lavandula* L. in Spain. *Proceedings ICEOFF 1989 New Delhi*, 4, 15–26.

Hegnauer, R. (1966). Labiatae. In: *Chemotaxonomie der Pflanzen*, 4, Birkhäuser Verlag, Basel and Stuttgart, pp. 289–346.

Hoffmann, W. (1979). Lavendel-Inhaltstoffe und ausgewachte Synthesen. *Seifen-Oele-Fette-Wachse*, 105, 287–91.

Jean, F. I. (1991). Extraction au four micro-ondes des diverses plantes cultivees et spontanitees. *Revisita Ital EPPOS (Numero Speciale)*, 504–10.

Karetnikova, A. I., Kustova, S. D., Fedulova, I. and Karpova, T. I. (1969). Composition of the alcohol part of lavender oil. *Maslo-Zhir. Prom.*, 35, 23–5.

Khalil, A. M., Ashy, M. A., El-Tawil, B. A. H. and Tawfiq, N. I. (1979). Constituents of local plants Part 5: The coumarin and triterpenoid constituents of *Lavandula dentata* L. plant. *Pharmazie*, 34, 564.

Lalande, B. (1984). Lavender, lavandin and other French oils. *Perf. Flav.*, 8, 117–21.

Lawrence, B. M. (1992). Chemical components of Labiatae oils and their exploitation. In: R. M. Harley and T. Reynolds (eds), *Advances in Labiate Science*, pp. 399–436. Royal Botanic Gardens, Kew.

Le Men, J. and Pourrat, H. (1953). Distribution of ursolic acid among the Labiatae. *Ann. Pharm. Franç.*, 11, 190–2.

Naef, R. and Morris, A. F. (1992). Lavender–lavandin-A comparison. *Revista Ital. EPPOS (Numero Speciale)*, 364–77.

Ognyanov, I. (1984). Bulgarian lavender oil and Bulgarian lavandin oil. *Perf. Flav.*, 8, 29–41.

Papanov, G., Bozov, P. and Malakov, P. (1992). Triterpenoids from *Lavandula spica*. *Phytochemistry*, 31, 1424–6.

Papanov, G. Y., Malakov, P. and Tomova, K. (1984). Triterpenoids from *Lavandula vera*. *Nauchni Tr.-Plovdivski*, 22, 213–20.

Papanov, G. Y., Gacs-Baitz, E. and Malakov, P. Y. (1985). 2-N-Phenylaminonaphthalene from *Lavandula vera*. *Phytochemistry*, 24, 3045–6.

de Pascual Teresa, J., Urones, J. G., Sanchez, A. and Basabe, P. (1978). *Lavandula pedunculata* components II triterpenes. *An. Quim. (Madrid)*, 74, 675–7.

Pedersen, J. A. (2000). Taxonomic implications of phenolics in the Lamiaceae determined by ESR spectroscopy. *Biochemical Systematics and Ecology*, 28, 229–53.

Prager, M. J. and Miskiewicz, M. A. (1979). Gas chromatographic-mass spectrophotometric analysis, identification and detection of adulteration of lavender, lavandin and spike lavender oils. *JAOAC*, 63, 1231–8.

Rabotyagov, V. D. and Akimov, Y. A. (1990). Inheritance of oil content and composition in tetra- and sesquidiploids of lavender. *Genetika (Moscow)*, 26, 283–92.

Saito, N. and Harborne, J. B. (1992). Correlations between anthocyanin type, pollinator and flower colour in the Labiatae. *Phytochemistry*, 31, 3009–15.

Skoula, M., Abidi, C. and Kokkabu, E. (1996). Essential oil variation of *Lavandula stoechas* L. ssp. *Stoechas* growing wild in Crete (Greece). *Biochemical Systematics Ecology*, 24, 255–60.

Tajuddin, S. A. S., Nigam, M. C. and Husain, A. (1983). Production of lavender oil in Kashmir Valley. *Indian Perfum.*, 27, 56–9.

Tomás-Barberán, F. A., Grayer-Barkmeijer, R. J., Gil, M. I. and Harborne, J. B. (1988). Distribution of 6-hydroxy, 6-methoxy- and 8-hydroxyflavone glycosides in the Labiatae, the Scrophulariaceae and related families. *Phytochemistry*, 27, 2631–45.

Touche, J., Derbesy, M. and Llinas, J. R. (1981). Maillettes et Lavandes fines françaises. *Revisita Ital. EPPOS*, 58, 320–3.

Upson, T. M., Grayer, R. J., Greenham, J. R., Williams, C. A., Al-Ghamdi, F. and Chen, F.-H. (2000). Leaf flavonoids as systematic characters in the genera *Lavandula* and *Sabaudia*. *Biochemical Systematics and Ecology*, 28, 991–1007.

Xaver, H. and Andary, C. (1988). Polyphenols of *Lavandula stoechas* L. *Bulletin Liaison Group Polyphenols*, 133, 624–6.

10 Distillation of the lavender type oils
Theory and practice

E.F.K. Denny

Preliminary review

Lavender plant characters for distillation

In the *Lavandula* genus, there are two plant species and one hybrid which produce aromatic oils of interest to the perfumery industry. In all three groups there are many thousands of different genotypes yielding individual oils which differ widely in commercial value. The oil is secreted in small globular glands between the ribs of the flowers' hairy calices (Figure 10.1). Traces of immature oil may also be found elsewhere in the plant material but if the flowers are ripe these traces do not affect the perfume; but with under-ripe flowers they may contribute to a 'green' note in the oil.

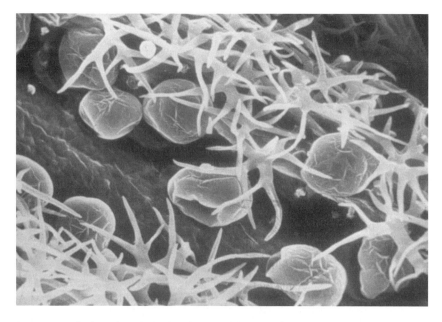

Figure 10.1 The lavender oil glands: oil glands on a calyx of ripe lavender.
Source: Electron microscopy by kindness of the John Innes Institute, Norwich, UK.
Note: Hairs in abundance.

Distillation of the lavender type oils 101

These lavender type oils can be extracted from the flowers by steam distillation. Being the products of distillation they are, by definition, 'essences' and are called 'essential oils'. For the purposes of distillation they belong to the class of *superficial* oils from *absorptive* surfaces because they are wholly secreted on the surface of the herb and the calices' hairy character gives them an absorptive surface. The oil's distillation proceeds differently from that for *subcutaneous* oils or those from less absorptive herbs.

Harvesting for distilling

In normal summer conditions the lavender is ready for distillation when about half the petals on an average flower head have wilted. Flowers that are harvested earlier than this may be sappy and require wilting in the sun if simple stills are to return an oil which attains the potential yield and quality of the particular cultivar. Irrespective of the method of harvesting, about 6 to 10 cm of bare stalk should be cut with the flower heads. This assists the steam's even distribution through the herb charge in the still.

Traditionally, the flowers were cut by hand with sickles and laid back on the top of the parent plant to wilt in the sun. In due course the flowers were picked up and carted to the distillery, where they were packed into stationary stills by hand.

In modern systems, mechanical harvesters load the flowers into special trailers which themselves become the distilling vessels (Figure 10.2). Handling loose flowers at the distillery becomes unnecessary and would actually be harmful. Mechanical harvesters usually break some of the superficial glands and the oil will evaporate and be lost from these if the herb is subsequently exposed to the open air. So the flowers must be fully ripe before mechanical harvesting is started, because they cannot be laid out in the sun to wilt.

Figure 10.2 Harvesting at about two tons lavender flowers per hour.
Note: Filling still cylinder in the field.

All heat losing surfaces of the still base, still cylinder and still lid must be insulated with at least a double thickness of good quality carpet under-felt.

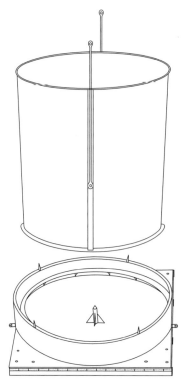

A. Lifting rope to lifting beam
B. Top flange for fitting with still lid
C. Tipping pivot with rope on dead eye
D. Cylinder wall of mild steel plate
E. Steel strap of suitable strength see (J) below
F. Cross brace of inverted angle iron
G. Rotation pivot bush to suit rotation spigot on trailer and/or locating pin in still base
H. Cylinder bottom plate with quadrants cut out for mesh
J. Tipping swivel spigot welded into steel strap (E)

Figure 10.3 Sketches illustrating the modern cylindrical still system which is the one best suited to lavender, although the American type box trailers are also good when used on flat ground.

However, this does not create a problem. If the boiler is big enough to enable the distillery to keep up with the harvester, the mechanized system is so rapid that harvesting can start later and finish sooner than is possible with the manual methods.

The distilling operation

In the past the best stills have been vertical cylinders with perforated grids to support the herb about 150 mm above the closed bottom. This creates a pressure equalizing chamber beneath the charge into which steam can be fed from a satellite boiler. Some of the modern trailer systems have retained the cylinder and grid principle but in an advanced form (Figure 10.3). Others, such as those copied from the American mint industry, are large rectangular boxes where the steam is introduced through an array of sparge pipes on the trailer floor.

The vertical height of the space above the grid or trailer floor, into which the flowers and stalk are packed, should be at least 1.3 m. It will be obvious that the upward travelling steam must pass sufficient herb to gather an economic quota of oil. Less obvious, but just as important, is the need for an adequate charge height to minimize the number of oil coated liquid cloud particles that escape from the still and are ultimately lost in the discarded distillate water.

The diameter of the still depends on the size of the boiler. For each 3 kg/min of steam supply available to the still, the charge's horizontal cross-section area should be close to 1 m^2.

If it is practical to do so, the herb should be tramped down firmly in the still to attain an even density of packing around 275 kg/m^3 and attention should be paid to ensuring that the herb is packed tightly against the still wall.

The still is closed by a steam-tight lid which often carries the outlet pipe. A flow of steam is then introduced at the bottom of the still so that it percolates upward through the charge. Starting at the bottom the steam condenses on the herb surfaces and, surrendering its latent heat, raises successive layers to boiling point. When the appropriate temperature reaches the top of the still, the moving steam will cause any oil that is exposed on the herb surface to start to vaporize. The on-coming steam will then drive a mixture of oil and water vapours off the top of the charge and from there it is led through a condenser. When they are reduced back to the liquid state, the oil and water are immiscible and separate spontaneously according to their densities.

Our key words are 'steam will cause oil to vaporize', because only oil that is actively vaporized inside the still will be recovered from the herb. So distillers need to understand the factors which promote this vaporization as well as those which impede it.

The theory of steam distillation

Properties of vapours

All vapours contain more energy than their parent liquids. So, to turn a liquid into a vapour we must apply energy to it in the form of heat. The actual amount of heat required to vaporize a unit mass of liquid without raising the temperature of the vapour above that of the liquid, is a precise characteristic of each individual compound and is called its 'latent heat of vaporization'. If a vapour is condensed back to liquid the energy of the molecules is reduced and the characteristic quantity of latent heat is given out.

Every liquid continually emits moving vapour molecules from its surface. If the vapour space were closed, the moving molecules would impinge on the enclosing walls and exert a pressure on them which increases with rising temperature. At each temperature, the magnitude of this so-called 'vapour pressure' is another precise characteristic of the particular liquid. When continuous heat is applied to a liquid its temperature will rise only until its vapour pressure becomes equal to the surrounding pressure, whatever that may be. Further heat will merely create more vapour. The liquid is then said to 'boil' and the temperature at which this occurs is its 'boiling point' under the prevailing pressure. If two immiscible liquids contribute molecules to the same vapour space, the mixture will boil when the temperature reaches the point where the sum of two the individual vapour pressures becomes equal to the ambient pressure.

Lavender stills work under virtual atmospheric pressure, the average of which at sea level is taken as being equal to that at the bottom of a column of mercury 760 mm tall. This is written as 760 mmHg or 1 atm abs, where abs stands for 'absolute'. At 99.6°C, linalyl acetate, a typical essential oil component, exerts a vapour pressure of 12 mmHg and water exerts 748 mmHg. So this mixture exerts $12 + 748 = 760$ mmHg at 99.6°C and it will boil away at that temperature under normal atmospheric pressure. On its own, linalyl acetate would boil at 220°C and tend to decompose. But in the presence of steam, this, and all other essential oils, can be boiled away from the surface of the herb at temperatures which are slightly below the boiling point of pure water.

Steam distillation for recovering herbaceous oils relies on bringing oil and water into contact at a temperature close to the boiling point of water. Then the addition of a small amount of vapour pressure from the oil will make the mixture's total pressure equal to the surrounding pressure. The oil and water mixture must then boil away from the herb at whatever rate the heat to support this vaporization can be applied to it.

This distilling operation looks so simple that, until recently, no one suspected that the factors which control it are quite complex. In particular, it was not realized that the way heat is applied to vaporize the oil needs special study and that a correct understanding of this factor makes the difference between an efficient operation and a very bad one.

The transference of heat

At the start of the process, when steam first contacts the lavender's oil glands, they burst and form numerous patches of homogeneous oil on the herb surface. At the perimeter of each virtually circular oil patch there is an oil-water *interface* (Figure 10.4) where oil is in contact with water from the steam that condensed on the herb to raise its temperature.

Inside the boiling still, the temperature distribution in each micro-system allows steam particles to condense on the herb surfaces in a way that transfers their surrendered latent heat to vaporize contiguous oil. Steam particles cannot condense directly onto the homogeneous oil patches from the broken glands because oil and water are immiscible. But they will readily condense into water surfaces in immediate contact with oil at the oil-water interface surrounding each surface patch, and the oil can be vaporized only from this perimeter. If that were just a thin circular line of contact, the area into which steam could usefully condense would be insignificant and very little oil would vaporize.

But if the herb surface is absorptive or hairy like that of lavender, oil and water will intermingle at the interface by capillary action. This vastly enlarges the target area into which we need the steam to condense and it ensures that the maximum amount of steam will be continually giving up heat to vaporize the oil. Then the radius of the surface patch shrinks rapidly as the oil is boiled away from its circumference, and the time taken by a given flow of steam to reduce the

(a)

(b)

(c)

Figure 10.4 Effects of steam moisture: oil and water on an absorptive surface. (a) Drops of oil (lavender) and water (dyed red) spreading on cartridge paper. (b) The spreading areas of oil and water meet and the two liquids intermingle. (c) A minute drop of water lodges on the area over which the oil drop in a and b above has spread. The intermingling of oil and water on this medium, at least, is clear. (See Color Plate XVII.)

radii of the surface oil patches to nil is the 'extraction' time for the oil (see 'Distillery throughput' on page 114).

Obviously we need to ensure that there is sufficient water on the herb surfaces to follow up the perimeters of the oil patches as they recede during distillation. Otherwise the vital heterogeneous mixing of oil and water at the interface will be much reduced and the oil's proportion of the distillate will decline prematurely. On the other hand, since this mixing of oil and water depends on capillary action, we must also make sure that there is never enough water present to flood the herb surface and overwhelm its absorptive capacity. This happy medium is achieved by adjusting the moisture content of the steam.

Matching the steam to the herb

Steam from ordinary commercial boilers consists of about 97 per cent dry saturated vapour and some 3 per cent by weight of liquid water particles in the form of cloud. Some of these cloud particles can lodge on herb surfaces inside the still and help to keep them moist. But if the steam is generated by a kettle in the bottom of the still itself, this 'wetness fraction' is much larger and the wetter steam can deposit much more water on the herb.

If the steam is generated under significant gauge pressure in a satellite boiler and expands to virtual atmospheric pressure on entering a lavender still, the surplus heat which is given out can vaporize some or all of the steam's normal wetness fraction. Then the relatively dry steam will deposit little or no water on the herb. Further, if all the equipment is well insulated against heat loss, a good boiler's expanding steam can take up a quantity of excessive moisture from the herb surface which will vary directly with the amount by which the generating pressure exceeds about 300 kPa.

Early in the distilling season the lavender is likely to be sappy. Then water from collapsing aqueous plant cells will be added to that which condenses on the herb during the initial heating up phase. This takes up much of the herb's absorptive capacity and reduces the intermingling of oil and water at the interface. The area into which steam particles can usefully condense will be further reduced by any moisture subsequently deposited from wet steam. Then distillation can become very slow and inefficient. At the same time, excess liquid may run off the flooded herb surfaces and set up a reflux flow that washes oil to the bottom of the still. This reflux shows as discoloured water gathered in the bottom of the still at the end of the run. That dirty water should not be there.

If it is essential to process the lavender when it has a large content of natural moisture, it will be advisable to generate the steam from a separate boiler capable of operating at pressures as high as 700 kPa. In case of extreme need, steam generated under that pressure could tend to take up nearly 1.5 per cent of its own weight in moisture from unduly wet herb surfaces. As the season advances and the lavender becomes less sappy, the generating pressure can be progressively reduced so that the expanded steam's moisture content gradually increases by amounts that are not enough to set up a reflux flow.

Late in the season the lavender becomes much drier and very absorptive. The water that condensed to heat the charge is stationary on the herb surface. It does not follow up the perimeters of the oil patches when they retreat as the oil is boiled away. Then only moisture deposited from wet steam can keep contact with the oil and maintain the heterogeneous mixture at the interface on which the vital transference of heat depends.

At this stage steam with a relatively large wetness fraction will be required for maximum efficiency. The steam may be supplied from a low pressure evaporator or a well designed kettle generating steam in the bottom of the still itself. The latter will require anti-splash baffles to prevent turbulent water wetting the bottom of the charge. The same effect can be had by injecting steam from any boiler into water in the bottom of the still. But in both these latter cases, care will be taken to ensure that the steam is not wet enough to flood the herb surfaces or set up a reflux flow.

Distillery equipment

The commercial scale still

Two of the sketches (Figure 10.5) depict orthodox schemes for stills that can each extract about half a tonne of true lavender oil per season for each cubic metre of their charge capacity. For these stationary stills, about $2\,m^3$ is a convenient size for lavender.

Distillation of the lavender type oils 107

Modern cylindrical stills however, exploit the fact that only the still base, below the grid level needs to be a fixture and the still lid, carrying the condenser, can be conveniently suspended on a crane and pulley system above it. The movable cylinder has a perforated bottom of which the perimeter flange seals by its own weight to a tubular neoprene gasket fixed to the still base. For handling at the distillery, either by fork lift or overhead gantry, this still body is carried on two pivotal lifting points which allow it to be inverted for emptying. It is slightly top heavy when full and self righting when empty. Both the lids and bases are common fits to all the cylinders. The latter are filled by the harvester and they shuttle between the field and the distillery on special trailers.

The American style box trailer stills can hold several tonnes of flowers and are, of course, independent movers on their own wheels. They have merely to be towed to a position under the suspended lid and connected to the steam supply when the lid has been clamped down. Some of these trailers have permanently built on lids to which only a condenser needs to be connected at the distillery. Usually they are equipped with a hydraulic hoist to tip the tray for emptying.

These modern systems, especially the cylinders, have the advantage that all operations are carried out quickly and conveniently at the one ground level. There is no need for any mezzanine floor. The choice between the two systems will depend on the nature of the terrain. The large box trailers cause problems in steering machinery across sloping ground where row crops lie along the lines of permissible slope derived from contours.

A still for small-scale work and test runs

A very satisfactory still for test runs can be made from a pair of 200 l drums, or equivalent cylinders, if they are fitted with flanges so that they can be clamped together with a gasket between them (Figure 10.6). It should also suit small-scale operators who aim to produce limited quantities of oil for strictly local sale. It will hold about 120 kg of ripe lavender. The upper drum should have an open arrangement of light steel bars in place of its original sheet metal bottom. The two drums can then be dealt with separately, which is convenient. A small manual trolley hoist on a short piece of overhead rail will facilitate handling if the drums are given suitable lifting points.

This still can be steamed by a simple evaporator, which can be put together from the details suggested on the diagram. It will produce steam that has a suitable wetness fraction for processing ripe lavender. The LP gas burner may be about 700 mm wide measured in a direction across the water tubes, but only 350–400 mm deep in a direction parallel with them. Take care that any hot water issuing from the $1\frac{1}{2}$ m safety stand pipe cannot scald the operators. Both the still and the kettle should be lagged against heat loss with a double thickness of good carpet under-felt. (In Figure 10.6 the abbreviation N.B. stands for Nominal Bore.)

The condenser

Essential oil distilleries must now use multi-tube condensers (Figure 10.7) where the vapour flow is distributed between a number of parallel stainless steel tubes which have a flow of cold water circulating round them. The restricted outlet of the old single tube, tapering coil type, causes excessive back pressure in stills using modern rates of steam flow.

The unit sketched for working with the test still shows the complete arrangement of a *simplified* multi-tube condenser (Figure 10.8). It differs from the more expensive *industrial* type in that

(a)

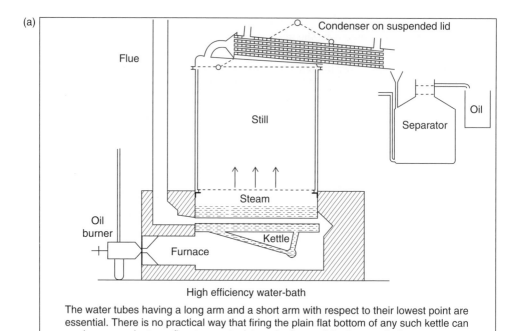

High efficiency water-bath

The water tubes having a long arm and a short arm with respect to their lowest point are essential. There is no practical way that firing the plain flat bottom of any such kettle can produce enough steam flow to counter internal reflux and the losses of oil.

(b)

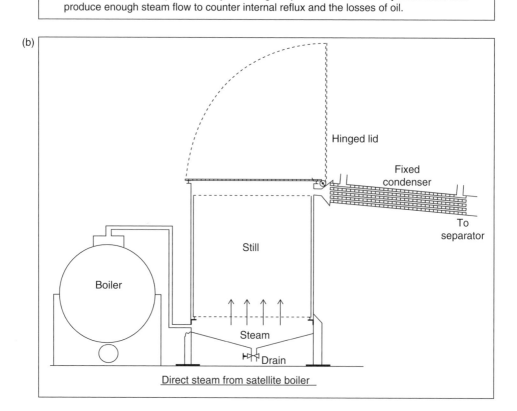

Direct steam from satellite boiler

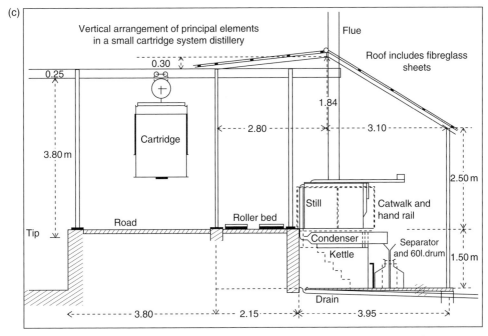

Figure 10.5 The two orthodox still schemes mentioned in distillation equipment. (a) Water-bath arrangement (b) Direct steam arrangement (c) Vertical arrangement of principle elements in a small cartridge system distillery.

it cannot be taken apart for cleaning the outsides of the tubes. But mud and loose dirt can be removed by agitation with washing soda. The construction of these simplified condensers is easy for professional fabricators and their price should be acceptable.

Industrial condensers are designed to be dismantled for cleaning and will be preferred where 'hard' cooling water may cause a calcareous encrustation of the tubes and seriously impair the transference of heat. Design details for these more complex units should be found in books dealing with the distillation of herbaceous oils.

The table (Figure 10.8) gives the number and length of the parallel tubes required in both simplified and industrial types, to condense a range of distillate flows in temperate climate areas. It allows for a connecting pipe from the still up to 3 m long, but having only one 90° bend. It is calculated for stainless steel tubes 19 or 20 mm in diameter, with wall thickness 1 mm, arranged on a hexagon pattern with centres 30 mm apart. The shell diameters are no more than 50 mm greater than that of the tube bundle and the coolant baffles are not more than $1\frac{1}{2}$ shell diameters apart. The coolant flow is fifteen times that of the distillate and the condensate emerges at a temperature some 22°C above that of the entering coolant water. If the cooling water fails or is not turned on, the maximum back pressure due to the connecting pipe and condenser is less than 1.4 kPa. (= 0.2 psi).

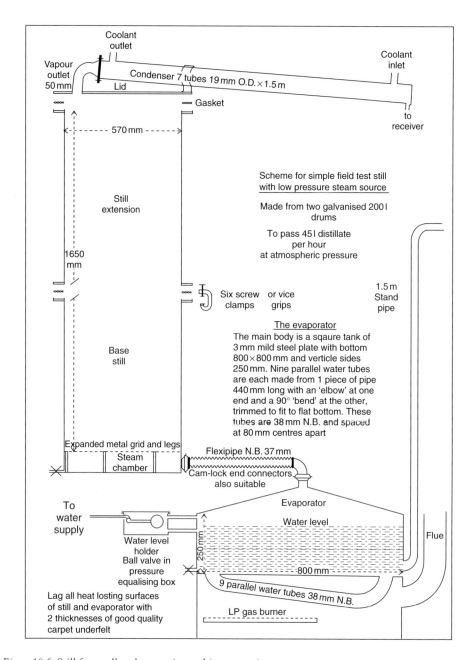

Figure 10.6 Still for small-scale operation and (parameter) tests.

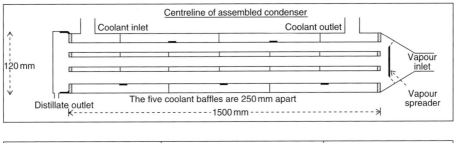

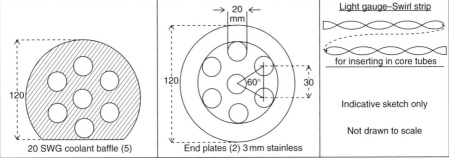

Figure 10.7 Multi-tube condenser: vapour-tube type.

The receiver–separator

When the liquid distillate is caught at the condenser outlet the oil floats on top of the water. If the receiver is suitably designed oil can be made to pour continuously from an outlet near its top while the water flows away from the bottom.

It was once believed that the oil and water should be separated at cool temperatures, to avoid a loss of oil due to its presumed solubility in warm water. But the main loss of oil susceptible to this solubility occurs at 100°C in the condenser and it is difficult to demonstrate any further significant loss of a condensed oil in water, even at 50°C.

The separation of oil and water is promoted by the difference in their densities and resisted by the viscosity of the water. In our case, the former increases and the latter decreases with rising temperature. So the conditions for the separation of lavender oil from water improve dramatically at higher temperatures. Tests have shown the small droplets of lavender oil rising at only 4.5 mm/min through water at 30°C, whereas they rose at 9 mm/min when the water was at 45°C.

Since the separator must have the water travelling downwards more slowly than the small oil droplets are rising through it, the cross-section area of the water's travel path at 45°C is only half what it would need to be at 30°C. Although modern separators that allow for the speeds of fluid travel are much larger in cross-section than the less effective types of former times, the higher temperatures do much to restrain their size.

The unit sketched for working with the small scale still shows the arrangement of a good receiver–separator (Figure 10.9). The inlet funnel and pipe leads distillate to the bottom of an open topped, inner cylinder, which can hold the first few minutes of distillate from a new

Distillate flow rate metric		Distillate flow rate pounds wt		Diameter connecting pipe		Outside diameter of condenser tubes		Number of tubes and pattern	Required length of tubes	
kg/hr	lit/min	lb/hr	lb/min	inch	mm	inch	mm	hex or Δ	feet	metre
30	0.5	66	1.1	1½	37	0.75	20	7 H	2.70	0.83
45	0.75	99	1.66	1½	37	0.75	20	7 H	4.05	1.23
60	1.00	132	2.20	2	50	0.75	20	7 H	5.40	1.65
100	1.67	220	3.66	3	75	0.75	20	15 Δ	4.21	1.30
150	2.50	330	5.50	3	75	0.75	20	19 H	4.98	1.52
200	3.33	440	7.33	3	75	0.75	20	27 Δ	4.68	1.42
250	4.17	550	9.17	4	100	0.75	20	31 H	5.09	1.55
300	5.00	660	11.00	4	100	0.75	20	31 H	6.10	1.86
350	5.83	770	12.83	4	100	0.75	20	37 H	5.97	1.82
400	6.66	880	14.67	4	100	0.75	20	42 Δ	6.01	1.83
500	8.33	1 100	18.33	5	125	0.75	20	55 H	5.74	1.75
600	10.00	1 320	22.00	5	125	1.00	25	37 H	8.20	2.50
800	13.33	1 760	29.33	6	150	1.00	25	42 Δ	9.20	2.80
1 000	16.66	2 200	36.66	6	150	1.00	25	61 H	8.20	2.50
1 250	20.83	2 750	45.83	8	200	1.00	25	69 H	9.10	2.75
1 500	25.00	3 300	55.00	8	200	1.00	25	85 H	9.00	2.74

Vapour in tube condensers
preferred tube arrangements
For cylindrical shells

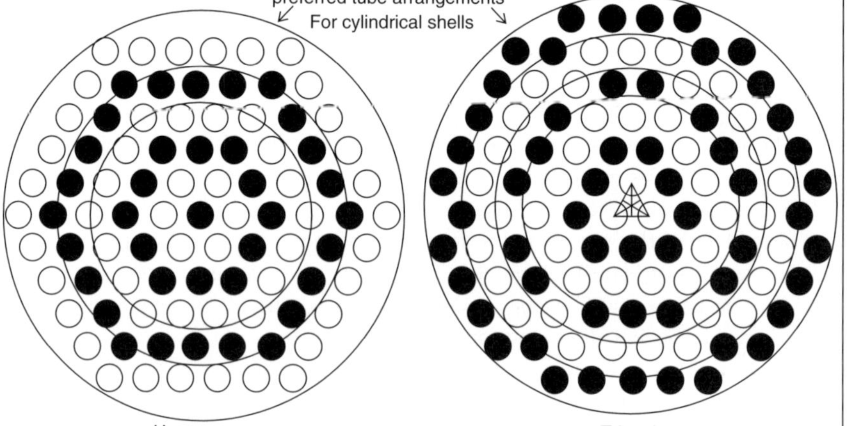

Hexagon
7 tube bundle diameter = 4 × tube O.D.
19 tube bundle diameter = 7 × tube O.D.
31 tube bundle diameter = 9 × tube O.D.
37 tube bundle diameter = 10 × tube O.D.
55 tube bundle diameter = 12 × tube O.D.
61 tube bundle diameter = 13 × tube O.D.
85 tube bundle diameter = 15 × tube O.D.
91 tube bundle diameter = 16 × tube O.D.

Triangle
27 tube bundle diameter = 9 × tube O.D.
42 tube bundle diameter = 10.5 × tube O.D.
69 tube bundle diameter = 13.5 × tube O.D.
102 tube bundle diameter = 16.5 × tube O.D.

Figure 10.8 Condensers for standard temperate zone conditions.

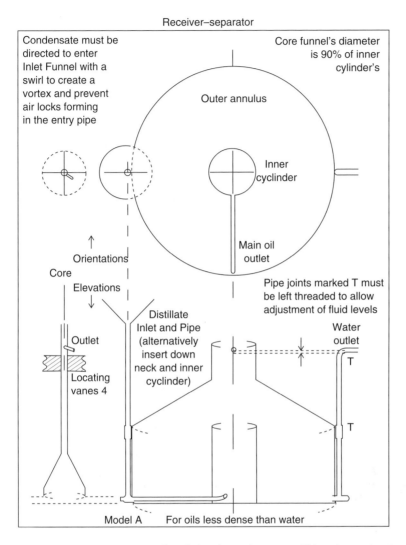

Figure 10.9 Receiver–separator for oils less dense than water. Although not clear in the sketch, the vessel's cylindrical neck continues upward beyond the low side of the main oil outlet by 30 mm.

charge. Often this oil rich early distillate is cooler and more dense than the hot water left in the vessel from the preceding run. It will go straight to the bottom and out the water discharge pipe if it is not caught and held until the temperatures have stabilized.

The separator's core is a column pipe on an inverted funnel which is inserted down the inner cylinder to rest on the inlet pipe. The latter is angled to make a swirl in the funnel. This reduces turbulence and effects a primary separation for which the outlet at the top of the column discharges oil just above the fluid level held in the neck of the main vessel.

The main cylinder encloses the outer annulus of which the cross-section area is such that the water will travel downward more slowly than the oil particles will rise through it. The water outlet pipe connects at the bottom of this annulus and is taken to a height which causes a depth of some 6 or 8 cm of oil to be retained in the separator's neck. This allows any suspended water droplets to sink away from the level of the main oil outlet pipe so that only oil that is relatively free of water will be delivered for collection. Even so, the oil should be dried over anhydrous sodium sulphate to ensure its keeping qualities.

It can be tested for dryness. One ml of oil should make a clear solution in 5 ml of chloroform. If the oil is not dry the solution will be milky.

Distillery throughput

To find how much plant material a prospective distillery will handle per day, it is simple to calculate the length of time each charge of known weight will have to occupy a still. This time falls into three periods.

First, the 'change time' is the interval between the steam being turned off from an exhausted charge and turned on again to the next one in the same still. With modern distilleries and handling systems it need not exceed 5 min.

The 'heating time' starts when the steam is turned on to the new charge and runs until the distillate is reckoned to start the normal rate of flow. The time taken to heat the charge depends on the supply of steam, which gives up 540 kilocalories of heat per kilo condensed, and the weights of steel and herb to be raised from day temperature to 100°C. The herb may be taken as having a mean specific heat close to 0.75 calories per gram.

The vapours which condense to heat the charge determine the size of the oil patches whose radii the steam must reduce to extract the oil. During the time, say 'x' seconds that each layer of herb, one centimetre thick, takes to attain boiling temperature, it receives a deposit of condensing oil equal to the amount vaporized in 'x' seconds from each and every similar layer below it in the still. The lavender's very absorptive surface enables it to receive this oil without becoming flooded.

Some of this oil will lodge in the natural oil patches formed by ruptured glands so the natural oil areas on the top layer of herb are increased by an equal amount for every centimetre of charged height. Then the time taken by the passing steam to reduce the radii of these enlarged oil areas to nil, is the extraction time for the oil.

Finally, the 'extraction time' lasts from the start of the normal distillate flow until the herb is virtually exhausted of its oil. For all steam distillations, this time can be calculated from two parameters which are derived from test distillations and are specific to the tested herb. For superficial oils from absorptive surfaces the parameters are:

1. 't' minutes. This is the time taken by the flow of suitably dried or moistened steam to exhaust the oil from one typical surface oil patch, regarded as having area 'a' on the lowest layer of herb in the still. For lavender, the numerical value of 't' is inversely proportional to the steam's rate of displacement over the herb surface.
2. 'S' is the ratio that the basic area 'a' bears to the increment, 'δa', in the area of the surface oil patches at the top of the charge, due to each cm of charge height. The value of 'S' is inversely proportional to the herb's oil content per kilo and to the mean overall density of charge packing.

Then if 't' and 'S' are both adjusted inversely for differences between the conditions of the test distillation and those of the industrial operation now being planned, the extraction time 'T' minutes for a charge of the tested herb having height 'H' cm is given by

Color Plate I (See Chapter 2, p. 13. Tim Upson)
Figure 2.6 *L. angustifolia* 'Hidcote' – cultivated at Cambridge University Botanic Garden.

Color Plate II (See Chapter 2, p. 14. Tim Upson)
Figure 2.7 *L. latifolia* – France, Col de Ferrier nr. Grasse. View of flower spike showing linear bracts and bracteoles.

Color Plate III (See Chapter 2, p. 15. Tim Upson)
Figure 2.8 *L. x intermedia* 'Grosso' – cultivated at Norfolk Lavender, UK. The most widely cultivated lavandin for oil production.

Color Plate IV (See Chapter 2, p. 16. Tim Upson)
Figure 2.9 *L. dentata* var. *dentata* – Morocco.

Color Plate V (See Chapter 2, p. 18. Tim Upson)
Figure 2.10 L. *stoechas* subsp. *stoechas*.

Color Plate VI (See Chapter 2, p. 21. Tim Upson)
Figure 2.11 L. *multifida* – close up showing flower spike, Morocco near Oued Laou.

Color Plate VII (See Chapter 2, p. 22. Tim Upson)
Figure 2.12 L. canariensis – plant in full bloom, Tenerife near Chio.

Color Plate VIII (See Chapter 2, p. 24. Tim Upson)
Figure 2.13 L. antineae – cultivated at Cambridge University Botanic Garden.

Color Plate IX (See Chapter 2, p. 28. Tim Upson)
Figure 2.14 L. *subnuda* – cultivated at Cambridge University Botanic Garden. Section Subnudae.

Color Plate X (See Chapter 2, p. 29. Tim Upson)
Figure 2.15 L. *aristibracteata* – cultivated at Cambridge University Botanic Garden. Section Subnudae.

Color Plate XI (See Chapter 2, p. 31. Tim Upson)
Figure 2.16 *L. bipinnata* – close up of flower, cultivated at Cambridge University Botanic Garden. Section Chaetostachys.

Color Plate XII (See Chapter 6, p. 60. Simon Charlesworth)
Figure 6.1 Out-buildings and sales area fronted by the display of lavenders.

Color Plate XIII (See Chapter 6, p. 70. Simon Charlesworth)
Figure 6.2 L. *stoechas* subsp. *Stoechas* 'Kew Red'.

Color Plate XIV (See Chapter 6, p. 71. Simon Charlesworth)
Figure 6.3 L. *stoechas* × *viridis* 'Willow Vale' L. *angustifolia* Hidcote.

Color Plate XV (See Chapter 6, p. 72. Simon Charlesworth)
Figure 6.4 L. *stoechas* × *viridis* 'Helmsdale'.

Color Plate XVI (See Chapter 6, p. 73. Simon Charlesworth)
Figure 6.5 L. *minutolii*.

(a)

(b)

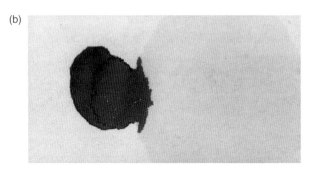

(c)

Color Plate XVII (See Chapter 10, p. 105. E.F.K. Denny)
Figure 10.4 Effects of steam moisture: oil and water on an absorptive surface.

the formula:

$$T = t(1 + H/S)^{1/2}$$

where 't' for lavender is 10.15 min when the steam's rate of displacement is 3 kg per minute for each square metre of the charge's cross-section area at the top of the still and 'S' has the numerical value 40.95, when the herb contains 7.73 g of oil per kilo and the charge packing density is 300 kilos/m^3. Then for any size of still, if the parameters are adjusted inversely for variations from these conditions and applied to the formula, the time 'T' for extracting the lavender oil will be accurate within 3 per cent.

Two other related items

Since lavender is a 'superficial' oil, that is, borne entirely on the surface of the herb, both the heating time and the extraction time are exactly inversely proportional to the rate of displacement of the steam. If the power of the boiler and the rate of flow of the distillate are doubled, these two times for each charge are halved. Conversely, if we double the diameter of a cylindrical still we increase the cross-section area of the charge by a factor of four. If the boiling rate is not changed, the steam's rate of displacement is then only one-quarter of what it was before and each charge will take four times as long to distil.

The amount of steam which must actually pass to exhaust the oil from any charge, depends on the amount of oil in that particular load. It is independent of the mass of the herb in the still. If a charge includes so much barren material that it contains only one-third of the normal amount of oil, its extraction time at the standard rate of flow will be the same as that taken by a charge of normal herb that has only one-third of the normal height.

Summary

Optimum recovery of all essential oils by steam distillation requires not only the elimination of reflux flow caused by the delivery of more water than the herb surfaces can hold, but also the existence of conditions which promote the maximum transference of latent heat from the steam to the oil. Since these advantageous circumstances do not occur naturally, the distiller must create them by matching the steam's wetness fraction to the absorptive property of the herb. The aim is to ensure that the herb surfaces remain moist during the whole distillation while retaining some residual absorptive capacity. They must not become flooded before all their oil has been removed.

With simple distilling equipment, where it is not easy to vary the moisture content of the steam, it is necessary to harvest the lavender at the stage of ripeness when its natural moisture content and absorptive capacity are suited to the quality of the steam available.

In more sophisticated distilleries, the judicious management of a good commercial boiler offers steam which ranges from water bath quality with its large wetness fraction for processing very absorptive herb towards the end of the harvest, to steam that is so dry that it can even take up a little excess moisture from herb surfaces that may be too wet early in the season. (Technically that steam would be slightly 'superheated' as it entered the bottom of the still and would instantly revert to being saturated steam near 100°C as it gathered moisture.)

If the lavender flowers are in good condition and the still is at least 1.3 m tall, the ratio of water to oil in the whole completed distillate should not exceed 20 to 1 by weight. If this is achieved, the steam cannot be too dry. Conversely, if blackened water does not accumulate beneath the charge, the steam is unlikely to be too wet. If the steam's moisture content is adjusted to give a result between those two undesirable extremes, the distillation will be at its most efficient and damage to the oil will be minimal.

References

Denny, E.F.K. (1991 updated to 2001) *Field Distillation for Herbaceous Oils*, 3rd edn, Denny, McKenzie Associates, PO Box 42, Lilydale, Tasmania 7268, Australia, E-mail <timdenny@southcom.com.au>.

11 Lavender essential oil

Standardisation, ISO; adulteration and its detection using GC, enantiomeric columns and bioactivity

Maria Lis-Balchin

Definition of lavender and *Lavandula* oils

The International Organisation for Standardization or ISO, defines Oil of French Lavender, ISO 3515 as 'The oil obtained by steam distillation of recently picked lavender flowers (*Lavandula angustifolia* P. Miller) either growing wild or cultivated in France'.

The ester value, expressed as linalyl acetata is given as 38 per cent minimum and 58 per cent maximum with a camphor content of 0.5 per cent minimum. The established chromatographic profile includes the main identifying components (Table 11.1).

Spike lavender (*L. latifolia* (L.) Medikus) has a separate ISO (4719:1992), as does oil of lavandin abrialis (*L. angustifolia* P. Miller × *L. latifolia* (L.) Medikus), France (Table 11.2a). The latter has a requirement for a minimum linalyl acetate content of 27/37 per cent maximum and linalool 28/38 per cent with camphor at 7/11 per cent maximum. Oil of lavandin grosso (*L. angustifolia* P. Miller × *L. latifolia* (L.) Medikus), France also has an ISO (Table 11.2b).

Lavender grown for oil production

L. angustifolia is mainly propagated by seed, sown in spring or autumn, depending on the severity of the winters in the region (Weiss). Sowing can be done directly into fields but more often the first stage is in nursery beds, where the plants remain for about a year. Clonal plants are made

Table 11.1 *L. angustifolia* P. Miller ISO 3515 1987

Component	Min	Max
Optical rotation −11 to −7; Ester min. 38%; max. 58% as linalyl acetate		
Trans-β-ocimene	2	6
Cis-β-ocimene	4	10
Octanone-3	—	2
1,8-cineole	—	1.5
Limonene	—	0.5
Camphor	—	0.5
Linalool	25	38
Linalyl acetate	25	45
Terpinen-4-ol	2	6
Lavandulol	0.3	
Lavandulyl acetate	2	
α-terpineol	—	1

Table 11.2a Lavandin abrialis ISO 3054 1987
           L. angustifolia P. Miller × L. latifolia
           (L.f) Medikus

Component	Min	Max
Ester (lin. acetate) min. 27%; max. 37%; Optical rotation −5 to −2°		
Trans-β-ocimene	3	7
Cis-β-ocimene	1.5	3.5
1,8-cineole	6	11
Camphor	7	11
Linalool	28	38
Linalyl acetate	20	28
Terpinen-4-ol	—	1
Borneol	1.5	3.5
Lavandulol	0.5	1.5
Lavandulyl acetate	1	2

Table 11.2b Lavandin grosso ISO 8902 1986
           L. angustifolia P. Miller × L. latifolia (L.f)
           Medikus light yellow

Component	Min	Max
1,8-cineole	4	7
Camphor	6	8
Linalool	25	35
Linalyl acetate	28	38
Terpinen-1-ol-4	2	4
Borneol	1.5	3
Lavandulol	0.3	0.5
Lavandulyl acetate	1.5	3
Caryophyllene oxide	tr	tr

via cuttings. Healthy mother plants are cut down near ground level and the branches can be stored for months before preparing the cuttings of 10–15 cm with 1–2 branchlets. These are also planted in a nursery, usually in the spring, for a year. Green cuttings can be used but these require tender care, growth hormones and misting. The plants are planted out in rows 1.5 m apart with 0.4–0.4 m in between rows; giving 10,000 plants per ha for *L. angustifolia* and about half for the hybrids (Weiss, 1997).

Husbandry has now improved the lavender crops and include fertilizers, often as ash (Chaisse and Blanc, 1990). The soil is loosened superficially 2–3 times a year to remove weeds or else weed killers are used. There are many lavender pests and diseases which have attacked crops since hundreds of years, and this reduces a possible 15–20-year lifespan to three years. The lavender fields in France and elsewhere are therefore declining as they become uneconomical. Root rot due to *Armillaria mellea* is very serious as a fungal disease; the insect *Thomasiniana lavandulae* (Diptera) is the most serious as its larvae feed under the bark, causing damage to the tops of branches. Other diseases are due to the fungus *Rosellinia necatrix*, the homoptera *Hyalesthes obsoletus, Cechenotettix martini, Eucarazza elegans*; coleoptera include *Arima marginata, Chrysolina americana* and *Meligethes subfumatus*; Lepidoptera include *Sophronia humerella*;

Argyrotaenia pulchellana, Pyterophorus spicidactyla, and many others (Chaisse and Blanc, 1990). The decline in lavender has been going on for over twenty-five years.

Harvesting was done by hand, especially in the mountains, using a sickle, but mechanical harvesters are now fully developed, cutting 7500 kg/day compared to hand harvesters cutting 500 kg. The yield of lavender oil is 40 kg/ha and lavandin is up to 120 kg. Spike lavender yields 50 kg/ha. The harvested lavender is left in the fields for a few days then steam distilled or extracted with CO_2 or other solvent.

Lavender oils and solvent extracts

Arctander (1960) lists several different lavender and lavandin oils as well as lavandin and lavender concretes and absolutes.

Lavender absolute and concrete

Lavender absolute and concrete (*L. angustifolia* P. Miller or *L. officinalis*) is produced from direct extraction of the herb with solvents (concrete) and thence extraction with absolute alcohol from the concrete or from the distillation water. The concrete is reasonably priced and can be used in soaps, fougeres, chypres, ambres, tabac perfumes etc. It blends well with bergamot, citrus oils, musks, eugenol etc.

Lavender oil

Lavender oil (*L. angustifolia* P. Miller or *L. officinalis*) is steam distilled from the freshly cut flowering tops of lavender. Most of the lavender plants were originally grown and distilled in the higher areas of Mediterranean France (600–1500 m), but nowadays a substantial amount comes from China. Exact figures for the production of the oil is difficult to obtain due to the immense amount of adulteration, mixing, cutting, and addition of synthetics or simply synthetic lavender oil itself.

The true oil is almost colourless and has a sweet, floral, herbaceous, refreshing odour with a pleasant, balsamic-wood undertone. It has a fruity-sweet top-note which is very transient, and the whole oil has low tenacity. It is used in colognes, lavender-waters, fougères, chypres, abres, floral and non-floral perfumes. It blends well with bergamot and other citrous oils, clove, patchouli, rosemary etc. It also blends with amyl salicylate, citronellol, geraniol, musks and numerous other aromachemicals (Arctander, 1960).

Adulterants

Adulterants include: acetylated lavandin, synthetic linalool and linalyl acetate, fractions of ho leaf oil and rosewood oil, terpinyl propionate, isobornyl acetate, terpineol, fractions of rosemary, aspic oil, lavandin etc. Lavandin oil, being produced in at least a ten-fold excess in comparison to lavender oil, is used to cut the oil as it is so much cheaper, as well as abundant (Arctander, 1960; Lis-Balchin, 1995).

Terpeneless lavender oil is produced by careful vacuum distillation, a 'topping off' of about 10 per cent of the oil is sufficient to make it mellower and softer and more soluble in dilute alcohol and of course with increased stability.

Lavandin oils and solvent extracts

Lavandin absolute and concrete

The lavandin absolute (from concrete) is produced by extracting the concrete with absolute alcohol, the alcoholic extracts are then chilled and then evaporated continuously under reduced vacuum towards the end of the distillation. The lavandin absolute is a viscous dark green liquid of herbaceous odour, resembling the flowering lavandin. It is sweeter than the essential oil. It is used for fougères, new-mown-hay types, herbaceous, floral fragrances, forest-notes and refreshing colognes. It blends well with patchouli, clove oil, bergamot and lime and rounds off rough ionones.

Lavandin absolute can also be produced from distillation, as some of the components are reasonably water-soluble and in a 20 ton still, a vast amount of lavandin can be extracted with benzene or petroleum ether and thence re-extracted with alcohol.

Lavandin oil

Lavandin oil was first produced in the late 1920s, but has since escalated well above that of true lavender. Lavandin is a hybrid, produced by crossing true lavender (*L. angustifolia* P. Miller) with spike lavender (*L. latifolia* L.). There are many such hybrids growing all over Europe and other parts of the world and due to the larger plants, the yield is much higher giving a greater monetary return per acre. The oil is pale yellow to almost colourless and has a strongly herbaceous odour with a distinctive top-note which is fresh camphene-cineole-like and disappears very rapidly from a blotter (Arctander, 1960).

Lavandin oil is used in large quantities for a fresh note in perfumes. It is well adapted for detergent products and needs no strong fixatives, but in soaps there is a necessity for good fixation.

Lavandin oil blends well with natural and synthetic products in perfumery including: clove oil, amyl salicylate, citronella and patchouli. Fixation is accomplished usually by sesquiterpene fractions from labdanum, nitromusks, coumarin and oakmoss (Arctander, 1960).

Natural changes in the oils due to different source or extraction process

Boelens (1995) reviewed in great depth the chemical and sensory evaluation of *Lavandula* oils. He recalled that lavandulol, one of the most important constituents was discovered in 1942. The main production countries nowadays are France and Bulgaria. In 1984, world production of lavender oil was 200 tons (Lawrence, 1984); Bulgaria produced 100–129 tons; France, 55 tons; USSR, 35 tons; Australia, 5 tons.

The chemical composition of *L. angustifolia*, *L.* 'Grosso' and *L. latifolia* extracted by steam distillation is very different in itself, with the spike lavender, *L. latifolia* having a much higher camphor content. The method of extraction also makes a difference to the composition (Table 11.3). Major components of lavender oil produced by hydrodiffusion and microwave extraction in hexane (Jean *et al.*, 1991) show variability to a very small extent here, but when the composition of lavandin 'Grosso' oil from steam distillation, oil from CO_2 extraction and an absolute (Pellerin, 1991) is compared (Table 11.4), substantial changes in linalool and linalyl acetate become apparent. There are also substantial changes in the main chemical components of natural *Lavandula* hybrids collected in Tanaro Valley, Italy (Peracino *et al.*, 1994) as seen in Table 11.5.

Table 11.3 Major compounds of lavender oil produced by hydrodiffusion and microwave extraction in hexane

Component	Hydrodiffusion	Microwave extraction
Linalyl acetate	27.3	29.8
Linalool	26.5	34.6
Lavandulyl acetate	8.2	5.0
(Z)-β-ocimene	8.0	4.7
(E)-β-ocimene	2.9	4.4
β-caryophyllene	7.1	1.6
Terpinen-4-ol	2.7	8.8
Lavandulol	1.1	0.6

Source: Jean et al., 1991.

Table 11.4 Composition of lavandin grosso oil from steam distillation, oil from CO_2 extraction and an absolute

Component	Steam distilled	CO_2 extract	Absolute
Myrcene	1.1	0	0
1,8-cineole	7.2	3.2	0
(Z)-β-ocimene	1.1	0.5	0
(E)-β-ocimene	0.6	0	0
Linalool	42.5	17.5	10.1
Camphor	7.8	4.5	1.7
Linalyl acetate	21.0	33.5	28.4
Lavandulyl acetate	2.5	1.9	1.8
β-caryophyllene	1.3	3.4	2.4

Source: Pellerin, 1991.

Table 11.5 Changes in main chemical components of natural L. hybrids collected in Tanaro Valley, Italy

Component	Hybrid no. (%)					
	1	2	3	4	5	6
Linalool	39.4	38.3	29.6	53.1	35.6	34.6
Linalyl acetate	22.9	18.0	28.0	15.5	19.5	29.1
Lavandulyl acetate	3.8	0.2	4.3	0.1	3.7	2.2
(Z)-β-ocimene	0.4	0.4	0.4	0.3	0.2	0.1
(E)-β-ocimene	0.6	0.7	0.7	0.5	0.7	0.7
β-caryophyllene	6.1	5.8	5.6	9.5	7.3	9.7

Source: Peracino et al., 1994.

Detection of adulteration

Identification of fixed oils

Lavender oil is frequently adulterated and the real or preferred oil can only be detected by experienced Noses and stringent chemical analysis. Adulteration or dilution with carrier oils (fixed oils like almond oil) is easily detected by putting a drop of the sample on blotting paper or a piece of cloth and looking for signs of a halo of grease remaining after a few hours (as pure essential oils would evaporate completely, leaving no residual mark).

Gas chromatography and enantiomeric columns

Ordinary gas chromatography can be used to detect solvents, when they are used in reasonable quantities, if one knows where to look for the solvents. Ordinary gas chromatography, however, with or without mass spectrometry (MS) or other identification facilities, like infra-red (IR) etc., are not sophisticated enough to find most adulterations when fractions of other oils or synthetic components are used. Adulteration could be detected using the presence of dehydrolinalool, dihydrolinalool, dehydrolinalyl acetate and dihydrolinalyl acetate, which indicated that the oil had been adulterated with synthetic linalool and/or linalyl acetate (Agnel and Teissiere, 1984).

The determination of such adulteration of essential oils was perfected by the use of special enantiomeric or chiral columns, mainly composed of an α-cyclodextrin phase (Ravid et al., 1992; Lis-Balchin et al., 1998, 1999). Ravid et al. (1986) showed that lavender oil had (3R)-(−)-linalyl acetate of an optical purity of 93 per cent. In 1990, Mosandl and Schubert showed that genuine lavender oil had 100 per cent®-(−)-linalyl acetate.

It is worth noting that chiral columns can also be used by synthetic chemists and those involved in adulteration of essential oils, as the same type of column can be used to separate out the enantiomers, which could then be added in the correct proportion for a given essential oil!

Bioactivity measurements

The bioactivity, as determined by the action of the oils against twenty-five different bacterial species, twenty different *Listeria monocytogenes* cultivars, three different fungi and also their antioxidant action (Lis-Balchin and Deans, 1997; Lis-Balchin et al., 1996, 1998) was not correlated with either the geographical source of the lavender oil specimens or their chemical composition (Table 11.6). There was no correlation between the percentage of the main components, linalool

Table 11.6 Correlation between high linalool or linalyl acetate content of commercial essential oils and bioactivity

Essential oil	Linalool (%)	Linalyl acetate (%)	Anti-bacterial activity*	Anti-fungal[a] activity		
				Aspergillus niger	Aspergillus ochraceus	Fusarium culmorum
Lavender (l)	29.7	42.8	19	82	90	79
Lavender Bulgarian	51.9	9.5	23	84	29	8
Spike lavender	43.1	4.0	19	93	58	31
Lavender French 2	29.1	43.2	13	57	44	77
Lavandin	28.7	39.4	17	93	86	69
Lavender French 1	26.1	47.9	16	93	58	31
Lavender Bulgarian[b]	2.3	79.8	22	74	84	89

Source: Lis-Balchin et al., 1998.

Notes
* Antibacterial activity tested against 25 different bacteria.
a Antifungal activity is high when numbers approach 100.
b Extracted with supercritical carbon dioxide.

and linalyl acetate, and the bioactivity, when seven different 'lavenders', including spike lavender were assessed.

Pharmacological evaluations using both the smooth muscle of guinea-pig ileum and striated skeletal muscle, showed considerable differences in activity for different 'lavenders' (Lis-Balchin et al., 1997a,b). This was a sensitive method for lavender and work using enantiomers (Lis-Balchin et al., 1996; 1999) have indicated that there is also scope for seeing more differences in activity due to individual enantiomers, which react differently in different tissues.

In conclusion, due to the high sensitivity, biological evaluation, using several different parameters, can be a useful tool in evaluating essential oils and checking for their adulteration.

References

Boelens, M.H. (1986) The essential oil of spike lavender, *Lavandula latifolia* Vill (*L. spica* DC). *Perf. Flav.*, 11, 43–63.
Boelens, M.H. (1995) Chemical and sensory evaluation of *Lavandula* oils. *Perf. Flav.*, 20, 23–51.
Chaisse, E. and Blanc, M. (1990) Les ravageurs de la lavande et du lavandin. *Phytoma*, 419, 45–6.
Jean, F.I. et al. (1991) Extraction au four micro-ondes des diverses plantes cultivées et spontanitées. *Riv. Ital EPPOS* (Numero Speciale), 504–10.
Lawrence, B.M. (1992) Progress in essential oils. *Perf. Flav.*, 17(2); 46–9 (6), 59–60.
Lawrence, B.M. (1994) Progress in essential oils. *Perf. Flav.*, 19(1), 40–2.
Lis-Balchin, M. (1995) *Aroma Science: The Chemistry and Bioactivity of Essential Oils*, Amberwood Pub. Ltd., Surrey.
Lis-Balchin, M. and Deans, S.G. (1997) Bioactivity of selected plant essential oils against *Listeria monocytogenes*. *J. Appl. Microbiol.*, 82, 759–62.
Lis-Balchin, M., Deans, S.G. and Eaglesham, E. (1998) Relationship between the bioactivity and chemical composition of commercial plant essential oils. *Flav. Fragr. Journal*, 13, 98–104.
Lis-Balchin, M. and Hart, S. (1997a) A preliminary study of the effect of essential oils on skeletal and smooth muscle *in vitro*. *J. Ethnopharmacology*, 58, 183–7.
Lis-Balchin, M. and Hart, S. (1997b) Pharmacological effect of essential oils on the uterus compared to that on other different tissue types. *Proc. 27th Int. Symp. Ess. Oils*, Vienna, Austria, 8–11 Sept. 1996, eds. Ch. Franz, A. Mathé and G. Buchbauer, Allured Pub. Corp., Carol Stream, Ill., pp. 24–8.
Lis-Balchin, M., Hart, S., Deans, S.G. and Eaglesham, E. (1996) Comparison of the pharmacological and antimicrobial action of commercial plant essential oils. *J. Herbs, Spices Med. Plants*, 4, 69–86.
Lis-Balchin, M., Ochocka, R.J., Deans, S.G. and Hart, S. (1999) Differences in bioactivity between the enantiomers of α-pinene. *J. Essent. Oil Res.*, 11, 393–7.
Naef, R. and Morris, A.F. (1992) Lavender-Lavandin-A comparison. *Riv. Ital. EPPOS*, 3, special issue, Feb., 365.
Pellerin, R. (1991) Supercritical fluid extraction of natural raw materials for the flavor and fragrance industry. *Perf. Flav.*, 16(4), 37–39.
Ravid,U., Putievsky, E., Katzir, I., Ikan, R. and Weinstein, V. (1992) Determination of the enantiomeric composition of citronellol in essential oils by chiral GC analysis on a modified γ-cyclodextrin phase. *Flav. Fragr. J.*, 7, 235–8.
Weiss, E. A. (1997) *Essential Oil Crops*. CAB International, Oxon.

12 Lavender oil and its therapeutic properties

Gerhard Buchbauer

Introduction

Today, one of the most appreciated essential oils (EO) used in phytotherapy and real aromatherapy as well as in esoteric alternative treatments, is the EO of lavender. Its popularity stems not only from the well known, fresh and pleasant floral-woody-herbaceous fragrance, but still more on account of its therapeutic properties for which this EO has been used in folk medicine since ancient times. The present review covers the literature on the biological – mainly therapeutic – properties of lavender, lavender oil, linalool, linalyl acetate and *Lavandula angustifolia* found in various databases, for example, MEDLINE®, from 1960 till 1999.

First, the effects on the nervous system are reviewed, effects for which this EO is famous, then allergic and antiallergic activities are discussed, followed by possible anticancer properties (mainly of linalool). The antibacterial, antimicrobial and antifungal properties which are common to most EOs, as well as repellent activities, are discussed in another part of this book. Finally, other EO-effects observed in humans and animals are explored. Not included in this review are mere cosmetic applications of the EO of lavender, such as in perfumes or eau de colognes and activities which cannot be related to therapeutic applications.

Relaxing, sedative, antistress, anticonvulsive and spasmolytic properties

CNS effect

Thirty years ago a Bulgarian research team had already reported on certain central neurotropic effects of the EO of lavender and found this oil to be relaxing, calming and stress relieving (Tasev *et al.*, 1969). In another paper the authors showed that linalool possesses central depressive effects similar to those of lavender oil, namely anticonvulsive, inhibitory to the spontaneous motor activity even if this had been stimulated by caffeine, or amphetamine pretreatment: at higher doses it was disturbing to the motor co-ordination and prolonged the narcotic effect of hexobarbital, ethyl alcohol and chloral hydrate. But, in toxic doses, linalool shows a narcotic effect. Even more effective than this prominent alcohol is terpinen-4-ol, a minor constituent of the EO of lavender (Atanassova-Shopova *et al.*, 1970; 1973; Imaseki *et al.*, 1962).

Mood-influencing effect and sedation

The EO of lavender, frequently used in aromatherapy, positively affected mood, EEG patterns of alertness and mathematical computations. This result of a test was assessed with 40 adults where the so-called lavender group subsequently had increased β-power in the EEG, suggesting increased drowsiness, and reported on a less depressed mood and of feeling more relaxed;

furthermore they performed some (simple) mathematical computations faster and more accurately upon aromatherapeutical treatment (Diego et al., 1998). These results are in agreement with the findings of Kerl (1997) who tested the efficiency of children's memory in an elementary school. Especially anxious and timid pupils, who had bad scores on account of their inability to recite a paragraph learnt by heart, increased their marks after inhalation of lavender oil, which was presented on common olfactory strips to the children. Pupils who were originally self-assured and/or lethargic showed no increase, rather even a decrease, in their marks. Therefore, the anxiety-relieving potential of this EO was shown. A similar goal was achieved by an investigation by Ludvigson et al. (1989) where the effects of the odours of lavender and cloves on cognitive skills of college students were studied. Lavender adversely influenced arithmetic reasoning. Discussions raised by the results included implications for the claim that lavender is physiologically relaxing, the surprising absence of an effect of odour change on memory and possible interactions of odour and personality factors.

Another study with the aim to link the effects of odorants with the emotional process through autonomic nervous system responses was performed and found that among the tested odorants the inhalation of lavender oil elicited mostly 'happiness'. More than 60 subjects showed similar autonomic responses which can be transcribed into basic emotions (Vernet-Maury et al., 1999). Similar results were obtained in an experimental study which was performed on 122 patients to evaluate the use of aromatherapy. Those patients who received an aromatherapeutical treatment with lavender oil reported a significantly greater improvement in their mood and perceived levels of anxiety compared to controls (Dunn et al., 1995). Changes in electroencephalogram measurements to different odours revealed that the EO of lavender significantly increased the regression coefficient of the power spectra of frequency fluctuation of α-waves. Thus, showing that this form of an experiment can be used for the evaluation of psycho-physiological responses (Lee et al., 1994).

Inhalation studies using lavender in animals

Evidence for the sedative properties of the EO of lavender after inhalation is also provided by Buchbauer and his team. In animal experiments, this oil significantly decreased the motility of 'normal' test mice as well as of animals rendered hyperactive or in other words 'stressed' by a intraperitoneal caffeine injection before the test (Buchbauer et al., 1991). The experiments were performed on young mice, housed under normal conditions, not under any stressful conditions for the 'normal' approach, therefore physiological patterns of behaviour would be expected. For the period under investigation using essential oil inhalation, specially designed plastic cages were used with adequate air volume provision and a light barrier installed. This light barrier would be interrupted when the animals crossed it and was a reflection of their locomotor activity. This in turn is directly related to their degree of sedation or the opposite, that is, agitation. The animals were allowed 1 h for adaptation and then the EO was applied in glass tubes and the vapour diffused out in a continuous air-supply for an hour. The activity of the animals was monitored over an hour and data taken of their actual activity at 15 min, 30 min and at 1 h. The results were compared with a group of control mice exposed to the same conditions in another cage at the same time. The hyperactivated or 'stressed' mice were even more sedated by lavender vapour than those under 'normal' physiological conditions (Table 12.1).

Furthermore, the main constituents of this oil, namely linalool and linalyl acetate, elicited a similar, although weaker effect. The sedative effect was clearly visible in the behaviour of the mice, which rested in a corner, did not show signs of social activity like grooming, sniffing each other, exploring the environment and generally exhibited a reduced motility.

Other EOs have been tested in these experiments as well, showing in many cases that the hyperactivity caused by a caffeine pretreatment could easily be overcome by the inhalation of

Table 12.1 Effect of lavender and other EOs and fragrance compounds on motility in mice after 1 h of inhalation

Compound	Effect on motility	Effect on motility after caffeine
Lavender oil	−78.4	−91.7
Linalool	−73.0	−56.7
Linalyl acetate	−69.1	−46.7
Neroli oil	−65.3	+1.9
Rose oil	−9.5	+4.3
Valerian root oil	−2.7	−12.0
Citronellal	−49.8	−37.4
Citronellol	−3.5	−13.7
Eugenol	+2.1	−38.7
Methyl salicylate	+16.6	−49.9

certain single fragrance compounds as well as of certain EOs. These included neroli oil, rosemary oil, sandalwood oil and also some common EO components: citronellal, benzaldehyde, linalool, linalyl acetate, α-terpineol and phenylacetate (Buchbauer et al., 1991; 1993; Jager et al., 1992).

The correlation of the motility of the animals to the linalool concentration in serum has also been proven (Buchbauer et al., 1991) as has the absorption of linalool from percutaneous application of lavender oil as a massage oil (Jager et al., 1992), thus furnishing evidence of the aromatherapeutical use of lavender, for example, in the form of herbal pillows, employed in folk medicine since ancient times in order to facilitate falling asleep or to minimise stressful situations. The experiments to determine percutaneous absorption from lavender massage was conducted on a male subject, the EO of lavender being diluted with peanut oil (2 per cent v/v) and one and a half grams of this mixture was spread over an area of the stomach and massaged gently into the skin for 10 min, after which the residue was removed (Jager et al., 1992). Blood samples were taken from the cubital vein at intervals for 90 min and later analysed. The maximal levels of both linalool and linalyl acetate were found in the blood after 20 min and there was virtually none left after 90 min. The experiment indicated that lavender components were very rapidly absorbed into the blood, but these experiments did not prove conclusively that this occurred from skin absorption, as the volatile oil could well have been absorbed by direct inhalation. Experiments, using complete occlusion of the nasal and oral passages under positive oxygen input using specially sealed masks have now added factual evidence to these experiments indicating that skin absorption did in fact occur.

A possible dependence of the calming effect on the lipophilicity of the applied odorant was also discussed (Buchbauer et al., 1993b). Qualitative analyses of the cortex of the test animals, treated with various odorants before sacrifice, showed that these could be found in this part of the brain, thus proving that the fragrance compounds must have passed the blood–brain barrier by means of a humoral transport (Buchbauer et al., 1993a).

One fragrance compound, 1,8-cineole was studied more intensively after inhalation using functional imaging in the brain (Nasel et al., 1994), its concentration in the blood after prolonged inhalation (Stimpfl et al., 1995) and its pharmacokinetics in humans (Jager et al., 1996). There are also studies on carvone after inhalation and massage (Fuchs et al., 1997) which indicate among other things that enantiomeric forms have a different metabolism and that they retain their enantiomeric quality after absorption into the blood (Jager et al., 2000).

Studies of chiral fragrances on the human autonomic nervous system parametres and self-evaluation have also been of interest (Heuberger et al., 2000) as has the influence of essential oils on human attentiveness (Ilmberger et al., 2001). Recent studies of sandalwood essential oil and α-santalol applied to man by inhalation and percutaneous absorption have shown contradictory effects: inhalation produced a stimulating effect, while the latter produced a relaxant effect, which may be due to the effect of the massage (Hongratanaworakit et al., 2000a,b). All these studies indicate a complex behaviour of essential oils and components and strongly caution against predicting similar effects of all essential oils or components on the basis of limited studies.

Interestingly, the relaxing effect of lavender could not only be observed in humans or in test animals, but also when applied to the treatment of stress and travel sickness of pigs (20 animals) which were transported by road for 2 h each day over a 2-day period. When lavender straw was used as bedding or mixed with wheat straw the incidence and severity of travel sickness was significantly decreased, as measured by concentrations of cortisol in the pigs saliva (Bradshaw et al., 1998).

A Japanese patent describes the method for evaluating the antistress effect of fragrances, by determination of cortisol in the saliva. The occurrence of this adrenocortical hormone in saliva is claimed to be related to the anti-stress effect of fragrances (Tanisawa et al., 1999). Another patent deals with hypnotics, foods and feeds containing the EO of lavender (Yano, 1998). A third patent proclaims the 'stress relaxation fragrances, their compositions and cosmetics containing them' which includes linalyl acetate or phenyl ethyl alcohol (Ishitoya et al., 1997).

In a review, Pahlow (1988) cited various forms of applications of lavender, for example, in a bath which is used to overcome the stress of the day and to gain a sort of calmness by means of relaxation and anticonvulsion; pillows filled with lavender flowers are also very common, and even, but to a lesser extent, the so-called sedative teas and sleep-promoting teas which have to be prepared as an infusion using dried lavender flowers.

Sedative properties of enantiomers of linalool

In other experiments, the sedative properties of linalool were examined using the optically active alcohols, (R)-$(-)$-, and (S)-$(+)$-linalool and its racemate. The subjects were healthy young volunteers, who were first required to evaluate thirteen impression items as a sensory evaluation before and after work. Mental work and physical activity was also included as well as listening to a recording. The effect of the racemic mixture of both alcohols showed a tendency towards a greater decrease of the β-wave, identical to the outcome observed for (R)-$(-)$-linalool, the genuine enantiomer of the EO of L. angustifolia. The effect of (S)-$(+)$-linalool, obtained from coriander oil, showed the reverse (Sugawara et al., 1998).

Potentiation of $GABA_A$ receptors

Recently, an important paper from another Japanese research team was published: Aoshima et al. (1999) discussed the potentiation of $GABA_A$ receptors expressed in Xenopus oocytes by perfumes and phytoncides. Essential oils, including various lavender oils and lavender perfumes as well – and such phytoncides as leaf alcohol, hinokitol, pinene, eugenol, citronellol and citronellal, potentiated the response in the presence of GABA at low concentrations (10 and 30 µM), possibly because they bound to the potentiation-site in $GABA_A$ receptors and increased the affinity of GABA to the receptors. Since it is known that the potentiation of $GABA_A$ receptors by benzodiazepine, barbiturate, steroids and anesthetics induces an anxiolytic, anticonvulsant and sedative activity, the results obtained by Aoshima et al., suggest the possibility that the intake of perfume or phytoncid by various ways modulates the neural transmission in the brain through ionotropic $GABA_A$ receptors.

Sleep promoter

The relief of stress and the relaxant property of lavender oil – a problem especially found in hospitals, hospices and homes for the aged, was emphasised again by Lis-Balchin (1997) as well as by Delaveau et al. (1989). In order to test the hypotheses that lavender oil has a sedative effect and that the resultant sleep promotes therapeutic activity, a pilot study was arranged with acutely ill elderly people as well as later on with long-term patients. The results showed a positive trend towards sleep improvement using lavender (Hudson, 1996). However, a somewhat contradictory statement was published by Uehleke (1996) who said that the EO of lavender elicits no sedative properties but exerts some 'harmonising' effects. A difference between stimulation of the olfactory system and pharmacodynamic effects after absorption through the skin is discussed.

To investigate the effects of lavender oil for insomnia, the hours of sleep were initially measured for 2 weeks, then measured again for another 2-week period after medication withdrawal, and then measured for a final 2 weeks during which lavender oil was diffused into the ward. The amount of time spent asleep was significantly reduced after withdrawal of medication, but returned to the same level with lavender oil as that under medication (Hardy et al., 1995).

Swiss mice, who had diluted lavender oil (1/60 in olive oil) orally administered, showed sedative effects, which were observed using various tests (hole broad test, four plates test, plus maze test, potentiation of barbiturate sleeping time). A significant interaction occurred with pentobarbital, because the sleeping time was increased and the time actually asleep was shortened (Guillemain et al., 1989).

Psychopharmacological effects

Relationships could be established between personality traits and the effect of odour on task performance using several odours, such as lavender (Knasko, 1992). A Japanese patent application claims the usage of several monoterpenes as brain stimulants and/or enhancers of brain activity. These compounds can be incorporated into food, such as chewing gums (Nakamatsu, 1995).

A psychopharmacological *in vivo* evaluation of one of the main components of lavender oil, namely linalool, proved its dose-dependent, marked sedative effect on the CNS, including hypnotic, anticonvulsant and hypothermic properties. The psychopharmacological effect could be caused by an inhibitory activity of this mono-terpene alcohol on glutamate binding in the cortex of the test animals, for example, rats (Elisabetsky et al., 1995b). The sedative properties of linalool could also be proved by Elisabetsky et al. (1995a). These effects revealed the usefulness of the traditional folk medicinal use of several plant species by the indigenous peoples in different continents. The effects of odorant inhalation on pentobarbital-induced sleeping time in rats were studied by Komori et al. (1997).

Spasmolytic and sedative effects on rats and fish

Various terpene alcohols possess spasmolytic and sedative properties as shown in animal experiments using mice and also on the rat duodenum (Imaseki and Kitabatake, 1962). Among these alcohols linalool had also been tested. The calming properties of at least some of these alcohols like linalool, could be shown by the inhibition of the fight aggressiveness of fishes, even if in this special case this alcohol exerted the weakest effect (Binet, 1972; Binet et al., 1972). This experimental design had already been used years ago by Laboratoires Meram (1966) who tested a series of terpene alcohols, linalool included. The results were similar to those mentioned above and additionally a potentiation of the barbiturate narcosis of the rats was found.

In a comprehensive study, using nine commercial EOs, the spasmolytic activity on the rat isolated phrenic nerve diaphragm was investigated. In comparison with the activity on field-stimulated guinea pig ileum preparations, the EO of lavender significantly reduced the twitch response to nerve stimulation without any hints of a contracture which was the case, for example, with the EO's of fennel or dill (Lis-Balchin et al., 1997a). Recently, these authors showed that the mechanism of this spasmolytic activity was post-synaptic and not atropine-like. It was most likely to be mediated through cAMP and not through cGMP. The tonus decrease of skeletal muscle preparations of the phrenic nerve diaphragm of rats was also shown (Lis-Balchin et al., 1999).

Linalool is not only one of the main components of the EO of lavender but also of some other oils, such as Hyssop oil from *Hyssopus officinalis* L. var. *decumbens* (Lamiaceae). Together with 1,8-cineole and limonene its spasmolytic potential was shown by the inhibition of acetyl choline- and $BaCl_2$-induced contractions of isolated guinea pig ileum by means of a concentration dependent, but non-competitive manner. The other two components showed only a weak, but spasmogenic action (Mazzanti et al., 1998).

Effects on the autonomic system

The effect of different odorants, among them lavender, on the emotional process was estimated in terms of autonomic nervous system activity. Parametres for the latter were: skin potential, skin resistance, skin blood flow and skin temperature, instantaneous respiratory frequency and heart rate. It seems that the autonomic response reflects the odour valence only through some parametres related to the main preferential channels. Thus a global autonomic pattern has to be considered (Alaoui et al., 1997). In another study the effects of two fragrance oils on human CNS were reported using neuro-physiological measurements. Lavender oil increased auditory reaction time and slowed critical flicker fusion frequency, irrespective of the subject's preference. Specific characteristic changes produced by the fragrance on quantitative EEG were noted. A decrease of 'fast' activity during lavender inhalation was observed as well. Thus, the effects of fragrance oils must be considered from the psychological as well as the physiological point of view (Yagyu, 1994). The physiological response to seven odours (among them lavender) was assessed by EEG recordings and topographical brain maps. Interestingly, subjects rated the odour of lavender oil as unpleasant. The odours that caused the greatest increase in theta waves, included the EO of lavender. Jasmine and lavender tended to induce theta wave increases sooner than birch tar. Widespread increase in theta waves occurred in most subjects during stimulation with the appropriate odour (Klemm et al., 1992).

In an interesting assay, the anxiety reduction potency of lavender odour was studied and the possibility of whether frequency fluctuation coefficients of alpha waves could be an indicator for this reduction was examined. As a result, anxiety decreased when the subject felt in a good mood and the anterior frequency fluctuation asymmetry also correlated with the anxiety level. The slope coefficient of frequency fluctuation of alpha waves was found to be a good indicator in estimating the state of anxiety (Yoshida et al., 1998).

In another experiment to investigate the effects of the odour of EOs, such as lavender oil, on human brain function, event-related potentials were recorded on the EEG in nine healthy adults, while they were working on a visual oddball task. As a result, differences between the conditions with and without the odour and between the EO's were observed in the distribution of current density in the anterior part and the deep areas of the brain (Kana et al., 1998). Schulz et al. (1997) mentioned among other psycho-phytopharmaceuticals lavender flowers as suitable to overcome many disorders, inconveniences of health, ailments, indispositions, states of restlessness, uneasiness and nervousness, then difficulties in falling asleep and also anxiety. Delta- and theta-waves showed an increase especially in a pharmaco-EEG two hours *post-application*, similar, but lesser effects were noted on the delta waves in comparison with a placebo and to diazepam.

Finally, the findings of Rovesti (1971; 1973; 1974) published in a series of papers, should be mentioned here as they are in agreement with a lot of results ascertained later by other scientists. Lavender oil as well as its constituent linalool, relieves anxiety and considerably lowers nervous excitability. Antidepressive and antispasmodic effects upon application of an essence of lavender have been noticed.

Effect on glutamate-binding sites

Another paper reports on the anticonvulsive effects of inhaling lavender oil vapour. Pentetrazol- and nicotine-induced convulsions as well as electroshock convulsions, but not strychnine-induced ones, in mice were blocked significantly and dose-dependently by inhalation of this oil. These effects may be based on an augmentation of GABA-ergic action (Yamada *et al.*, 1994). Elisabetsky *et al.* (1999) investigated the pharmacokinetic basis of the anticonvulsant properties of linalool by examining its effects on behavioural and neurochemical aspects of glutamate expression. Specifically, linalool inhibits by means of a competitive antagonism the L-[^{3}H] glutamate binding at CNS-membranes as well as it delaying NMDA- and blocks quinolinic acid induced convulsions. Finally, there was a report on mineral baths which have antiseptic, antispasmodic and tranquilizing properties as well as analgesic, bactericidal, phytocidal, tonic or other hygienic actions. The EOs can be collected from fresh flowers and from cultivated plants, such as lavender (Biskys *et al.*, 1996).

Allergic and antiallergic activities

Recently, a study was published where the effects of lavender oil on mast cell-mediated immediate-type allergic reactions in test animals were investigated. This EO inhibits dose-dependently the mast cell-dependent ear swelling response induced by an irritant administered either topically or intradermally. The same effect can be observed on passive cutaneous anaphylaxis as well as by studying the histamine release from the peritoneal mast cells. Furthermore, lavender oil exerted a significant inhibitory effect on anti-dinitrophenyl-IgE-induced tumour necrosis factor-α-secretion from these mast cells. These results show the versatility and usefulness of the EO of lavender in skincare preparations for all skin types (Kim *et al.*, 1999).

A series of EOs belonging to the Lamiaceae family were investigated as to their systemic allergic reactions using the prick-by-prick technique with dried commercial plants and prick tests with extracts. Skin tests with inhalants were positive to grasses as well as to ones with plants of the Lamiaceae family with the exception of basil and lavender. Plants belonging to the Lamiaceae seem to show a cross-sensitivity on the basis of clinical history and *in vitro* and *in vivo* test results (Benito *et al.*, 1996). In contrast, contact allergy reactions to various EOs used in aromatherapy, such as lavender oil, jasmine and rosewood oil, were found. Laurel, eucalyptus and pomerance (bitter orange) produced positive skin reactions, thus showing an allergic airborne contact dermatitis. A similar dermatitis was reported on inhalation of lavender fragrance in Difflam gel (Schaller *et al.*, 1995; Rademaker, 1994; Brandao, 1986). A facial 'pillow' dermatitis due to lavender oil allergy was also reported (Coulson *et al.*, 1999).

A toxicity profile study by the BIBRA-working group (1994) stated that lavender oil applied to human and animal skin, produced little or no irritation, but it has caused sensitization, photosensitization and pigmentation in humans. Its principal effect following administration by oral, injection or inhalation routes to rodents, was sedation. Lavender oil was rapidly absorbed through intact human skin, facts which already have been reported elsewhere (see Jäger et al., 1992; Fuchs et al., 1997). It is said that airborne sesquiterpene lactones, especially emitted from Asteraceae plants, cause allergic reactions. This should also be the case with feverfew (*Tanacetum parthenium*, Asteraceae), but headspace analysis of the air around the intact plant revealed mainly (~88 per cent) volatile monoterpenes (in total 41 volatiles), among them also linalool. No sesquiterpene lactones were detected, thus only the other airborne volatiles could be responsible for the 'compositae dermatitis' (Christenson et al., 1999). Twenty-two volatile compounds have been detected in the headspace of oilseed rape (*Brassica napus* spp. *oleifera*, Brassicaceae), among them linalool. These volatiles were made responsible for respiratory mucosal and conjunctival irritations. But, since only between 50 and 87 per cent of the total volatiles, emitted in all of the entrainments carried out with flowering oilseed rape plants, belong to the monoterpene fraction also other volatiles, such as dimethyl sulfide, or sabinene, or isomyrcenol, or (*E*)-3-hexene-1-ol could be the cause for the above mentioned irritation (Butcher et al., 1994). Also in another case, linalool and hydroxy citronellal were made responsible for a facial psoriasis caused by contact allergy to fragrance compounds in an after shave. In the meantime, according to the knowledge of the author of this review on lavender, hydroxycitronellal possesses this allergenic potential (de Groot et al., 1983). In an exhaustive test with mice linalool-oxidation products, such as linalyl hydroperoxide and linalyl epoxide, could be detected as real sensitisers. The furanoid form gives rise to the formation of a carbon centred reactive radical as intermediate in the skin sensitisation (Bezard et al., 1997). A toxicity profile study performed by the BIBRA working group (1995) found that linalool was irritant to the skin of various species of laboratory animals. In man, this monoterpene alcohol has shown only a small ability to cause skin irritation and sensitisation.

Anticancer properties

Perillyl alcohol, a component of the EO of lavender as well as the most important metabolite of d-limonene, is currently under investigation as a chemo-preventative and chemo-therapeutic agent. A pharmacokinetic study has been performed using stable-isotope labelled internal standards. Two new major metabolites besides intact perillyl alcohol, perillic acid and *cis*- and *trans*-dihydro perillic acid have been found in human plasma (Zhang et al., 1999). Another study, dealing with the same monoterpene alcohol, was designed to test its chemo-preventative potential especially in a lung tumour-bioassay. Perillyl alcohol is an inhibitor of farnesyl transferase. In the early development stages of mouse lung carcinogenesis the *ras*-protein undergoes a series of modifications, and farnesylation at the cysteine is one of these, which leads to the anchoring of *ras*-p 21-gen to the plasma membrane in its biologically active state. Perillyl alcohol administered to test mice showed a 22 per cent reduction in tumour incidence and a 58 per cent reduction in tumour multiplicity (Lantry et al., 1997). Perillyl alcohol also inhibited significantly the incidence (percentage of animals with tumours) and multiplicity (tumour/animals) of invasive adenocarcinomas of the colon and exhibited increased apoptosis of the tumour cells. Therefore consumption of diets containing fruits and vegetables rich in monoterpenes, such as d-limonene, as well as using also the EO of lavender, for example, as a flavour compound in the provençal kitchen (Frohn et al., 1997), reduces the risk of developing cancer of the colon, mammary gland, liver and lung (Reddy et al., 1997). Not only perillyl alcohol, but also other terpene alcohols, such as nerolidol, β-citronellol, linalool and menthol, showed inhibitory activities on induced neoplasia of the large bowel and duodenum. Nerolidol, especially, has an impact on the protein prenylation and is able to reduce the adenomas from 82 per cent in the controls to 33 per cent in rats fed with these compounds. Also the number of tumours/rat decreased from 1.5 in the controls to 0.7 in the nerolidol group (Wattenberg, 1991).

Other activities and effects

A protective effect by aerial parts of lavender flowers against enzyme-dependent lipid peroxidation is described by Hohmann et al. (1999). Plant volatile oils have been found to exert certain beneficial effects on the human body in maintaining the level of polyunsaturated acids (PUFAs), to protect them from becoming oxidised (Deans et al., 1995). These EOs show a greater antihydrolytic effect than commercial preservatives, such as BHT, on butter, thus reducing its oxidation (Singh et al., 1998). Linalool, on the contrary, showed only marginal inhibitory effects, even at high concentrations, on lipid peroxidation of PUFA's (Reddy et al., 1992). Natural concentrations of some EOs were examined for effects on the system lipid-peroxidation-antioxidant-defense and lipid metabolism in 150 patients with chronic bronchitis. Lavender oil promoted normalisation of the level of total lipids and the ratio of total cholesterol to its α-fraction (Siurin, 1997). Inhalation of lavender oil volatiles, had no effect on the content of cholesterol in the blood, but reduced its content in the aorta and reduced also the incidence of atherosclerotic plaques in the aorta, that is these EOs are said to produce an angioprotective effect (Nikolaevskii et al., 1990).

Linalyl acetate, a prominent constituent of the EO of *Hedyosmum brasiliense*, is used on account of its analeptic and febrifuge characteristics (Gabriel et al., 1998). Anti-inflammatory properties and peripheral analgesic effects are also attributed to this monoterpenic ester (Moretti et al., 1997). The local anaesthetic activity of the EO obtained from *L. angustifolia* Mill. was studied by *in vivo* tests on the rabbit conjunctiva and *in vitro* in a rat phrenic nerve-hemidiaphragm preparation (Ghelardini et al., 1999). The authors found that the EO, as well as the main constituents, linalool and linalyl acetate, were able to reduce drastically in a dose-dependent manner, the electrically evoked contractions of the rat phrenic-hemidiaphragm. In the rabbit conjunctival test these odorants allow a dose-dependent increase in the number of stimuli necessary to provoke the reflex, thus confirming *in vivo* the local anaesthetic activity.

As a potent radical scavenger, lavender oil is used as a component in topical formulations to relieve the pain associated with rheumatic and musculo-skeletal disorders (Billany et al., 1995). The influence of inhalation of terpenic volatiles on the blood pressure in human subjects was investigated and found that immediately after the end of the jogging tour, these EO-volatiles lessened the raised systolic blood pressure rapidly to normal values (Suzuki et al., 1994; Romine et al., 1999). A hypoglycemic effect caused by linalool in normal and streptozotocin-diabetic rats was described by Afifi et al. (1998). Linalool leads also to a hepatic peroxysomal and microsomal enzyme induction in rats (Roffey et al., 1990; Chadba et al., 1984). A choleretic and cholagogic activity of Bulgarian lavender oil as well as of a mixture of linalool and α-terpineol was found (Peana et al., 1994; Gruncharov, 1973).

Periodontal diseases can be treated with a mixture of EO's, among them lavender oil (Yamahara et al., 1994; Sysoev et al., 1990). An immunotoxicity assessment of food flavouring ingredients indicated that the majority of the compounds tested did not modulate the cell-mediated or humoral immune response (Gaworski et al., 1994). Linalool was shown to penetrate the *ex-vivo* porcine oral mucosa in Franz' cells very easily (Ceschel et al., 1997). Lavender oil is suitable for prevention and treatment of decubitus ulcers as well as of insect bites, athletes foot and skin rash (Hartwig, 1996; Karita, 1996). This oil can also be used for the topical treatment of acne and dilated pores and for the prevention of facial scarring and blemishes of the face and body (Anon, 1997a,b). The EO of lavender is also a component of a mixture which is suitable as a hair growth stimulant and for the treatment of Alopecia areata (Betourne, 1995; Hay et al., 1998). Lavender oil also reduced somewhat the perineal discomfort after childbirth when added to the bath water, probably on account of its antiseptic and healing properties, but side effects were recorded (Cornwell et al., 1995; Dale et al., 1994).

A series of papers describe the use of EOs as skin penetration enhancers, especially for the transdermal absorption of various drugs and medicaments: Nifedipine and lavender oil (Thacharodi et al., 1994), metoprolol and (+)- /(−)-linalool (Kommuru et al., 1999), propanolol (Kunta et al., 1997), ketoprofen and linalool (Kommuru et al., 1998) and azidothymidine and linalool (Kararli et al., 1995).

Enhancement of the growth of plants, for example, cultures of vegetables such as radish, kidney beans and lettuce, by applying linalool vapours is claimed in a Japanese patent application (Sato et al., 1995). Some papers deal with the olfactory sense in behavioural discrimination of volatiles by the honey bee *Apis mellifera* (Lozano et al., 1996; Sandoz et al., 1995; Pham-Delegue et al., 1993; Akas et al., 1992).

A review, especially on Oleum spicae (=the EO of *lavandula latifolia*), reports on the treatment of bronchial diseases as well as of chronic arthritis and also as an adjuvant treatment with antibiotics (Fröhlich, 1968). Another review about lavender oil covers all interesting facts from chemistry to uses of this natural product (Tewari et al., 1987). Toxicological aspects of linalool are reviewed by Powers et al. (1985). A very informative review mainly on the *Lavandula* species cultivated in the southern part of France, the Provence, furnishes valuable details on the harvesting and EO-composition but nearly nothing about its use in medical treatments (Galle-Hoffmann, 1997).

The best review so far – at least to the opinion of the author of this article – is given by Schmidt (1996). The mere title of his paper *'a lavender oil is a lavender oil – or perhaps not at all?'* already points in the direction of this review. Schmidt covers the history, lavender species, cultivation, quality, the analysis of this EO, the problem of the linalool enantiomers and adulteration of the oil, toxicity, market and usage in perfumery. Finally, two other articles on lavender and lavender oil should be mentioned, published in the same issue of the above-mentioned journal (Collin, 1996; Häringer, 1996).

Lavender oil increased the response rate during the alarm period in a dose-dependent manner in the same way as diazepam, indicating that it has an anticonflict effect in ICR mice using the 'Geller-type-conflict-test' (Umezu, 2000). Enantiomeric stereospecificity of (R)-(−)-linalool or (S)-(+)-linalool evoked different odour perception and responses not only with chiral dependence but also with task dependence when administered to subjects both before and after 10 min of physical work. The (R)-(−)-linalool gave a more positive response to the sensory test and also produced a greater decrease in beta waves after work in comparison with that before work. This is in contrast to the case of mental work, which resulted in a tendency for agitation accompanied by an increase in beta waves of the EEG (Sugawara et al., 2000). Inhalation of lavender oil furnished evidence for a possible correlation between alpha 1 activity (alpha 1 (8–10 Hz) of EEG at parietal and posterior temporal regions) and subjective evaluation as it was significantly decreased under odour conditions in which subjects felt comfortable and showed no significant change when subjects felt uncomfortable (Masago et al., 2000). As a decrease of alpha-wave activity means higher attentiveness, these results seem to be contradictory to the many findings of relaxation, sedation, etc., discussed in this chapter. In experiments to clarify the anticonvulsive mechanisms of linalool, its effects on binding of [^{3}H]MK801 (NMDA-antagonist) and [^{3}H]muscimol (GABA$_A$-agonist) to mouse cortical membranes showed a dose dependent non-competitive inhibition on the antagonist binding but no effect on the muscimol binding suggesting that the anticonvulsant mode of action of linalool includes a direct interaction with the NMDA receptor complex and no direct interaction with the GABA$_A$-receptors, although changes in GABA-mediated neuronal inhibition or effects on GABA release and uptake cannot be ruled out (Silv-Brum et al., 2001a). Linalool also inhibits the binding of [^{3}H]glutamate and [^{3}H]dizocilpine to brain cortical membranes and significantly reduces potassium-stimulated glutamate release as well as glutamate uptake, but not interfering with basal glutamate release (Silva-Brum et al., 2001b). An interesting diagnostic method in brain research uses linalool to determine how odour processing is altered in patients with unilateral supratentorial brain tumours. Patients with right-sided lesions showed distinct deficits in a discrimination task after stimulation of the right and left nostril. In contrast, patients with left-sided lesions only had an attenuation of correct reactions after left-sided stimulation. The first patients group showed bilateral impairment, thus supporting the importance of the right hemisphere in olfaction (Daniels et al., 2001). Linalool showed not only marked sedative effects at the CNS, including hypnotic, anticonvulsant and hypothermic properties, but also an inhibitory effect on the acetylcholine release and on the channel open time in the mouse neuromuscular junction which demonstrates a local anaesthetic action either on the voltage or on the receptor-activated channels (Re et al., 2000).

Concluding remarks

The EO of *L. angustifolia* (Lamiaceae) has been used since ancient times on account of its healing properties in the treatment of various diseases, ailments, disorders and discomforts as well as in cosmetic formulations and in perfumery. By means of these properties lavender flowers as well as their oil are even accepted in various pharmacopöeas (see also Wichtl, 1997). Aromatherapy as well as aromachology make use of this EO to a great extent and also nearly all alternative healing treatments (scientific and esoteric and unscientific ones) know of its beneficial activities. The characteristic woody-herbaceous odour with green-floral notes creates a sort of 'well' feeling in the user by which it can be used as a mood tonic, antidepressant and psychological remedy, for example, against shock of injury (Worwood, 1994). In that way the aromachological usage of this EO is equally important as its use in (scientific) aromatherapy (see Buchbauer et al., 1994; Buchbauer, 1998). As can be seen by the above compilation of therapeutic properties, this oil as well as its main constituents, linalool and linalyl acetate, show a lot of beneficial effects which can already be explained by scientific methods, thus rendering this oil a very important and valuable treasure in phytotherapy. Since scientific research has only recently started, a lot of valuable medicinal information can be expected in the future.

References

Afifi, F.U., Saket, M. and Jaghabir, M. (1998), Hypoglycemic effect of linalool in normal and streptozotocin diabetic rats. *Acta Tecnol. Legis Medicam*, 9, 101–6.

Akas, R.P. and Getz, W.M. (1992), A test of identified response classes among olfactory receptor neurons in the honey bee worker. *Chem. Senses*, 17, 191–209.

Alaoui, I.O., Vernet, M.E., Dittman, A., Delhomme, G. and Chanel, L. (1997), Odor hedonics: connection with emotional response estimated by autonomic parametres. *Chem. Senses*, 22, 237–48.

Aoshima, H., and Hamamoto K. (1999), Potentiation of $GABA_A$ receptors expressed in *Xenopus* oocytes by perfume and phytoncid. *Biosci. Biotechnol. Biochem.*, 63, 743–8.

Atanassova-Shopova, S. and Roussinov, K.S. (1970), On certain central neurotropic effects of lavender essential oil. *Izvest. Inst. Fiziol. Sofiia*, 13, 69–77.

Atanassova-Shopova, S., Roussinov, K.S. and Boycheva, I. (1973), On certain central neurotropic effects of lavender essential oil. II. communication: Studies on the effects of linalool and of terpineol. *Izv. Inst. Fiziol. Bulg. Akad. Nauk*, 15, 149–56.

Benito, M., Jorro, G., Morales, C., Pelaez, A. and Fernandez, A. (1996), Labiatae allergy: systemic reactions due to ingestion of oregano and thyme. *Ann. Allergy, Asthma, Immunol.*, 76, 416–18.

Betourne, M. (1995), Hair growth stimulants containing glycyrrhizic acid ammonium salt. *Fr. Demande* 2709952; *Chem. Abstr.*, 123, 20, 17450 c.

Bezard, M., Karlberg, A.T., Montelius, J. and Lepoittevin, J.P. (1997), Skin sensitization to linalyl hydroperoxide: support for radical intermediates. *Chem. Res. Toxicol.*, **10**, 987–93.

BIBRA working group (1994), Lavender oil: BIBRA toxicity profile of lavender oil. *Govt. Reports Announcements & Index* (GRA & I), Issue 19, 1996.

Billany, M.R., Denman, S., Jameel, S. and Sugden, J.K. (1995), Topical antirheumatic agents as hydroxyl radical scavengers. *Int. J. Pharm.*, **124**, 279–83.

Binet, P. (1972), Action de quelques alcools terpéniques sur le système nerveux des poissons. *Ann. Pharm. Franc.*, **30**, 653–8.

Binet, L., Binet, P., Micque, M., Roux, M. and Bernier, A. (1972), Recherches sur les proprietés pharmacodynamiques (action sédative et action spasmolytique) de quelques alcools terpéniques aliphatiques. *Ann. Pharm. Franc.*, **30**, 611–16.

Biskys, V., Mikelskas, R., Lozinskij, I., Taran, V., Biskys, M. and Butkus, R. (1996), Composition for therapeutic baths. *Lith.* 4080; *Chem. Abstr.* **126**, 15, 203781 y.

Bradshaw, R.M., Marchant, J.N., Meredith, M.J. and Broom, D.M. (1998), Effects of lavender straw on stress and travel sickness in pigs. *J. Altern. Complement. Med.*, **4**, 271–5.

Brandao, F.M. (1986), Occupational allergy to lavender oil. *Contact Dermatitis*, **15**, 249–50.

Buchbauer, G. (1998), Aromatherapie, naturwissenschaftlich betrachtet. *Z. Phytother.*, **19**, 209–12.

Buchbauer, G. and Jirovetz, L. (1994), Aromatherapy – use of fragrances and essential oils as medicaments. *Flav. Fragr. J.*, **9**, 217–22.

Buchbauer, G., Jáger, W., Našel, B., Ilmberger, J., Dietrich, H. (1994), The biology of essential oils and fragrance compounds, *Proceedings from the Aromatherapy Symposium Essential Oils, Health & Medicine*, pp. 69 – 76, New York.

Buchbauer, G., Jirovetz, L., Czejka, M. and Nasel, C. (1993a), New Results in Aromatherapy Research. *24th Intern. Symp. Essent. Oils*, Berlin, July 21–24, 1993, abstract, p18.

Buchbauer, G., Jirovetz, L., Jäger, W., Plank, C. and Dietrich, H. (1993b), Fragrance compounds and essential oils with sedative effects upon inhalation. *J. Pharm. Sci.*, **82**, 660–4.

Buchbauer, G., Jirovetz, L., Jäger, W., Dietrich, H., Plank, C. and Karamat, E. (1991), Aromatherapy: evidence of sedative effects of the essential oil of lavender. *Z. Naturforsch.*, **46c**, 1067–72.

Butcher, R.D., MacFarlane-Smith, W., Robertson, G.W. and Griffiths, D.W. (1994), The identification of potential aeroallergen/irritant(s) from oilseed rape (*Brassica napus* spp. *oleifera*): volatile organic compounds emitted during flowering progression. *Clin. Exp. Allergy*, **24**, 1105–14.

Ceschel, G.C., Maffei, P., Moretti, M.D.L., Peana, A.T. and Demontis, S. (1997), Ex vivo permeation through porcine oral mucous membrane of Salvia desoleana Atzei & Picci essential oil from topical formulations. *Farm. Vestn.*, **48**, 240–1.

Chadba, A. and Madyastha, K.M. (1984), Metabolism of geraniol and linalool in the rat and effects on liver and lung microsomal enzymes. *Xenobiotica*, **14**, 365–74.

Christensen, L.P., Jakobsen, H.B., Paulsen, E., Hodal, L. and Andersen, K.E. (1999), Airborne compositae dermatitis: monoterpenes and no parthenolide are released from flowering *Tanacetum parthenium* (feverfew) plants. *Arch. Dermatol. Res.*, **291**, 425–31.

Collin, P. (1996), Speiklavendel. *Forum Aromatherapie & Aromapflege*, **1**, 5–6.

Cornwell, S. and Dale, A. (1995), Lavender oil and perineal repair. *Mod. Midwife*, **5**, 31–3.

Coulson, I.H. and Khan, A.S.A. (1999), Facial 'pillow' dermatitis due to lavender oil allergy. *Contact Dermatitis*, **41**, 111.

Dale, A. and Cornwell, S. (1994), The role of lavender oil in relieving perineal discomfort following childbirth: a blind randomized clinical trial. *J. Adv. Nurs.*, **19**, 89–96.

Daniels, C., Gottwald, B., Pause, B.M., Sojka, B., Mehdorn, H.M. and Ferstl, R. (2001), Olfactory event-related potentials in patients with brain tumors. *Clin. Neurophysiol.*, **112**, 1523–30.

Deans, S.G. and Noble, R.C. (1995), Beneficial effects of plant volatile oils in maintenance of levels of polyunsaturated fatty acids, prevention or mitigation of deleterious changes in nervous tissue, elevation of protein levels and prevention or mitigation of retinal degeneration. *PCT. Intern. Appl.* 95 05838; *Chem. Abstr.* **122**, 21, 256419 k.

De Groot, A.C. and Liem, D.H. (1983), Facial psoriasis caused by contact allergy to linalool and hydroxycitronellal in an after-shave. *Contact Dermatitis*, **9**, 230–2.

Delaveau, P., Guillemain, J., Narcisse, G. and Rousseau, A. (1989), Sur les proprietes neuro-depressives de l'huile essentielle de lavande. *C.R. Seances Soc. Biol. Ser. Fil.*, **183**, 342–8.

Diego, M.A., Jones, N.A., Field, T., Hernandez-Reif, M., Schauberg, S., Kuhn, C., MacAdam, U., Galamaga, R. and Galamaga, M. (1998), Aromatherapy positively affects mood, EEG patterns of alertness and math computations. *Int. J. Neurosci.*, **96**, 217–24.

Dunn, C., Sleep, J. and Collett, D. (1995), Sensing an improvement: an experimental study to evaluate the use of aromatherapy, massage and periods of rest in an intensive care unit. *J. Adv. Nurs.*, **21**, 34–40.

Elisabetsky, E., Brum, L.F. and Souza, D.O. (1999), Anticonvulsant properties of linalool in glutamate-related seizure models, *Phytomedicine*, **6**, 107–13.

Elisabetsky, E., Coelho de Souza, G.P., dos Santos, M.A.C., Siquieira, I.R., Amador, T.A. and Nunes, D.S. (1995a), Sedative properties of linalool. *Fitoterapia*, **66**, 407–14.

Elisabetsky, E., Marschner, J. and Souza, D.O. (1995b), Effects of linalool on glutamatergic system in the rat cerebral cortex. *Neurochem. Res.*, **20**, 461–5.

Fröhlich, E. (1968), Oleum spicae: Übersicht über klinische, pharmakologische und bakteriologische Untersuchungen. *Wr. Klin. Woschenschr.*, 345–50.

Frohn, B. and Joas, A. (1997), *Ätherische Öle, Aromen und Düfte*, Südwest Verlag, München, pp 152–4.

Fuchs, N., Jäger, W., Lenhardt, A., Böhm, L., Buchbauer, I. and Buchbauer, G. (1997), Systemic absorption of topically applied carvone: influence of massage technique. *J. Soc. Cosmet. Chem.*, **48**, 277–82.

Gabriel, M.M., Moreira, E.A., Miguel. O.G., Nakashima, T. and Lopes, M. (1998), Phytochemical study of essential oil from *Hedyosmum brasiliense* Mart. ex Miq. (Chloranthaceae). *Rev. Brasil. Farm.*, **79**, 65–8.

Galle-Hoffmann, U. (1997), Lavendel in der Provence. *Dtsche. Apoth.-Ztg.*, **137**, 3986–9.

Gaworski, C.L., Vollmuth, T.A., Dozier, M.M., Heck, J.D., Dunn, L.T., Ratajczak, H.U. and Thomas, P.T. (1994), An immunotoxicity assessment of food flavoring ingredients. *Food Chem. Toxicol.*, **32**, 409–15.

Ghelardini, C., Galeotti, N., Salvatore, G. and Mazzanti, G. (1999), Local anaesthetic activity of the essential oil of *Lavandula angustifolia*. *Planta Med.*, **65**, 700–3.

Gruncharov, V. (1973), Clinico-experimental study on the choleretic and cholagogic action of Bulgarian lavender oil. *Vutr. Boles.*, **12**, 90–6.

Guillemain, J., Rousseau, A., Delaveau, P. (1989), Effets neurodepresseurs de l'huile essentielle de *Lavandula Angustifolia* Mill. *Ann. Pharm. Fr.*, **47**, 337–43.

Häringer, E. (1996), Balance von Physis und Psyche, Historisches, Pharmakologisches, Medizinisches zu Lavendel. *Forum Aromatherapie & Aromapflege*, **1**, 16–18.

Hardy, M., Kirk-Smith, M.D. and Stretch, D.D. (1995), Replacement of drug treatment for insomnia by ambient odor. *Lancet*, **346**, 701.

Hartwig, G. (1996), Essential oils for prevention and treatment of decubitus ulcers. *Germany Offen.* 19518836; *Chem. Abstr.*, **126**, 5, 65430 e.

Hay, I.C., Jamieson, M. and Ormerod, A.D. (1998), Randomized trial of aromatherapy. Successful treatment for alopecia areata. *Arch. Dermatol.*, **134**, 1349–52.

Hohmann, J., Zupko, I., Redei, D., Csanyi, M., Falkay, G., Mathe, I. and Janicsak, G. (1999), Protective effects of the aerial parts of *Salvia officinalis*, *Melissa officinalis* and *Lavandula angustifolia* and their constituents against enzyme dependent and enzyme-independent lipid peroxidation. *Planta Med.*, **65**, 576–8.

Hongratanaworskit, T., Heuberger, E. and Buchbauer, G. (2000a), Effects of sandalwood oil and α-santalol on humans. 1. Inhalation. Poster: *Int. Essent. Oil Symp.* Hamburg, Sept. 10–13.

Hongratanaworskit, T., Heuberger, E. and Buchbauer, G. (2000b), Effects of sandalwood oil and α-santalol on humans. 2. Percutaneous administration. Poster: *Int. Essent. Oil Symp.* Hamburg, Sept. 10–13.

Hudson, R. (1996), Nursing: the value of lavender for rest and activity in the elderly patient. *Complement. Ther. Med.*, **4**, 52–7.

Ilmberger, J., Heugerger, E., Mahrhofer, C., Dessovic, H., Kowarik, D. and Buchbauer, G. (2001), The influence of essential oils on human attention: 1. Alertness. *Chem. Senses*, **26**, 239–245.

Imaseki, I. and Kitabatake, Y. (1962), Effect of essential oils and their components on the isolated intestine of mice. *Yakugaku Zasshi*, 82, 1326–32; *Chem. Abstr.* (1963), 58, 7279 a.

Ishitoya, T. and Okada, M. (1997), Stress relaxation fragrances, their compositions, and cosmetics containing them. *Japan Kokoai Tokkyo Koho*, 97 302377; *Chem. Abstr.*, 128, 5, 53075 w.

Jäger, W., Buchbauer, G., Jirovetz, L. and Fritzer, M. (1992), Percutaneous absorption of lavender oil from a massage oil. *J. Soc. Cosmet. Chem.*, 43, 49–54.

Jager, W., Nasel, B., Nasel, Ch., Stimpfl, T., Vycudilik, W. and Buchbauer, G. (1996), *Chem. Senses*, 21, 477–80.

Jager, W., Mayer, M., Platzer, P., Reznicek, G., Dietrich, H. and Buchbauer, G. (2000), Stereoselective metabolism of the monoterpene carvone by rat and human liver microsomes. *J. Pharm. Pharmacol.*, 52, 191–7.

Kana, I., Yoshihiko, K., Kenichi, F., Kazumi, N., Tatsumara, M. and Kenichi, T. (1998), Effects of the odor of essential oils on brain function reflected in the results of the new EEG analysis program LORETA. *Chem. Senses*, 23, 215.

Kararli, T.T., Kirchhoff, C.F. and Penzotti, Jr., S.C. (1995), Enhancement of transdermal transport of azidothymidine (AZT) with novel terpene and terpene like enhancers: in vivo-in vitro correlations. *J. Control. Release*, 34, 43–51.

Karita, T. (1996), Preparation of capsules, storage thin sheets, bag-type dosage forms for volatile drugs and topical administration of bag-type forms for treatment of skin diseases. *Japan Kokai Tokkyo Koho*, 96 109137; *Chem. Abstr.*, 125, 8, 96069 c.

Kerl, S. (1997), Zur olfaktorischen Beeinflussbarkeit von Lernprozessen. *Dragoco Report*, 44, 45–59.

Kim, H.M. and Cho, S.H. (1999), Lavender oil inhibits immediate-type allergic reaction in mice and rats. *J. Pharm. & Pharmacol.*, 51, 221–6.

Klemm, W.R., Lutes, S.D., Hendrix, D.V. and Warrenburg, S. (1992), Topographical EEG maps of human responses to odors. *Chem. Senses*, 17, 347–61.

Knasko, S.C. (1992), Ambient odor's effect on creativity, mood, and perceived health. *Chem. Senses*, 17, 27–35.

Kommuru, T.R., Khan, M.A. and Reddy, I.K. (1998), Racemate and enantiomers of ketoprofen: phase diagram, thermodynamic studies, skin permeability, and use of chiral permeation enhancers. *J. Pharm. Sci.*, 87, 833–40.

Kommuru, T.R., Khan, M.A. and Reddy, I.K. (1999), Effect of chiral enhancers on the permeability of optically active and racemic metoprolol across hairless mouse skin. *Chirality*, 11, 536–40.

Komori, T., Tamida, M., Kikuchi, A., Shoji, K., Nakamura, S. and Nomura, J. (1997), Effects of odorant inhalation on pentobartbital-induced sleep time in rats. *Human Psychopharmacol.*, 12, 601.

Kunta, J.R., Goskonda, U.R., Brotherton, H.O., Khan, M.A. and Reddy, I.K. (1997), Effect of menthol and related terpenes on the percutaneous absorption of propanolol across excised hairless mouse skin. *J. Pharm.* Sci., 86, 1369–73.

Laboratoires Meram (1966), Nouveaux sedatifs et spasmolytiques. *Brevet Special de Medicament, Fr. Demande* 4055 MI.

Lantry, L.E., Zhang, Z., Gao, F., Crist, K.A., Wang, Y., Kellof, G.J., Lubet, R.A. and You, M. (1997), Chemopreventive effect of perillyl alcohol on 4-(methylnitrosamino)-1-(3-pyridyl)-1-butanone induced tumorigenesis (C3H/HeJ X A/J)F1 mouse lung. *J. Cell. Biochem.*, 27 (Suppl.), 20–5.

Lee, C.F., Katsuura, T., Shibata, S., Ueno, Y., Ohta, T., Hagimoto, S., Sumita, K., Okada, A., Harada, H. and Kikuchi, Y. (1994), Responses of electroencephalogramm to different odors. *Ann. Physiol. Anthropol.*, 13, 281–91.

Lis-Balchin, M. (1997), Essential oils and 'aromatherapy': their modern role in healing. *J. Royal Soc. Health*, 117, 324–9.

Lis-Balchin, M. and Hart, S. (1997), A preliminary study of the effect of essential oils on skeletal and smooth muscle in vitro. *J. Ethnopharmacol.*, 58, 183–7.

Lis-Balchin, M. and Hart, S. (1999), Studies on the mode of action of the essential oil of lavender (*Lavandula angustifolia* Miller). *Phytother. Res.*, 13, 540–2.

Lozano, V.C., Bonnard, E., Gauthier, M. and Richard, D. (1996), Mecamylamine-induced impairment of acquisition and retrieval of olfactory conditioning in the honeybee. *Behav. Brain. Res.*, 81, 215–22.

Ludvigson, H.W. and Rottmann, T.R. (1989), Effects of ambient odors of lavender and cloves on cognition, memory, affect and mood. *Chem. Senses*, 14, 525–36.

Masago, R., Matsuda, T., Kikuchi, Y., Miyazaki, Y., Iwanaga, K., Harada, H. and Katsuura, T. (2000), Effects of inhalation of essential oils on EEG activity and sensory evaluation. *J. Physiol. Anthropol. Appl. Human Sci.*, 19, 35–42.

Mazzanti, G., Lu, M. and Salvatore, G. (1998a), Spasmolytic action of the essential oil from *Hyssopus officinalis* L. var. *decumbens* and its major components. *Phytother. Res.*, 12 (Suppl.), S92–4.

Moretti, M.D.L., Peana, A.T. and Satta, M. (1997), A study on anti-inflammatory and peripheral analgesic action of Salvia sclarea oil and its main components. *J. Essent. Oil Res.*, 9, 199–204.

Nakamatsu, Y. (1995), Brain stimulants containing terpenes. *Japan Kokai Tokkyo Koho*, 95 258113; *Chem. Abstr.*, 124, 2, 15522 z.

Nasel. B., Nasel, Ch., Samec, P., Schindler, E. and Buchbauer, G. (1994), Functional imaging effects of fragrances on the human brain. *Chem. Senses*, 19, 359–64.

Nikolaevskii, U.V., Kononova, N.S., Pertsovskii, A.I. and Shinkarchuk, I.F. (1990), Effect of essential oils on the course of experimental atherosclerosis. *Patol. Fiziol. Eksp. Ter.*, 52–3.

Anon (1997a), Ointment for prevention of facial scarring and blemishes of the face and body. *Fr. Demande*, 2743720; *Chem. Abstr.*, 128, 1, 7221 t.

Anon (1997b), Topical pharmaceuticals for the treatment of acne and dilated pores. *Fr. Demande*, 2743722; *Chem. Abstr.*, 128, 1, 7307 a.

Pahlow, M. (1988), Hausmittel in der Apotheke. *Dtsche. Apoth.-Ztg.*, 128, 2462–3.

Peana, A., Satta, M., Moretti, M.D.L. and Orecchioni, M. (1994), A study on choleretic activity of salvia desoleana essential oil. *Planta. Med.*, 60, 478–9.

Pham-Delegue, M.H., Bailez, O., Blight, M.M., Masson, C., Picard-Nizou, A.L. and Wadhams, L.J. (1993), Behavioral discrimination of oilseed rape volatiles by the honeybee Apis mellifera L. *Chem. Senses*, 18, 483–94.

Powers, K.A. and Beasley, V.R. (1985), Toxicological aspects of linalool: a review. *Vet. Human Toxicol.*, 27, 484–6.

Rademaker, M. (1994), Allergic contact dermatitis from lavender fragrance in Difflam gel. *Contact Dermatitis*, 31, 58 9.

Re, L., Barocci, S., Sonnino, S., Mencarelli, A., Vivani, C., Paolucci, G., Scarpantonio, A., Rinaldi, L. and Mosca, E. (2000), Linalool modifies the nicotinic receptor-ion channel kinetics at the mouse neuromuscular junction. *Pharmacol. Res.*, 42, 177–182.

Reddy, A.C. and Lokesch, B. (1992), Studies on spice principles as antioxidants in the inhibition of lipid peroxidation of rat liver microsomes. *Molec. Cell. Biochem.*, 111, 117–24.

Reddy, B.S., Wang, C.X., Samaha, H., Lubet, R., Steele, V.E., Kelloff, G.J. and Rao, C.V. (1997), Chemoprevention of colon carcinogenesis by dietary perillyl alcohol. *Cancer Res.*, 57, 420–25.

Roffey, S.J., Walker, R. and Gibson, G.G. (1990), Hepatic peroxisomal and microsomal enzyme induction by citral and linalool in rats. *Food & Chem. Toxicol.*, 28, 403–8.

Romine, I.J., Bush, A.M. and Geist, C.R. (1999), Lavender aromatherapy in recovery from exercise. *Percept. Mot. Skills*, 88, 756–8.

Rovesti, P. (1971), L'aromaterapia dell' essenza di lavanda. *Aromi, Saponi, Cosmetici, Aerosol*, 53, 251–69.

Rovesti, P. (1974), Tolerance et pharmacologie de l'essence de lavande. *Aromi, Saponi, Cosmetici, Aerosol*, 56, 71–7.

Rovesti, P. and Colombo, E. (1973), Aromatherapy and aerosols. *Soap, Perf. & Cosmet.*, 46, 475–7.

Sandoz, J.C., Roger, B. and Pham-Delegue, M.H. (1995), Olfactory learning and memory in the honeybee: comparison of different classical conditioning procedures of the proboscis extension response. *Compt. R. Acad. Sci. Paris, Serie III*, 318, 749–55.

Sato, T., Hata, H., Ishii, H. and Watanabe, S. (1995), Plant growth enhancement. *Japan Kokai Tokkyo Koho*, 95 87845; *Chem. Abstr.*, 123, 3, 27820 n.

Schaller, M. and Korting, H.C. (1995), Allergic airborne contact dermatitis from essential oils used in aromatherapy. *Clin. Exp. Dermatol.*, 20, 143–5.

Schmidt, E. (1996), Ein Lavendelöl ist ein Lavendelöl, ist ein Lavendelöl – oder doch nicht? *Forum Aromatherapie & Aromapflege*, 1, 7–15.

Schulz, V., Hübner, W.D. and Ploch, W. (1997), Clinical studies with phyto-psychopharmaceuticals. *Z. Phytother.*, 18, 141–54.

Silva-Brum, L.F., Elisabetsky, E. and Souza, D. (2001a), Effects of Linalool. *Phytother. Res.*, 15, 422–55.

Silva-Brum, L.F., Emanuelli, T., Souza, D.O. and Elisabetsky, E. (2001b), Effects of linalool on glutamate release and uptake in mouse cortical synaptosomes. *Neurochem. Res.*, 26, 191–4.

Singh, G., Kappor, I.P.S. and Pandey, S.K. (1998), Studies on essential oils. Part 14. Natural preservatives for butter. *J. Med. Aromat. Plant. Sci.*, 20, 735–9.

Siurin, S.A. (1997), Effects of essential oil on lipid peroxidation and lipid metabolism in patients with chronic bronchitis. *Klin. Med. Moskau*, 75, 43–5.

Stimpfl, T., Nasel, B., Nasel, Ch., Binder, R., Vycudilik, W. and Buchbauer, G. (1995), Concentration of 1,8-cineole in human blood after prolonged inhalation, *Chem. Senses*, 20, 349–50.

Sugawara, Y., Hara, C., Tamura, K., Fujii, T., Nakamura, K., Masujima, T. and Aoki, T. (1998), Sedative effects on humans of inhalation of essential oil of linalool: sensory evaluation and physiological measurements using optically active linalools. *Anal. Chim. Acta*, 365, 293–9.

Sugawara, Y., Hara, C., Aoki, T., Sugimoto, N. and Masujima, T. (2000), Odor distinctiveness between enantiomers of linalool: difference in perception and responses elicited by sensory test and forehead surface potential wave measurement. *Chem. Senses*, 25, 77–84.

Suzuki, M. and Aoki, T. (1994), Effects of volatile compounds from leaf oil on blood pressure after exercising. *Mokuzai Gakkaishi*, 40, 1243–50. *Chem. Abstr.*, 122, 17, 204810 m.

Sysoev, N.P. and Lanina, S.I. (1990), The results of sanitary chemical research into denture base materials coated with components from essential oil plants. *Stomatologia*, 59–61.

Tanisawa, S. and Suga, C. (1999), Method for evaluating the anti-stress effect of fragrance. *Japan Kokai Tokkyo Koho*, 99 19076; *Chem. Abstr.*, 130, 12, 148677 z.

Tasev, T., Toleva, P. and Balabanova, V. (1969), Neurophysical effect of Bulgarian essential oils from rose, lavender and geranium. *Folia Medica*, 11, 307–17.

Tewari, R. and Sharma, A. (1987), Chemistry of lavender oil, a review. *Current Res. Med. & Aromatic Plants*, 9, 92–104.

Thacharodi, D. and Rao, D.P. (1994), Transdermal absorption of nifedipine from microemulsions of lipophilic skin penetration enhancers. *Int. J. Pharm.*, 111, 235–40.

Uehleke, B. (1996), Phytobalneology. *Z. Phytother.*, 17, 26–43.

Umezu, T. (2000), Behavioral effects of plant-derived essential oils in the geller type conflict test in mice. *Jpn. J. Pharmacol.*, 83, 150–153.

Vernet-Maury, E., Alaoui-Ismaeli, O., Dittmar, A., Delhomme, G. and Chanel, J. (1999), Basic emotions induced by odorants: a new approach based on autonomic pattern results. *J. Autonom. Nerv. Syst.*, 75, 176–83.

Wattenberg, I.W. (1991), Inhibition of azoxymethane-induced neoplasia of the large bowel by 3-hydroxy-3,7,11-trimethyl-1,6,10-dodecatriene (nerolidol). *Carcinogenesis*, 12, 151–2.

Wichtl, M. (1997), *Teedrogen und Phytopharmaka*, Wissenschaftl. Verlagsgesellsch., Stuttgart.

Worwood, V.A. (1994), *The Fragrant Pharmacy*, Bantam Books, Toronto-New York-London-Sydney-Auckland.

Yagyu, T. (1994), Neurophysiological findings on the effects of fragrance: lavender and jasmine. *Integrative Psychiatry*, 10, 62–7.

Yamada, K., Mimaki, Y. and Sashida, Y. (1994), Anticonvulsive effects of inhaling lavender oil vapour. *Biol. Pharm. Bull.*, 17, 359–60.

Yamahara, J., Mik, A. and Yamaguchi, Y. (1994), Herbal extracts containing d-borneol for treatment of periodontal diseases. *Japan Kokai Tokkyo Koho*, 94 247864; *Chem. Abstr.*, 122, 2, 17180 u.

Yano, H. (1998), Hypnotics, foods and feeds containing lavender essential oil. *Japan Kokai Tokkyo Koho*, 98 25246; *Chem. Abstr.*, 128, 10, 119700 e.

Yoshida, T. and Okazaki, Y. (1998), Anxiety reduction using odors and frontal lobe function. *Chem. Senses*, 23, 215.

Zhang, Z., Chen, H., Chan, K.K., Budd, T. and Ganapathi, R. (1999), Gas chromatographic-mass spectrometric analysis of perillyl alcohol and metabolites in plasma. *J. Chromatogr., B, Biomed. Sci.-Appl.*, 728, 85–95.

13 Pharmacology of *Lavandula* essential oils and extracts *in vitro* and *in vivo*

Stephen Hart and Maria Lis-Balchin

Introduction

There are large numbers of *Lavandula* species, hybrids and cultivars growing all over the world. The main uses for them is based on their aroma as they are used in a variety of perfumes, soaps, creams as well as for their decorative and odorant qualities (as dried lavender bunches) or in scented cushions to ensure relaxation and promote better sleep. For the perfumery and other odorant products, true lavender, *Lavandula angustifolia* is favoured, closely followed by its hybrids *L. x intermedia*, which includes cultivars like 'Grosso' and 'Abrialii' in the commercial essential oil market. All these contain linalool and linalyl acetate as their main components. The more camphoraceous *L. latifolia*, or spike oil, is not widely used in the products listed above, but can be used in some food products and household cleaners and has had a greater medicinal usage than *L. angustifolia* in the past.

Folk-medicinal usage of *Lavandula* species

The Abbess Hildegarde (1098–1179) wrote about lavender and its ability to delouse and to clear eyes and also drive away evil spirits. Lavender was used in many medicines in medieval Wales and England in conjunction with numerous other herbs. Gerard (1636) mentions spike lavender as being effective against catalepsy, migraine and fainting. Psychological healing was also possible, for example, for passions of the heart and inward and outward grief (Culpeper, 1653). He mentioned the usage of lavender for the stomach, liver and obstructions in the spleen, for the kidneys and against colic and wind (suggesting a relaxant effect). He, however, also stated that it brought on menses and caused abortion of the dead child and after-birth (i.e. a contractile effect).

Culpeper stated that true sweet lavender (*L. angustifolia*) was not considered good for anything medicinal, only as a perfume, and, he also warned against excessive use of the 'oil of spike'. However, these statements have been forgotten or completely ignored and all the lavenders are now considered equal, in the absence of scientific proof.

Study of *in vivo* pharmacological activity

There are three appropriate approaches to the study of the effects of lavender in the whole animal which differ by their route of administration. The classical approach is parenteral administration into the conscious or anaesthetised animal. Results from such experiments will demonstrate the pharmacology of the compound, but may be difficult to relate to their normal use. Exposure of animals to the aroma of an essential oil, or the application of the diluted oil to depilated skin, obviously mimics in animals the normal use of the compounds in humans. The ideal investigation will use human volunteers and two approaches are possible. In the first the purpose of the experiment is to determine the pharmacology of the essential oil and a dose-related effect would be expected while in the second one is interested in whether, at the concentrations normally used, the oil has any significant effects. The latter experiments require careful design, which will differ depending on whether the aroma alone, or aroma plus massage plus absorption across skin, is being investigated.

Study of *in vitro* pharmacological activity

The preparation chosen to assess the *in vitro* pharmacological activity of an essential oil or its components on smooth muscle must respond to spasmogenic agents, that is, those causing a contraction and spasmolytic agents, which will relax smooth muscle. Preparations of intestinal smooth muscle are robust and, although spasmogenic and spasmolytic activity can be identified on those from rabbit and the large intestine of the guinea-pig, the field stimulated guinea-pig ileum allows quantitative experiments to be performed readily. The field stimulated guinea-pig ileum also enables a neurogenic response to be distinguished from a myogenic response and the receptors involved to be elucidated. When the reported use of the essential oil indicates a targeted organ such as the uterus or lungs then appropriate preparations of these smooth muscles can be studied. For a more complete study of the *in vitro* pharmacology it would be necessary to include vascular and cardiac tissues and to examine activity on skeletal muscle such as the rat phrenic nerve hemi-diaphragm preparation. It must of course be remembered that essential oils contain many components and the observed pharmacological action of the oil may be due to several active principles acting by different mechanisms or, indeed, by opposing mechanisms which may give rise to biphasic responses. It is also important to verify that the solvent for an essential oil or its components, in the concentration that will be in contact with the tissue, does not itself have a spasmogenic or spasmolytic action. Methanol is satisfactory for most smooth muscle preparations at a dilution of not less than 1:125, but it has been reported that hexane extracts may give misleading results (Lis-Balchin *et al.*, 1997). It is questionable whether aqueous emulsions are appropriate: Gamez *et al.* (1990) emulsified their oils in Tween 80 and other vehicles include Arlatone 285 (Reiter and Brandt, 1985) and dimethylsulphoxide, DMSO, (Aqel, 1991).

Alternative *in vitro* approaches include the use of cell cultures or binding studies in which the ability of the oil to displace a radio-labelled ligand from a particular receptor on a membrane preparation is measured. Such binding studies are especially useful in the investigation of the possible mode of action of an essential oil within the brain.

Investigations of *Lavandula* essential oils

The aroma of lavender must be one of the best known and the biological activity of the vapour from the essential oil has been investigated in several species including man. As early as 1920, Plant looked at the actions of 'waters' of lavender on intestinal activity in the dog and a year later Macht and Ting (1921) reported on the sedative action of the vapour from tincture of lavender on rats. More recently, Buchbauer's group in Vienna have performed detailed experiments on mice and humans (Buchbauer *et al.*, 1993).

Triebs (1956) and Sticher (1977) have reviewed the results from early experiments on some of the constituents of lavender, which is one of the oils reviewed by Tisserand and Balacs (1995) in their guide on the safety of essential oils.

Experiments *in vivo*

Cardiovascular system

Lavender oil and its component, linalool, produce a fall in blood pressure in experimental animals probably due to peripheral vasodilatation (Tisserand and Balacs, 1995).

Central nervous system

Macht and Ting (1921) trained rats to negotiate a maze to find food and then determined the effect of exposure to the vapour from essential oils on the time taken to reach the food and the number of errors occurring. Most of their experiments were on valerian and various incense, but three rats were exposed to a tincture of lavender and in each case the time to reach the food doubled, with two animals also making errors in the maze. From these results is was concluded by the authors that lavender had a slight sedative action and it was suggested that the vapour from essential oils might be stimulating olfactory sense organs directly. Delaveau *et al.* (1989) administered lavender essence (*L. angustifolia* P. Miller) orally to mice and observed changes in activity to electrical stimulation, which were interpreted as an anxiolytic effect. In addition, it was shown that lavender essence enhanced the hypnotic action of pentobarbitone. Such an effect on barbiturate sleeping time may indicate an effect on the brain but could also be due to inhibition of liver enzymes which metabolise barbiturates. The sedative action of lavender is confirmed by the work of Buchbauer *et al.* (1991) who, like Macht and Ting (1921), studied the effects of the vapour rather than the oil administered orally. Buchbauer *et al.* (1991) exposed mice to the vapour from lavender (*L. angustifolia* Mill.), and the components linalool and linalyl acetate, and found a reduction in overall activity, as monitored in activity boxes. Scores in the activity box are increased when mice are pre-treated with caffeine given by the intraperitoneal route and this increase in activity is very sensitive to inhibition by lavender oil, linalool and linalyl acetate (Buchbauer *et al.*, 1993). Interestingly, exposure of mice to the vapour from linalool led to measurable levels of the compound in their plasma. The authors considered that the sedative action is due to the absorption of linalool and its subsequent transport to the brain by the blood, rather than a direct stimulation of olfactory receptors. Results from experiments in mice and rats thus confirm that lavender is a sedative.

Elisabetsky *et al.* (1995) report that linalool produces a dose-dependent inhibition of the binding of glutamate (an excitatory neurotransmitter in the brain) to its receptors on membranes prepared from the cerebral cortex of the rat, which is a possible explanation for the observed sedative effects. More recently, Elisabetsky *et al.* (1999) have related this action to an anticonvulsant activity of linalool in rats. Lavender is one of the few essential oils that have been studied in well-controlled experiments in humans. Buchbauer *et al.* (1993) measured two types of reaction times in male and female students breathing either air or vapour from the essential oil of lavender. Reaction times were increased by lavender and it was possible to show that this was not due to a peripheral effect but to central sedation. Attempts have also been made to monitor the effects of inhaled essential oils on blood flow through the brain using computerised tomography. Buchbauer *et al.* (1993) observed no changes with lavender oil or linalyl acetate but a decrease with 1,8-cineol. It has also been reported that the vapour of lavender essential oil reduces certain brain waves (contingent negative variation, CNV) in humans (Torii *et al.*, 1988; Manley, 1993).

Experiments on *in vitro* guinea-pig ileum preparations

Method

Segments of ileum (2 cm) from a freshly killed guinea-pig are mounted in an organ bath (25 ml) containing Krebs solution (composition, mM: NaCl 118.1, KCl 4.7, $NaHCO_3$ 25.0, Glucose 11.1, $MgSO_4$ 1.0, KH_2PO_4 1.0 and $CaCl_2 \cdot 2H_2O$ 2.5) maintained at 37°C and gassed continually with 95 per cent oxygen in carbon dioxide. The preparation is placed under a tension of 1 g and contractions recorded through an isometric force transducer connected to a pen recorder or MacLab. The method of Paton (1954) is used to stimulate nerves within the wall of the intestine. Two platinum electrodes, attached to a stimulator, are placed on either side of the intestine and a square wave (width, 0.5 msec) delivered every 10 sec (0.1 Hz) at an appropriate voltage (about 50 V) produces a regular and reproducible contraction of the intestine. This contraction is not affected by the presence of a ganglion-blocking agent, but is inhibited by atropine indicating that it is due to the stimulation of post-ganglionic parasympathetic nerves leading to the release of acetylcholine which acts on muscarinic receptors on smooth muscle. The addition of an extract with spasmogenic activity during field stimulation is recognised as a rise in the base-line and/or an increase in the size of the electrically induced contraction. A reduction in the size of the contraction indicates spasmolytic activity, which can be further analysed as described below (Figure 13.1).

General action on intestinal smooth muscle

Plant (1920) applied waters of lavender to the intestine of dogs *in vivo* and reported increased activity, which was sometimes followed by relaxation and decreased peristaltic activity. Linalool

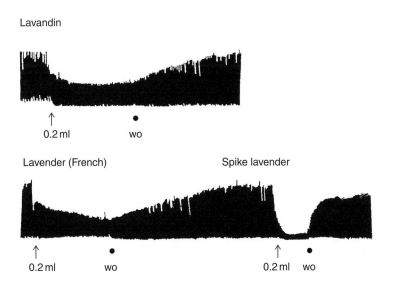

Figure 13.1 The spasmolytic effect of commercial *Lavandula* oils on the electrically-stimulated guinea-pig ileum *in vitro*. Note the slight contraction produced initially with lavender (French).

Notes
wo = wash-out.
Arrows indicate when the essential oil was added.

was reported to relax the small intestine of the mouse (Imaseki and Kitabatake, 1962) while Shipochlier (1968) observed a spasmolytic action on rabbit and guinea-pig gut by the essential oil of lavender (*L. spica* L.) which he judged to be a direct action on the muscle. Reiter and Brandt (1985) in a comprehensive study of several oils and their components report that linalool relaxes the longitudinal muscle of guinea-pig ileum.

Gamez *et al.* (1990) examined the oil of *L. dentata* L., and its components 1,8-cineole and α- and β-pinene, on rat duodenum and observed a spasmolytic activity. Izzo *et al.* (1996) differentiated between the longitudinal and circular muscle of the guinea-pig ileum and found that the essential oil of *L. angustifolia* Mill. relaxed both muscles. There appears therefore to be good agreement that the oils of lavender are spasmolytic on intestinal muscle but Lis-Balchin *et al.* (1996a) report that with some commercial samples the spasmolytic action is preceded by a contraction on guinea-pig ileum (Figure 13.1).

In addition to the European species of lavender mentioned above, two from New Zealand, namely a particular lavender (labelled TP) and another lavender (labelled C), the letters denoting different areas of production as well as *Lavandin* 'Grosso', have been shown to be spasmolytic on guinea-pig ileum (Lis-Balchin *et al.*, 1996b).

Recent experiments using three different extracts of several *Lavandula* species, that is, a cold methanolic extract, a tea, made with boiling water and a hydrosol, that is, the water remaining after steam/water distillation showed that the methanolic extracts of *L. angustifolia* dried flowers, *L. angustifolia* fresh flowers and fresh leaves, assessed separately, *L. stoechas* leaves and *L. viridis* leaves have a spasmolytic action on the guinea-pig ileum (Table 13.1). All the teas and hydrosols except for *L. angustifolia* dried flowers and *L. angustifolia* fresh leaves were only spasmolytic while the water-soluble tea extract of *L. angustifolia* dried flowers and the leaves of *L. angustifolia* showed an initial spasmogenic action.

Neurogenic or myogenic

A concentration of *Lavandula* essential oil which produces a significant inhibition of the contraction of the field stimulated guinea-pig ileum *in vitro*, also reduces the size of the

Table 13.1 Evidence for contraction, C and or relaxation, R for extracts and essential oils in electrically-stimulated guinea-pig ileum *in vitro*

Plant	Extract	Tea	Hydrosol	EO
L. angustifolia (dried flowers)	R	C/R (+base-line C)	C/R	R
L. angustifolia (fresh flowers)	R	R	R	R
L. angustifolia (leaves)	R	R(low) C/R(high)	*C/R	R
L. stoechas (leaves)	R	R	R	R
L. stoechas 'pedunculata' (leaves)	R	R	R	R
L. viridis (leaves)	R	N/A	R	R

Note: (+base-line C) = there is a distinct contraction rather than an increase in tone as shown by the other samples where C is shown; * strong contraction.

contraction of the smooth muscle induced by exogenous acetylcholine (Izzo et al., 1996; Lis-Balchin and Hart, 1999). This shows that the oil is not affecting nerve conduction (local anaesthetic action) or the release of acetylcholine (as seen with opioids such as morphine), but is acting directly on the smooth muscle. All samples studied have a myogenic action.

Atropine-like action

The concentration of *Lavandula* oil, which reduces the size of the contraction of the smooth muscle to acetylcholine, has a similar effect on the contraction due to exogenous histamine (Lis-Balchin and Hart, 1999) or barium chloride (Izzo et al., 1996). If the essential oil contains an atropine-like compound the response to histamine or barium would be unaffected by a concentration of the oil which reduces the response to acetylcholine: as this did not occur it can be concluded that the oil does not contain an atropine-like compound and that it is relaxing the smooth muscle by some other mechanism. None of the extracts studied have been reported to possess atropine-like activity (Figure 13.2).

Adrenoceptor mediated response

The presence of an adrenoceptor agonist in the essential oil would explain the relaxation of the intestinal smooth muscle and such activity would be sensitive to inhibition by a combination of alpha- and beta-adrenoceptor antagonists. A concentration of phentolamine and propranolol, which inhibits the spasmolytic action of exogenous noradrenaline, has no effect on the inhibition of the contraction of the field stimulated guinea-pig ileum produced by the *Lavandula* oil (43.6 per cent before the antagonists and 50.8 per cent in the presence of the antagonists, Lis-Balchin and Hart, 1999). The *Lavandula* essential oils do not, therefore, appear to contain a substance which stimulates adrenoceptors.

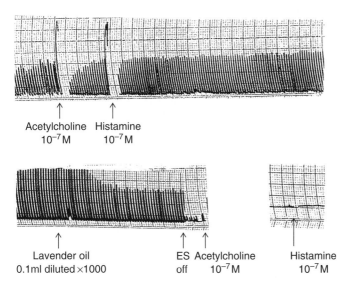

Figure 13.2 Mode of action of lavender oil on the guinea-pig ileum *in vitro*: effect of exogenous acetylcholine and histamine, indicating a similar effect and thus a post-synaptic action.

Involvement of cyclic adenosine monophosphate (cAMP)

The relaxation of intestinal smooth muscle, which occurs as a result of adrenoceptor activation, is mediated by an increase in the level of cAMP as a result of the stimulation of adenylate cyclase. The duration of action of the cAMP so produced is limited by its metabolism by phosphodiesterase. It is thus possible that the essential oil contains a substance which is capable of stimulating adenylate cyclase in which case the spasmolytic action would be expected to be potentiated by a phosphodiesterase inhibitor.

Several phosphodiesterase inhibitors are available which may be classified by their selectivity in inhibiting the seven or so different isoenzymes of phosphodiesterase which have been identified. Theophylline and trequinsin are non-selective inhibitors while rolipram (Banner *et al.*, 1995) and CDP 840 (Hughes *et al.*, 1996) inhibit the type IV isoenzyme.

Lavender oil and one of its components, linalool, appear to mediate a spasmolytic action on intestinal smooth muscle via a rise in cAMP (Lis-Balchin and Hart, 1999). Thus, a concentration of trequinsin, which potentiates the response of isoprenaline from 22.6 to 41.8 (per cent inhibition of twitch response to electrical stimulation), potentiates the spasmolytic activity of lavender oil from 18.2 to 31.1 per cent (Figure 13.3). The response to linalool was potentiated from 17.1 to 34.6 per cent. Similarly, a concentration of CDP840 which potentiates the response of isoprenaline from 23.6 to 36.6 per cent, also potentiates the response to lavender oil from 26.7 to 46.0 per cent.

Involvement of cyclic guanosine monophosphate (cGMP)

Sodium nitroprusside, and other compounds which lead to the local production of nitric oxide, produce a spasmolytic action which is mediated by an increase in cGMP as a result of the

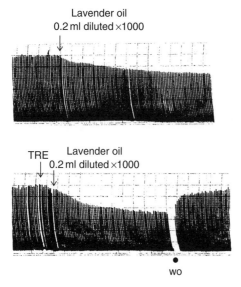

Figure 13.3 Mode of action of lavender oil on the guinea-pig ileum *in vitro*: the enhancement of spasmolysis following the application of Trequensin, a phosphodiesterase inhibitor, suggesting that cAMP is involved.

Note: TRE = trequensin addition.

stimulation of guanylate cyclase by nitric oxide. A further possible mechanism of action for the *Lavandula* oils is therefore via the stimulation of guanylate cyclase, a mechanism which is sensitive to inhibition by ODQ (1H-(1,2,4)oxadia-zolo(4,3-a)quinoxalin-1-one. Garthwaite *et al.*, 1995). A concentration of ODQ which blocks the spasmolytic action of sodium nitroprusside has no effect on the activity of lavender essential oil (Lis-Balchin and Hart, 1999).

Calcium channel blockade

Intestinal smooth muscle contraction involves the influx of calcium ions through L-type channels and calcium channel blockers, such as verapamil and nifedipine, which are used therapeutically for their ability to block calcium channels in vascular and cardiac muscle, have a spasmolytic effect on intestinal smooth muscle. It is therefore possible that essential oils with spasmolytic activity contain components capable of blocking calcium channels.

There are several methods by which possible calcium channel blockade can be identified and in each method one uses an accepted calcium channel blocker as a positive control. Such screening methods have been reviewed by Neuhaus-Carlisle *et al.* (1997) and Vuorela *et al.* (1997), but neither group mention lavender oil.

When normal Krebs solution is replaced by calcium-free Krebs solution the contraction due to field stimulation is lost, but returns gradually when the tissue is returned to normal Krebs solution. The rate of recovery of the twitch is delayed by the presence of a calcium channel blocking agent, thus in the presence of nifedipine (10^{-6} M) recovery of the twitch is delayed from 8.6 ± 1.9 to 30.0 ± 5.3 min. Lavender oil, at a concentration which inhibits the twitch and is potentiated by a phosphodiesterase inhibitor, does not delay the recovery (9.0 ± 3.7 min, Lis-Balchin and Hart, 1998).

When guinea-pig ileum is bathed with a depolarising-Krebs solution the addition of calcium chloride solution produces dose-related contractions that are sensitive to inhibition by calcium channel blockers. Nifedipine (4×10^{-5} M) inhibits the calcium contraction by 84 per cent while lavender oil, at a concentration which produces a significant inhibition of the twitch response (8×10^{-6} w/v), has no effect (Figure 13.4). However, at higher concentrations of lavender oil there is evidence of calcium block with 4×10^{-5} giving 93 per cent inhibition with a complete block occurring at 8×10^{-5} (Lis-Balchin and Hart, 2000). When the dose-response curve for lavender oil against inhibition of the electrically induced contraction of the guinea-pig ileum is plotted on the same graph as that for inhibition of the calcium contraction the two lines are significantly different. This indicates that although lavender oil is capable of blocking calcium movement into the smooth muscle cell this is not the primary mechanism of action in the guinea-pig ileum. Gamez *et al.* (1990), in their study on the rat duodenum of the essential oil from *L. dentata* L., showed that it, and the component 1,8-cineole, inhibited calcium-induced contractions and concluded that a block of calcium channels was contributing to the spasmolytic activity. There is recent evidence to show that the methanolic extracts of *L. angustifolia* (dry flowers, fresh flowers and fresh leaves) are calcium channel blockers as are the leaves of *L. viridis* and *L. stoechas* (Table 13.2).

Potassium channel opening

Some drugs which are used therapeutically to relax vascular smooth muscle act by opening potassium channels and thereby hyperpolarise the cell. Cromokalim is such a compound and it has a spasmolytic action on intestinal smooth muscle thus emphasising another possible mode of action for the spasmolytic action of *Lavandula* essential oils. In the presence of a potassium

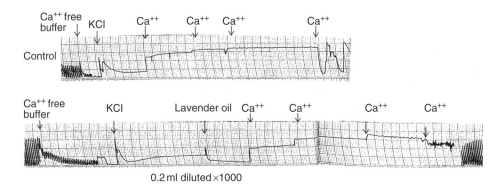

Figure 13.4 Mode of action of lavender oil on the guinea-pig ileum *in vitro*: investigating the possibility of calcium channel involvement using calcium-free buffers: the effect of depolarisation with KCl on added calcium with and without the addition of lavender oil at low concentration, showing that this is an unlikely mode of action at such concentrations.

Table 13.2 Evidence for Ca^{++} channel blocking by different extracts of *Lavandula* species

L. angustifolia (dry flowers)	MeOH extract
L. angustifolia (fresh flowers)	MeOH extract
L. angustifolia (fresh leaves)	MeOH extract and EO
L. viridis leaves	MeOH extract and hydrosol
L. stoechas leaves	Hydrosol
L. stoechas 'pedunculata'	MeOH extract

channel opener, guinea-pig ileum will not respond to a high concentration of potassium (60 mM) with a contraction. If tissue is contracted with a high concentration of potassium then a relaxation can be obtained with a calcium channel blocker but not with a potassium channel opener. There is no experimental evidence to suggest that lavender oil contains components capable of opening potassium channels (Lis-Balchin and Hart, 1999).

Spasmogenic action on intestinal smooth muscle

An initial contraction is sometimes observed with lavender essential oil before the spasmolytic action occurs (Lis-Balchin *et al.*, 1996a) and this is probably due to the presence of 1,8-cineole and α- and β-pinene, which have been shown to contract guinea-pig ileum (Lis-Balchin and Hart, 1997). Preliminary results indicate that α-pinene is not acting via muscarinic cholinoceptors or histamine receptors and that the two enantiomers of α-pinene do not have identical pharmacological activity (Lis-Balchin *et al.*, 1999). On the duodenum of the rat, Gamez *et al.* (1990), on the other hand, found 1,8-cineole to have spasmolytic activity. Recent experiments using methanolic and water-soluble extracts of *L. angustifolia* dried flowers, *L. angustifolia* fresh flowers and fresh leaves, assessed separately, *L. stoechas* leaves and *L. viridis* leaves have indicated that the water-soluble tea extract of *L. angustifolia* dried flowers and the leaves of *L. angustifolia* have some spasmogenic action, while all their essential oils have only a spasmolytic action.

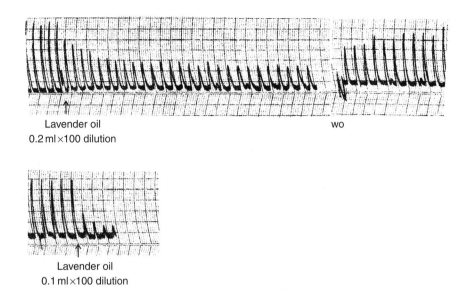

Figure 13.5 The effect of lavender oil on the spontaneously contracting uterus of the rat *in vitro*, showing the inhibition of contractions, which at higher lavender oil concentrations actually cease altogether.

Experiments on *in vitro* uterine preparations

Method

Uterus from a freshly killed rat is mounted in an organ bath containing Krebs solution maintained at 37°C and gassed continually with 95 per cent oxygen in carbon dioxide. Activity of the tissue is monitored with an isometric force transducer connected to a pen recorder. Uterine tissue may be quiescent but usually exhibits regular contractions and relaxations, after a short period of equilibration, depending upon the oestrus cycle of the rat. The ability of an essential oil to increase or decrease the overall activity of the uterus is readily demonstrated, but the elucidation of the mode of action is not as straightforward as with intestinal tissue.

General action on uterus

Linalool, linalyl acetate, α- and β-pinene and 1,8-cineole reduce uterine activity at concentrations which are spasmolytic on intestinal muscle, but the response to French lavender oil depends upon the oestrus state of the rat from which the uterus is removed (Lis-Balchin and Hart, 1997). In proestrus, lavender reduces spontaneous activity while in oestrus there is a slow increase in the amplitude of the spontaneous activity (Figure 13.5).

Experiments on *in vitro* bronchial preparations

Method

Activity on bronchial muscle is usually studied on guinea-pig tracheal muscle which is cut transversally to produce rings which are then tied together and mounted in an organ bath (Kitchen, 1984). An alternative preparation, which reduces the involvement of cartilage, is favoured by Reiter and Brandt (1985).

General actions on bronchial muscle

Reiter and Brandt (1985) reported the actions of several essential oils and their components on tracheal and intestinal smooth muscle including some components of *Lavandula* oils but not the oil itself. Linalool is spasmolytic, with intestinal muscle being more sensitive than tracheal. Linalool is one of many components and oils studied by Brandt (1988) and found to be spasmolytic on guinea-pig ileum and trachea. Alpha- and β-pinene, which are major components of the oil of *L. dentata* L., also relax guinea-pig tracheal muscle (Lis-Balchin and Hart, 2000).

Experiments on *in vitro* cardiac preparations

Method

A heart from a freshly killed rat or rabbit can be perfused with Krebs solution through the aortic arch by the method of Langendorf (1895) and the overall activity recorded by attaching a thread from the ventricles to a transducer. The perfused heart remains viable for several hours and the effect of essential oils on rate and force can be assessed on addition to the perfusion fluid and the mode of action investigated by the use of standard antagonists.

Alternatively, the auricles can be dissected from the heart and set up in an organ bath in a manner similar to that used for intestinal tissue.

General action on cardiac tissue

No literature has been found on this topic.

Experiments on *in vitro* skeletal muscle preparations

Method

It is more difficult to study skeletal muscle than smooth muscle *in vitro* because preparations of the former are usually thicker and do not remain viable in the organ bath. Two preparations which are successful are the chick biventer cervices and the rat phrenic nerve hemi-diaphragm. The diaphragm with attached phrenic nerves is dissected from a freshly killed rat and the diaphragm cut in half to give two preparations that are mounted in a 50 ml organ bath on a support which keeps the diaphragm secure and enables the stimulation of the phrenic nerve (Bulbring, 1946). Stimulation of the phrenic nerve (0.1 Hz, 5 msec, 20 V) causes regular and reproducible contractions of the diaphragm which are recorded via a transducer and pen recorder. The introduction of an essential oil into the organ bath during stimulation can raise the tone of the tissue and/or affect the size of the nerve-induced contraction. It is also possible to stimulate the muscle directly, which allows differentiation between compounds acting via the nerve or directly on the muscle.

General action on skeletal muscle

The essential oil of *L. angustifolia* Miller (2×10^{-4}) and also linalool and linalyl acetate reduce the size of the contraction in response to stimulation of the phrenic nerve and also when the muscle is stimulated directly (Lis-Balchin and Hart, 1994) (Figure 13.6). Thus the action would appear to be myogenic, however, Ghelardini *et al.* (1999) interpret their results with the essential oil of *L. angustifolia* and its components linalool and linalyl acetate, on the same preparation,

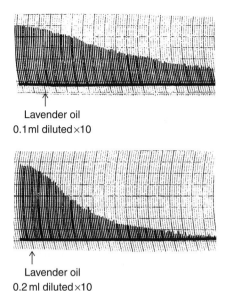

Figure 13.6 The action of lavender oil on skeletal muscle using the chick biventer muscle preparation *in vitro*. This shows an inhibition of electrically-stimulated contractions.

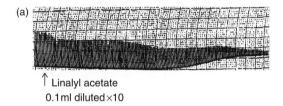

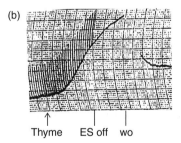

Figure 13.7 (a) The action of linalyl acetate on the skeletal muscle showing a spasmolytic action with a rise in the baseline at the same concentration as that of lavender oil (b) the action of thyme oil for comparison, showing a rise in baseline but no decrease in the size of the contractions.

to show local anaesthetic action. Similarly, Re *et al.* (2000) conclude from experiments on mouse neuromuscular junction that linalool has local anaesthetic action. Linalyl acetate also caused an increase in base-line or resting tone (Lis-Balchin and Hart, 1994), while limonene caused a rise in tone, with a decrease in the size of the contractions (Figure 13.7).

Conclusions

Essential oils prepared from *L. angustifolia* Mill., *L. dentata* L. and *L. spica* L. are spasmolytic on intestinal smooth muscle and this activity is probably due to the linalool and linalyl acetate. The mechanism of action is myogenic involving cAMP at low concentrations with block of calcium channels also being involved at high concentrations. When there is also some spasmogenic activity this is probably due to the presence of α- and β-pinene and 1,8-cineole. Linalool has been shown to have a local anaesthetic action on motor nerve skeletal muscle preparations and it is surprising that a similar activity is not seen with the field stimulated guinea-pig ileum preparation.

In the whole animal, the oils of lavender cause sedation which may be explained by the observation that linalool is capable of displacing glutamate from its receptor. Lavender oil also depresses skeletal muscle activity in the rat but the work of Buchbauer *et al.* (1993) suggests that this is not important in humans.

The fact that some extracts of *L. angustifolia* have a strong spasmogenic action in the dried flowers and fresh leaves is somewhat disturbing as so many modern herbal and aromatherapy books state that the teas are sedative and are often prescribed for upset stomachs. The results suggest that all the information has been mistakenly transcribed from early herbals like those of Culpeper (1653) where *L. spica*, a more camphoric lavender was used medicinally and not the very floral, *L. angustifolia*, which has always been used, as it is used today, mainly in perfumery and cosmetic products (except by aromatherapists who have no knowledge of plant taxonomy or past herbal literature). The spasmolytic results shown for the water soluble extracts of the more camphoraceous *L. stoechas* suggests that this mistake is probably the reason for the differences in the well-quoted action of the camphoraceous spike lavender and should not be confused with that of the non-canmphoraceous *L. angustifolia*.

References

Aqel, M. B. (1991) Relaxant effect of the volatile oil of *Rosmarinus officinalis* on tracheal smooth muscle. *J. Ethnopharmacol.*, 33, 57–62.

Banner, K. H., Marchini, F., Buschi, A., Moriggi, E., Semeraro, C., and Page, C. P. (1995) The effect of selective phosphodiesterase inhibitors in comparison with other anti-asthma drugs on allergen-induced eosinophilia in guinea-pig airways. *Pulm. Pharmacol.*, 8, 37–42.

Brandt, W. (1988) Spasmolytische wirkung atherischer Ole. *Zeitschrift fur Phytotherapie*, 9, 33–9.

Buchbauer, G., Jirovetz, L., Jager, W., Dietrich, H., and Plank, C. (1991) Aromatherapy: evidence for sedative effects of the essential oil of lavender after inhalation. *Z. Naturforsch*, 46, 1067–72.

Buchbauer, G., Jager, W., Jirovetz, I., Ilmberger, J., and Dietrich, H. (1993) Therapeutic properties of essential oils and fragrances. In: *Bioactive Volatile Compounds from Plants*, 159–65. R. Teramishu, R. G. Buttery and H. Sugisawa (eds). ACS symposium series 525. Washington DC: American Chemical Society.

Bulbring, E. (1946) Observations on the isolated phrenic nerve diaphragm preparation of the rat. *Br. J. Pharmac.*, 1, 38–61.

Delaveau, P., Guilleman, J., Narcisse, G., and Rousseau, A. (1989) Sur les proprietes neuro-depressives de l'huile essentielle de Lavande. *C.R. Soc Biol.*, 183, 342–8.

Elisabetsky, E., Marschner, J., and Souza, D. O. (1995) Effects of linalool on glutamatergic system in the rat cerebral cortex. *Neurochem. Res.*, 20, 461–5.

Elisabetsky, E., Brum, L. F., and Souza, D. O. (1999) Anticonvulsant properties of linalool in glutamate-related seizure models. *Phytomedicine*, 6, 107–13.

Garthwaite, J., Southam, E., Boulton, C. L., Nielsen, E. B., Schmidt, K., and Mayer, B. (1995) Potent and selective inhibition of nitric oxide-sensitive guanylyl cyclase by 1H-[1, 2, 4] Oxadiazolo[4, 3-a] quinoxalin-1-one. *Mol. Pharmacol.*, 48, 184–8.

Ghelardini, C., Galeotti, N., Salvatore, G., and Mazzanti, G. (1999) Local anaesthetic activity of the essential oil of *Lavandula angustifolia*. *Planta Med.*, 65, 700–3.

Hughes, B., Howat, D., Lisle, H., Holbrook, M., James, T., Gozzard, N., Blease, K., Highes, P., Kingaby, R., Warrellow, G., Alexander, R., Head, J., Boyd, E., Eaton, M., Perry, M., Wales, M., Smith, B., Owens, R., Catterall, C., Lumb, S., Russell, A., Allen, R., Merriman, M., Bloxham, D., and Higgs, G. (1996) The inhibition of antigen-induced eosinophilia and bronchoconstriction by CDP840, a novel stereo-selective inhibitor of phosphodiesterase type 4. *Br. J. Pharmacol.*, 118, 1183–91.

Imaseki, I. and Kitabatake, Y. (1962) Studies on effect of essential oils and their components on the isolated intestines of mice. *J. Pharmaceut. Soc. Japan.*, 82, 1326–8.

Izzo, A. A., Capasso, R., Senatore, F., Seccia, S., and Morrica, P. (1996) Spasmolytic activity of medicinal plants used for the treatment of disorders involving smooth muscle. *Phytother. Res.*, 10, S107–08.

Langendorff (1895). *Pflugers Arch ges Physiol.*, 190, 280.

Lis-Balchin, M. and Hart, S. (1997) Pharmacological effect of essential oils on the uterus compared to that on other tissue types. *Proc. 27th Int. Symp. Ess. Oils, Vienna, Austria*, Ch. Franz, A. Mathé and G. Buchbauer (eds), Allured Pub. Corp., Carol Stream, Ill. 29–32.

Lis-Balchin, M. and Hart, S. (1999) Studies on the mode of action of the essential oil of lavender (*Lavandula angustifolia* P. Miller). *Phytother. Res.*, 13, 540–2.

Lis-Balchin, M. and Hart, S. (2000) Unpublished observations.

Lis-Balchin, M., Hart, S., Deans, S. D., and Eaglesham, E. (1996a) Comparison of the pharmacological and antimicrobial action of commercial plant essential oils. *J. Herbs, Spices Med. Plants*, 4, 69–86.

Lis-Balchin, M., Deans, S., and Hart, S. (1996b) Bioactivity of New Zealand medicinal plant essential oils. *Proc. Int. Symp. Medicinal and Aromatic Plants*, L. E., Craker, L. Nolan, K. Shetty (eds), 13–27.

Lis-Balchin, M., Ochocka, R. J., Deans, S. G., Asztemborska, M., and Hart, S. (1999) Differences in bioactivity between the enantiomers of α-Pinene. *J. Essent. Oil Res.*, 11, 393–7.

Macht, D. I. and Ting, G. C. (1921) Experimental inquiry into the sedative properties of some aromatic drugs and fumes. *J. Pharmacol. Exp. Therap.*, 18, 361–72.

Manley, C. H. (1993) Psychophysiological effect of odor. *Crit. Rev. Food Sci. Nutr.*, 33, 57–62.

Neuhaus-Carlisle, K., Vierling, W., and Wagner, H. (1997) Screening of plant extracts and plant constituents for calcium-channel blocking activity. *Phytomedicine*, 4, 67–9.

Paton, W. D. M. (1954) The response of the guinea-pig ileum to electrical stimulation by coaxial electrodes. *J. Physiol.(Lond.)*, 127, 40–41P.

Pappe, L. (1868) *Florae Capensis Medicae*, Prodromus, 3rd ed., Cape Town.

Re, L., Barocci, S., Mencarelli, A., Vivani, C., Paolucci, G., Scarpantonio, A., Rinaldi, L., and Mosca, E. (2000) Linalool modifies the nicotinic receptor-ion channel kinetics at the mouse neuromuscular junction. *Pharmacol Res.*, 42, 177–82.

Reiter, M. and Brandt, W. (1985) Relaxant effects on tracheal and ileal smooth muscles of the guinea-pig. *Arzneim.-Forsch/Drug Res.*, 35, 408–14.

Shipochliev, T. (1968) Pharmacological investigation into several essential oils. First communication. Effect on the smooth musculature. *Vet. Med. Nauki*, 5, 63–9.

Sticher, O. (1977) Plant mono-, di- and sesquiterpenoids with pharmacological or therapeutical activity. In: *New natural products and plant drugs with pharmacological, biological or therapeutic activity*, H. Wagner and P. Wolf (eds), Springer-Verlag, Berlin.

Tisserand, R. and Balacs, T. (1995) *Essential oil safety*, Churchill Livingstone, London.

Torii, S., Fukuda, H., Kanemoto, H., Miyanchio, R., Hamauzu, Y., and Kawasaki, M. (1988) Contingent negative variation and the psychological effects of odor. In *Perfumery: The Psychology and Biology of Fragrance*. S. Toller and G. H. Dodds (eds) Chapman and Hall, New York.

Triebs, W. (1956) *Die Atherischen Öle*, Akademie-Verlag, Berlin.

Vuorela, H., Vuorela, P., Tornquist, K., and Alaranta, S. (1997) Calcium channel blocking activity: screening methods for plant derived compounds. *Phytomedicine*, 4, 167–81.

14 The psychological effects of lavender

Michael Kirk-Smith

Introduction

Scientific research into the psychological effects of lavender has only recently begun. However, there is a long tradition in folklore of its use by ordinary people (rather than clinical use) to affect psychological states. Presumably, this traditional use is founded on consistent observations of the effects of lavender on people's psychological states, though we should note that although such consistency may be suggestive of a real effect, it is always possible that the effect may still be due to placebo and expectation rather than some property of the lavender itself.

So how has lavender been commonly used? To get an idea of how 'ordinary' people, rather than clinicians, understood how lavender was used in everyday life, this chapter starts by looking at the references to lavender in fiction as represented in literature and plays. The speech of people is not available to us, but it is likely that any appearance in literature will be in terms, and of uses, commonly understood and known by the 'ordinary person'. This may even be more so in plays, since they present only speech. Moreover, main elements in literature and plays, and without which it is doubtful that they would be of interest, are accounts of people's feeling, thinking and doing, and these are psychology's objects of study.

Following these examples from literature and plays, recent scientific research will be reviewed to see why lavender might have psychological effects, and what these effects might be.

Psychology uses of lavender from literature and plays

To find how lavender has been used in literature and plays, the 'e-texts' available on Literature Online (www.lion.chadwyck.co.uk) were searched. These databases cover over 260,000 works in the English language. Ideally, mentions of lavender in works in other languages and cultures should be covered and contrasted, however, this was beyond the scope of the present chapter.

In the available literatures and plays it was found that lavender was mentioned for three purposes, first, as a calming or 'restorative' agent, second as a fragrance when storing bedsheets and clothes, and finally, as a pleasant scent by itself or to block out unpleasant smells.

To calm, that is, to relieve anxiety and stress

In an early text on herbalism, Culpeper ('The Complete Herbal', 1653) comments that lavender: 'Being an inhabitant almost in every garden, it is so well known, that it needs no description'. He views it as having definite psychological effects: 'Lavender is of a special good use for all the griefs and pains of the head and brain that proceed of a cold cause, as the apoplexy, falling-sickness, the dropsy, or sluggish malady, cramps, convulsions, palsies, and often faintings'. This is further emphasised in the prescription: 'Two spoonfuls of the distilled water of the flowers taken, helps them that have lost their voice, as also the tremblings and passions of the heart, and faintings and swooning, not only being drank, but applied to the temples, or nostrils to be smelled unto.' This early description of lavender's calming properties is reflected in both later literature and plays. Here is a selection of quotes from novels.

There are several mentions in Jane Austen's 'Sense and Sensibility' (1811) to the use of lavender. In the first, lavender water is used to bathe a wound, although there seems to be a suggestion that it might be calming too: '… a pin in her ladyship's head-dress slightly scratching the child's neck, produced from this pattern of gentleness such violent screams … and everything was done by all three, in so critical an emergency, which affection could suggest, as likely to assuage the agonies of the little sufferer. She was seated in her mother's lap, covered with kisses, her wound bathed with lavender-water …'.

The second mention relates to calming after a shock: 'Her face was crimsoned over, and she exclaimed, in a voice of the greatest emotion, 'Good God! Willoughby, what is the meaning of this? Have you not received my letters? … 'But have you not received my notes?' cried Marianne in the wildest anxiety … Marianne, now looking dreadfully white, and unable to stand, sunk into her chair; and Elinor, expecting every moment to see her faint, tried to screen her from the observation of others, while reviving her with lavender water.'

The last mention is similarly for calming: 'no attitude could give her ease; till growing more and more hysterical, her sister could with difficulty keep her on the bed at all, and for some time was fearful of being constrained to call for assistance. Some lavender drops, however, which she was at length persuaded to take, were of use; and from that time till Mrs. Jennings returned, she continued on the bed quiet and motionless.'

She also writes in 'Northanger Abbey' (1817): 'Catherine, supposing some uneasiness on Captain Tilney's account, could only express her concern by silent attention, obliged her to be seated, rubbed her temples with lavender-water, and hung over her with affectionate solicitude'.

Wilkie Collins presents a detailed account of the use and effects of 'red lavender' in this series of quotes from 'Man and Wife' (1870):

> 'Nerves, Lady Lundie. Repose in bed is essentially necessary. I will write a prescription.' He prescribed, with perfect gravity: Aromatic Spirits of Ammonia – 16 drops. Spirits of Red Lavender – 10 drops. Syrup of Orange Peel – 2 drams. Camphor Julep – 1 ounce. On a table at her side stood the Red Lavender Draught – in colour soothing to the eye; in flavour not unpleasant to the taste.'
>
> 'Her ladyship was feebly merry (the result, no doubt, of the exhilarating properties of the red lavender draught)'.
>
> '… her ladyship was free to refresh herself with another dose of the red lavender draught, and to sleep the sleep of the just who close their eyes with the composing conviction that they have done their duty.'
>
> 'He found his patient cured by the draught! It was contrary to all rule and precedent; it savoured of quackery – the red lavender had no business to do what the red lavender had done – but there she was, nevertheless, up and dressed …'

In the 'Black Robe' (1881) Wilkie Collins again comments on the use of red lavender: '"I declare I am agitated myself!" she exclaimed, falling back into her customary manner. "Such a shock to my vanity, Stella – the prospect of becoming a grandmother! I really must ring for Matilda, and take a few drops of red lavender."'

Arguably the greatest Victorian novelist, Charles Dickens, describes how lavender was delivered commercially in 'Little Dorrit' (1855): '… she explained that she put seventy-five thousand drops of spirits of lavender on fifty thousand pounds of lump sugar, and that she entreated Little Dorrit to take that gentle restorative …'

Finally, in 'Don Quixote' (1804) Miguel De Cervantes uses lavender as a metaphor for something gentle and relaxing: 'It is not to a wedding we are bound, but to go round the world, and play at give and take with giants and dragons and monsters, and hear hissings and roarings and bellowings and howlings; and even all this would be lavender, if we had not to reckon with Yanguesans and enchanted Moors.'

In plays, the speech must reflect 'real-life' uses of language as spoken in everyday life in order to be easily and immediately comprehensible, and so the following quotes from plays may reflect even more accurately how lavender was talked about and used over the last two centuries.

First, the earliest quotes are from two plays by Joanna Baillie: 'Oh they are such savages! I'm sure if I had not put lavender on my pocket handkerchief, like Mama, I should have fainted away' ('The Election', 1798), and '… my head has been put into such a confusion! La, ma'am! said my millener, do take some' lavender drops, you look so pale. Why, says I, I don't much like to take them, Mrs. Trollop, they a'nt always good.' ('The Tryal', 1798). Quotes from four other plays are similar:

> 'My dear, have some lavender, or you'd best have a thimble full of wine, your spirits are quite down, my sweeting.'
>
> (O'Keeffe, John, 'A beggar on horseback', 1798)

> Lady D. 'What is the cause of all this outcry?' Davy. 'Cause enough, my lady; this two – handed son of the church has kill'd your nephew.' Lady D.: The Lord be good unto me. Help me to lift him up; here, chafe his temples with lavender water; I don't see any bruises he has about him.'
>
> (Cumberland, Richard, 'The Walloons', 1813)

> 'The misfortune is, my dear fellow, that I've lost my chapeau bras'. 'Never mind, lean on my arm: we'll retire to my attic; and I've no doubt that a little sal volatile and red lavender will set all to rights again'.
>
> (Dibdin, Charles, 'Life in London', 1822)

> Polly. [Assuming a tragedy air.]: 'Dead! oh, dreadful tidings! Dead! our best friend, our patron, dead! Where is my pocket-handkerchief, and my spirits of lavender? [Takes both out of her reticule, affecting grief.]
>
> (Somerset, Charles, 'A day after the fair', 1828)

Taken together, these quotes from literature and plays suggest that lavender, especially 'red lavender', has been a well-recognised calming treatment for 'agitated' states.

'Laying up in lavender'

The bedchamber, bedsheets and linen

Brewer's 'Dictionary of Phrase and Fable' (1984) claims that the name lavender comes from the Spanish lavandera (a laundress), the plant used by laundresses for scenting linen. Indeed, the scenting of bed sheets is commonly mentioned in literature. This may have significance because of lavender's association with calming effects, as observed above, and/or moth repellent activity.

As regards the latter, Brewer (1984) gives the meaning of 'laid up in lavender' as 'taken great care of, laid away, as women put things away in lavender to keep off moths'. However, the following three quotes from plays suggest that the preparation of bedsheets with lavender relates to a hedonic quality given to them rather than in relation to storage:

> Stat: 'While in the Garret of Simiramis I make your Bed, lay on clean Sheets, Scented with Lavender, And sweep the Room out for your coming …'.
> (Cibber, Colley, 'The Rival Queans', 1729)

> Jenny: 'Pick … the lavender, I must be tying that, or you'll have linen, And naught to sweeten it!'.
> (Pfeiffer, Emily, 'The Wynnes of Wynhavod', 1882)

> Mrs. M. Cosey: 'Rosalie, lavender sheets, aired snow, snug supper, every thing what I call comfortable, eh?'.
> (Morton, Thomas, 'Town and Country', 1807)

One reference is to the smell of the bedroom in general: Heron: 'Your chamber hath been furnished hard by mine, And by my hand sweetened with lavender'. (Austin, Alfred, 'Flodden Field', Poet Laureate, 1897–1913).

A poem makes this positive quality clear: 'Crowd her chamber with your sweets, Not a flower but grows for her!, Make her bed with linen sheets, That have lain in lavender' (Duclaux, Agnes Mary Frances, 'Celia's Home-Coming', 1857).

In 'Mill on the Floss' (1860), George Eliot uses the term with some black humour in conjunction with the similar phrase of 'laying out of the dead' in preparation for burial: '… as for them best Holland sheets, I should repent buying 'em, only they'll do to lay us out in. An' if you was to die to-morrow, Mr Tulliver, they're mangled beautiful, an' all ready, an' smell o' lavender as it 'ud be a pleasure to lay 'em out'. Other mentions are more relevant to the storage aspect, although the hedonic aspect is still prominent:

> 'Rocking cradles, and covering jams, knitting socks for baby feet, Or piecing together lavender bags for keeping the linen sweet …'.
> (Austin, Alfred, 'Grandmother's Teaching' 'Soliloquies in Song', 1882)

> '… their display of the whitest linen, and their storing-up, wheresoever the existence of a drawer, small or large, rendered it possible, of quantities of rose-leaves and sweet lavender'.
> (Dickens, Charles, 'Bleak House', 1853)

> 'Then he pulled open the drawers, filled with his mother's things, and looked at them: there were lavender bags among the linen, and their scent was fresh and pleasant.'
> (Maugham, Somerset, 'Of Human Bondage', 1915)

> 'See, sir,' and here he opened a door and ushered Otto into a little white-washed sleeping-room, 'here you are in port. It is small, but it is airy, and the sheets are clean and kept in lavender'.
> (Stevenson, Robert Louis, 'Prince Otto', 1885)

Clothes

An early mention of the use of lavender in storing clothes relates to repelling moths: 'A Citie pestilence, A moath that eates up gownes, doublets and hose, One that with Bills, leades smocks and shirts together To linnen close adultery, and upon them Strowes lavender, so strongly, that the owners Dare never smell them after; hee's a broaker.' (Dekker, Thomas 'The Wonder of a Kingdome', 1636). A selection of other quotes illustrate that the storage is long-term or for rarely used items:

> 'When the gay triumph ceases, and the treasure Divided, all the Offices laid up, And the new cloathes in Lavender, what then?'.
> (Shirley, James, 'The Sisters', 1653)

> Ind: 'The onely sute you wear smels of the chest that holds in Limbo Lavender all your rest'.
> (Brome, Richard, 'The English Moor', 1659)

> Joe: 'That dress, mum, was made in France, so I've heard, Parish make, and it's got a Parish lining in the buzzum too. It's been laid up in lavender nigh upon twenty years among my old woman's relics.'
> (Palmer, T. A. 'Among the relics' (no date))

> 'In a wardrobe, fresh as spring with yearly lavender'.
> (Davidson, John, 'The Theatrocrat', 1905)

> 'A dingy brown coat, with vellom button-holes, to be sure, speaks an excellent taste; but then I would advise you to lay it by in lavender, for your Grandson's christening ...'.
> (Cowley, Hannah, 'Who's the dupe', 1779)

Since clothes are usually bought to be worn regularly, those that are kept in storage are necessarily used only for unusual occasions, for example, during mourning. Two quotes illustrate this association:

> 'As this remembrance came upon Mrs Greenow she put her handkerchief to her eyes, and Alice observed that that which she held still bore the deepest hem of widowhood. They would be used, no doubt, till the last day, and then put by in lavender for future possible occasions.'
> (Anthony Trollope, 'Can You Forgive Her?', 1865)

> Post Obit.: 'But come, Georgiana, I must go change my cloathes, and put by my mourning in lavender...'
> (Reynolds, Frederick, 'Folly as it flies', 1802)

Other items

In 'Mr. Lirriper's Lodgings', Charles Dickens comments on a place to put important papers: 'which I mean always carefully to keep in my lavender drawer as the first printed account of his'. Perhaps this is associated with moth repellent activity. One intriguing mention is in relation to claret: 'When we askt him for Claret, he vow'd not a drop, for he had in Lavender laid it all up …' (Ames, Richard, 'The Search after Claret', 1691). It is not clear whether this is an actual or synonymous use of the term.

Metaphorical usage

'Laying up in lavender' was apparently a well-accepted and known practice so that it was able to be used metaphorically as meaning 'putting away in storage' in a general sense. This could include lovers and others, here possibly with the additional meaning of 'and forgotten about', for example:

> Bertha: 'If she has fancies – and all girls have some – She knows her duty and will lay them by In lavender with other childish gauds; When the right royal lover clatters in.'
> (Warren, John Byrne Leicester, Baron de Tabley, 'The Soldier of Fortune', 1876)

> 'She lay'd him neatly in her Lap, And carried him to a House that stood Upon an Hill in an old Wood: And when she had the Urchin there, She laid him up in Lavender.'
> (Cotton, Charles, 'Scarronides', 1667)

> 'No, no; Joe Bangles' work is done. Shelve him, superannuate him, lay him up in lavender, he's only in the way.'
> (Gilbert, William Schwenck, 'Randall's Thumb', 1911)

The term could also be applied to the abstract: 'Take off your tragic airs, my dear friend, and fold them up and put them away in lavender. You'll never need them again.' (Montgomery, Lucy Maud, 'Anne's House of Dreams', 1917).

Following Brewer's comment that the term implies 'taking care of', it was seen as appropriate for a treasured object: Dumont: 'Oh, you darling instrument! (Hugging the violin.) If poeans ought to be sung to thee, for I owe thee every thing! You have made me happy, by enabling me to make two young creatures so, and shall be laid up in lavender, as the first fiddle of love!' (Webster, Benjamin, 'The Modern Orpheus', 1837).

Brewer also cites two other meanings of the term, notably, that persons who are in hiding are said to be 'in lavender', and for someone who is 'in pawn', that is, they have become so poor that they have had to pawn their belongings:

> 'The poor gentleman paies so deare for the lavender it is laid up in, that if it lies long at the broker's house he seems to buy his apparel twice.'
> (Greene: 'A Quip from an Upstart Courtier', 1592)

However, apart from his assertions and this one quote, no other citations for these usages were found in the literature searched, so perhaps these were colloquial at some time, but did not make it into literature.

Hedonic uses

As well as for calming and storing sheets and clothes, lavender was directly used to counter the stenches of older times: 'Yet the smell of a Yahoo continuing very offensive, I always keep my nose well stopped with rue, lavender, or tobacco leaves.' (Dean Swift, 'Gullivers' Travels', 1729). Two plays also comment on this use: 'Here is the Lavender for your Nose'. Sir.' Furrs. 'Stop, stop those holes James, there are more stinks than sweet smells.' (Newcastle, William Cavendish, Duke of, 'The Humorous Lovers', 1677). Baron: 'And why, I should like to know, must you drench your clothes and my sofa with lavender water?' Count: 'Pardonnez, mon Colonel; the smoke of tobacco is quite insupportable.' (Thompson, Benjamin, 'Lovers' Vows', 1801).

Lavender also seemed to be used in combination with rose-leaves as an ambient fragrance. George Eliot mentions this purely hedonic use in 'Silas Marner', 1861: 'Mrs Osgood's inclination to remain with her niece gave them also a reason for staying to see the rustic beauty's toilette. And it was really a pleasure from the first opening of the bandbox, where everything smelt of lavender and rose leaves ...'.

Similarly, Hans Christian Andersen (1872) refers to the combination in 'The Shepherdess and the Sheep': 'Let us get into the great pot-pourri jar which stands in the corner; there we can lie on rose-leaves and lavender', and Charles Dickens (1970) in 'The Mystery of Edwin Drood': '... he would quietly swallow what was given him, merely taking a corrective dip of hands and face into the great bowl of dried rose-leaves, and into the other great bowl of dried lavender ...'

It was also carried around on handkerchiefs, as commented by one itself: 'Julia never looked more lovely than she did that night. She anticipated much pleasure, and her smiles were in proportion to her anticipations. When all was ready, she took me from the drawer, let a single drop of lavender fall in my bosom ... asked Betts Shoreham ... 'What CAN there be in that pocket-handkerchief to excite tears from a mind and a heart like yours?''. (James Fenimore Cooper, 'Autobiography of a Pocket-Handkerchief', 1843).

This apparently amorous association is mentioned by Brewer, who states that the giving of lavender is a token of affection: ' He from his lass him lavender hath sent. Showing his love, and doth requital crave' (Drayton: Eclogue, IX), although no other mentions of such a use were cited in the searched literature.

In the following quotes the carrying and wearing of lavender is represented as somewhat foppish, and perhaps effete, when worn by men:

'No more the gaudy beau, With handkerchief in lavender well drench'd'
(Fergusson, Robert, 'The Canongate Play-house in Ruins. A Burlesque Poem', 1750–74).

Third footman: 'Damned hard, my friends: I have not been able to throw a drop of lavender into my handkerchief, yet!'
(Pratt, Samuel Jackson, 'Fire and frost: A comic drama', 1805).

Mrs S. 'Oh, I did, Mr. Gradgrind, I did; but the unhappy woman told me to mind my own business; and her guilty paramour – that lavender-scented parliament fellow – threatened to kill me if I gave the alarm.'
(Nation, William Hamilton Codrington, 'Under the Earth: A Romantic Drama', 1871).

Met: 'I have seen the ruffian somewhere ... Owed me half-a-crown for seven years, and wears lavender water!'
(Jerrold, Douglas William, 'Time Works Wonders', 1845).

Other applications

Shakespeare (1623) links lavender with middle age in a Winter's Tale, when Perdita quotes a courting youth: 'Here's flow'rs for you: hot lavender, mints, savory, marjoram; The marigold, that goes to bed wi' th' sun, and with him rises weeping; these are flow'rs of middle summer, and I think they are given to men of middle age. Y'are very welcome.'

There is a unique mention of lavender being used in cooking: Drug.: 'A finer sucking pig in lavender, with sage growing in his belly, was never seen!' (Murphy, Arthur, 'What We Must All Come To', 1764).

Finally, Beatrix Potter (1904) in 'Children's Stories' has a whimsical use: 'Old Mrs. Rabbit ... also sold herbs, and rosemary tea, and rabbit-tobacco (which is what WE call lavender).'

Research findings on psychological effects of lavender

The above review of literature and plays suggests common and consistent uses of lavender, certainly in England and the United States, and the primary psychological effect seems to be that of calming. This section examines the scientific research to see if there is a basis to this effect, that is, why lavender and its components might be calming, how it affects the brain and psychological states, and discusses issues related to the odour perception and, finally, examines whether these effects can be seen in practice. One important proviso is that lavender oils can vary in composition, and this may affect the results obtained in studies, yet many of the following studies do not specify the composition of the lavender oil, so that results may not be comparable (or repeatable).

Doing research on the psychological effects of odour on people is difficult and this is one reason why research on the psychological effects of odour has been limited. Many mechanisms may be involved in producing psychological effects. Once an effect is found, it is very difficult to know why it is occurring, for example, an odour may exert its effects through a mixture of pharmacological and various psychological mechanisms, for example, hedonic, learning and social mechanisms; all possibly operating at the same time and maybe even in different directions.

More specifically, psychological responses to odour may involve an interaction of an individual's experience, the particular situation, and the individual's current state of mind and this combination makes both interpretation of results and prediction difficult. A change in one of these may change the psychological response to an odour, even though the odour remains the same. These issues will be discussed in more detail later, although research does show that there may be a primary pharmacological mechanism involved in the calming effect of lavender.

Pharmacological mechanisms

Unlike psychological mechanisms, the pharmacological mechanism does not involve any perception of lavender odour, and the psychological effects are due to the odour compounds entering the body and acting directly on the brain, that is, via the blood stream through absorbtion through the lungs or olfactory mucosa. These effects depend on the specific molecular properties of lavender, and are therefore likely to give similar effects across all people.

The main terpenoid components of lavender oil, for example, linalool and linanyl acetate, are lipophilic and similar terpenes are found to suppress cell action potentials, possibly through direct steric interaction with cell membranes, thereby increasing their volumes and thus slowing the inflow of calcium ions and reducing sodium ion permeability (Teuscher et al., 1990). Such general suppression of cell electrical activity suggests the possibility of a light sedative or anaesthetic effect.

Elisabetsky et al. (1995) noted that plant extracts with a high linalool content are used as a home-made anticonvulsant remedy in Brazil. More specifically, they found that linalool competitively inhibited glutamate binding in the cells in the cortex of rats and suggested that this may cause the sedative effect. Linalool also inhibited glutamate-related epilepsy, and gave protection against pentylenetrazol (PTZ), quinolinic acid and transcorneal electroshock-induced convulsions (Elisabetsky et al., 1999).

However, the reduction in the number of convulsions caused by PTZ induced kindling, was not linked to changes in the glutamate binding levels that accompany kindling, so linalool may act by a variety of mechanisms to prevent convulsions other than glutamate binding inhibition. For example, Yamada *et al.* (1994) found that inhaled lavender oil reduced pentetrazol, nicotine and electroshock induced convulsions in mice, and suggested that these effects may be due to an augmentation of GABA-ergic action. Evidence for this mechanism comes from Aoshima and Hamamoto's (1999) investigation of the potentiation of $GABA_A$ receptors expressed in Xenopus oocytes by lavender oil components. These increased response in the presence of $GABA_A$ at low concentrations, possibly through binding to the potentiation-site in $GABA_A$ receptors, thus increasing the affinity of $GABA_A$ to the receptors.

This effect resembles the potentiation of $GABA_A$ receptors by benzodiazepine, barbiturate, steroids and anesthetics, which is associated with their anxiolytic, anticonvulsant and sedative effects.

These reductions in cell activity due to lavender oil and linalool have also been observed in non-CNS cells and tissues, reflecting a general spasmolytic or 'calming' effect. Mazzanti *et al.* (1998) found that linalool inhibited acetyl choline and barium chloride induced contractions of isolated guinea-pig ileum, and similarly, Lis-Balchin and Hart (1999) showed that lavender oil and linalool reduced the twitch response to nerve stimulation in isolated rat diaphragm phrenic nerve preparations, and found that the mechanism was post synaptic, not atropine-like and likely to be mediated through cAMP and not through cGMP.

Reflecting this pharmacological rather than psychological action, Buchbauer *et al.* (1993) found that a 78 per cent decrease in the motility of mice resulted from inhalation of ambient lavender oil for one hour, with similar effects for the main components, linalool (73 per cent) and linalyl acetate (63 per cent); these three being the most effective sedatives of forty-two oils and their components tested. When mice were first agitated by caffeine injection, inhaled lavender oil reduced motility by 92 per cent. They also note that the ability to sedate related to increasing lipophilic properties. This is consistent with Teuscher *et al.*'s (1990) suggestion that the effect is due to steric interaction with cell membranes. Inhalation of lavender oil led to a serum level comparable to an intravenous injection – this may be brought about by rapid absorption through both the nasal and lung mucosa (Buchbauer *et al.*, 1991). As might be expected, given the highly lipophilic nature of the brain, the components were transported across the blood brain barrier, being found in the cortex of test animals (Buchbauer *et al.* 1993a).

Paralleling the quotes in literature and plays, these findings suggest that lavender odour may act as a light intranasally-introduced sedative. Speculatively, such effects may also account for the use of lavender to scent sheets, that is, to ensure a good night's sleep as well as any moth repellent properties.

More significantly, these studies also identify how the effect may be caused and this knowledge should now allow its optimisation, for example, though breeding of high linalool lavenders and perhaps through chemical modification. Assuming that the active components in lavender can enter the brain, the next question is whether lavender actually affects brain activity, and this is addressed in the next section.

Brain responses to lavender

There have been several studies of the brain's responses to odour. These have used the changing electrical activity picked up by scalp electrodes in response to lavender odours as a measure of brain activity (EEG). EEGs have the advantage of being objective rather than subjective measures, although interpretation of the results may be complex (reflecting the trade-off of reliability and validity between the two types of measure).

The CNV measure (contingent negative variation) has been used to investigate the effect of lavender. The CNV is an upward shift in EEG waves which occurs when people are expecting something to happen, for example, between a signalling tone followed by a light, which the subject then switches off. Since the CNV is found to be reduced by sedatives and increased by stimulants, Torii *et al.* (1988) investigated the effect of lavender on CNV in trained Japanese perfumers. They found that it reduced CNV, but unlike sedatives, had no effect on reaction time or heart rate, possibly due to the low amounts absorbed.

These results were replicated by Lorig and Roberts (1990) in untrained American subjects. They comment that this suggests that CNV is robust across cultures. However, they also suggest that the effects may also involve cognitive mediation or mood induction due to previous experience with the odour as well as due to the direct physiological or pharmacological effects on the brain. This was because they found that the CNV also changed depending on what subjects' thought the odour was, even though the odour presented was different. They also used concentrations of lavender that the subjects found 'arousing and distracting' and suggest that this may also have led to the lowered CNV.

Changes in different frequency bands of the EEG to lavender and other odours have also been measured. Diego *et al.* (1998) examined how lavender oil affected the spectra of the EEG compared to rosemary oil. A group breathing lavender showed increased frontal alpha- and beta-2 waves, suggesting an increase in drowsiness, and this group also rated themselves as more relaxed. The study lacked a no-odour control group so perhaps taking part on the study alone without odour might have given similar results. Diego *et al.* (1998) did not report changes in theta waves. In contrast, Klemm *et al.* (1992) reported that lavender and other odours were associated with a general increase in theta waves, which is associated with increased drowsiness. However, the typical alpha wave increase during drowsiness was not observed, making these results hard to interpret.

A main finding by Klemm *et al.* (1992), which also contrasts with Diego *et al.*'s (1998) results, was that EEG responses to the odours varied greatly between subjects. They suggest that the difference observed in EEG responses may be due to subjects visualising the object or situations that the odours evoked, since these are likely to be different across subjects.

This evocation of specific objects or situations relates to the suggestion that all olfactory percepts are, in fact, associates (Booth, 1979). Certainly, odour words are virtually always names of the typical source object, thus, odours may be nothing more than carriers of the meaning of their associates, by evoking recognition of the source and by evoking any affect that may be associated with it (Kirk-Smith and Booth, 1987). As EEG measures neocortical activity, it can be considered to reflect people's affect and thoughts, and these will be influenced by individual associations with each odour. Klemm *et al.* (1992) comment that unless such individual thoughts and affect are accounted for, it may not be appropriate to look for consistent EEG responses across subjects.

Overall, these studies suggest that brain responses to lavender are in the direction of increased calmness rather than excitement. However, to investigate the pharmacological effect alone, rather than to have a confounding with perceptually-related effects, would require levels of lavender below awareness to be given for long enough for sufficient amounts to enter the brain.

Factors such as subject expectation, the experimental situation and its interpretation and personal experience of lavender may all influence objective EEG responses to lavender. In other words, EEG responses to lavender also depend on psychological factors unrelated to the chemical characteristics of lavender. These are especially important in understanding the psychological effects of lavender in laboratory conditions, and these are the subjects of the next section.

The effects of lavender on psychological responses

As a stimulus, the fragrance of lavender might be expected to have some effect on behaviour, cognition, affect, memory and mood. Several studies have investigated how lavender affects such psychological factors.

Ludvigson and Rottman (1989) compared the effects of lavender and cloves compared to a no-odour control on various tasks; cognitive – using arithmetical tests, vocabulary using synonyms, and analogies to test reasoning skills; memory – where twenty-four words were presented, half of the subjects were asked if each word was pleasant or not, the other half were asked if the word contained repeated letters, then all were asked to recall the words after 5 min; affect – whether they liked doing the tasks and would return to do similar studies and mood. Their study was somewhat complex with two sessions, where the three groups received any of the three other conditions in a second session a week later.

They found that cognitive functioning, particularly for arithmetic, was reduced by lavender in the first session, but not the second. There was no effect on mood or memory in either session. People with lavender in the first session found the session more enjoyable than the other groups; however, people who had no odour in the first session and lavender in the second session said they would not return. Ludvigson and Rottman (1989) comment that the intensity of the lavender odour was quite strong, so this may have caused a distraction in the first session and the decreased performance, and been more of an annoyance to those in the second session rather than first, where it was novel. The total time taken to do the various tasks was at least 25 min, so it may be that initial perceptual effects were mixed with later pharmacological effects as the lavender entered the brain.

In complete contrast, Diego et al. (1998) found that subjects performed arithmetical tests faster and more accurately following exposure to lavender odour for 3 min than rosemary odour. Since there were no control groups and no assessments of the perceived strengths and pleasantness of either odour were given it is hard to interpret these results.

Knasko (1992) studied the influence of odour pleasantness by comparing the effect of ambient lavender, lemon and dimethyl sulphide on mood and a standard creativity test in two 1 h sessions 1 week apart, counterbalanced as to control for one experimental condition. She found that lavender had no effects on either mood or creativity, though DMS did tend to lower mood ratings. She comments that the lavender concentration was moderate compared to Ludvigson and Rottman (1989) where a reduced performance was observed.

Degel and Koster (1999) compared the effects of lavender and jasmine on creativity and on arithmetical and concentration tasks over 45 min, but used imperceptible or unnoticed levels of odour, thus removing distracting and expectational cues. They found that lavender reduced the number of errors made in the arithmetical and concentration tasks compared to the control and jasmine groups, that is, in this study the improvement is likely to be an effect of the lavender odour rather than due to some general expectation effect. They suggest that doing these tasks is stressful and that the sedative effect of lavender may have improved performance by reducing arousal; also that the effect would therefore not be shown if the subjects had not been stressed.

In interpreting these results, several methododological points need to be borne in mind. In many studies of the effects of odour on human response, a main variable that frequently arises is whether the fragrance is perceived as pleasant or unpleasant, which also tends to correlate with the perceived strength of the odour (i.e. too much of a good thing is not liked).

The importance of the hedonic aspect is not surprising. Pleasantness or unpleasantness is related to the value of an odour in evoking recognition of the source, and approach or avoidance of the source. As such, hedonic responses play a fundamental part in an organism's survival behaviour. As well as whether the odour is pleasant or not, performance also may be affected depending on how the odour is attended to, particularly if the odour is perceived as an annoying distraction or as incongruent to the task.

In general, the issue behind many studies on the psychological effects of odour is how or whether a response or performance is changed when people are in an unpleasant or pleasant condition; and this will depend on the particular task, the particular conditions and people's perceptions of them. Since all these parametres can be changed (and inevitably differ between studies), this means that psychological responses to odours are malleable, and highly dependent on the particular situation.

For example, in several of the studies mentioned above lavender was used in concentrations high enough to be considered distracting or unpleasant by subjects, that is, at levels which subjects would not choose to use themselves. As reflected in the quotes from literature and plays, lavender, in self-administered amounts, is seen as a particularly pleasant odour and worthy of carrying around on handkerchiefs.

The level used also relates to arousal. Any sensory stimulus will cause some arousal by definition, and whether this arousal improves or decreases performance on a task will depend on the level of the stimulus, for example, too much may decrease performance, and too little will have little effect, and the pre-existing level of arousal, for example, if someone is underaroused, the stimulation may improve performance, but if they are already optimally aroused, the same stimulation may reduce it.

Given these factors, it is quite possible that lower and pleasanter levels of lavender might have given very different results. Olfaction is also different from the other senses in that the sense adapts, that is, after a while a smell is not perceived. This suggests that its perceptual effect may only last until it has adapted, and that intermittent exposure or varying concentrations might be needed to maintain an effect.

The impact of factors independent of the odour itself on the response to the odour can be both dramatic and counter-intuitive. For example, Rotton et al. (1978) hypothesised that the presence of an unpleasant odour (butyric acid) would create a negative mood, which would, in turn, cause a reduction in people's rated attractiveness of others. This follows the well-researched result that putting people in a bad mood makes them rate others worse. Contrary to their hypothesis, subjects in the presence of the bad odour actually rated (fictitious) others as more attractive compared to subjects who rated without the odour. However, it was discovered that the subjects had thought that the rated person was also suffering the bad odour, that is, was next door doing the same study. This perception may have increased positive feelings towards a 'fellow sufferer'. They repeated the study with the changed instructions that the imaginary rater was not present, and found that subjects now rated them lower. The conclusion of this study is that the psychological response caused by an odour was completely reversed simply because of a different belief about where a person was.

In other studies, the different responses are due to the different tasks imposed. Baron (1993) found that when women job applicants wore fragrance for a mock job interview, male interviewers liked them less and rated them lower in intelligence and friendliness than female applicants not wearing perfume. Baron suggests that the men may have felt that the perfume distracted them from objective assessment. In contrast, Kirk-Smith and Booth (1990) found that in the presence of perfume, men rated photographed women as sexier and softer compared to a no-perfume situation. In this experiment, there was no pressure on subjects to perform a task, and the low levels of odour were unlikely to be distracting.

Knasko et al. (1990) also examined how peoples' expectations about odours influences their psychological responses in a test in which three groups were told that either an unspecified neutral, pleasant or unpleasant odour was present. In fact, no odours were present at all. Even so, people in the 'pleasant odour' condition reported the least ill-health symptoms and those in the 'unpleasant odour' condition reported the most symptoms. Those in the 'pleasant odour' condition also reported a better mood, presumably because people have an implicit expectation that pleasant odours improve mood. In this case, merely thinking that a pleasant odour was present had the same effect as if the pleasant odour had been present (Knasko, 1992).

Individual experience can also affect response. Lavender was the favourite fragrance of an occupational therapist and an acquaintance of the author. She had begun a routine job that she found stressful, so she decided to make her life more pleasant by introducing the smell of lavender into her workplace. She later changed jobs, but now found that every time she smelt lavender it reminded her of the stressful job, even though she still found the smell pleasant. In this case, the hedonics are positive, but the affective associations are now negative.

The conclusion that can be drawn from examples such as these is that it is very difficult to make simple generalisations concerning the effects of any fragrance, such as lavender, on psychological responses which are based on the immediate perceptual effects, rather than the longer term pharmacological effects. A pharmacological effect is more likely to affect people similarly, but the additional psychological mechanisms will create complex effects at the individual level.

As for the EEG responses, psychological responses to a particular odour may depend not only on its pleasantness and congruity with a situation, but also on the demands of the situation and the perceptions, state of mind and the experience of the person. As stated above, it is quite possible that simply changing any one of these many variables will change the psychological response to the odour.

These considerations suggest that the time course of effects needs to be examined in lavender since initial perceptual effects may be replaced or accompanied by later pharmacological effects. The hedonic aspects also need to be carefully accounted for since variations in concentration may account for many of the effects observed. Furthermore, in real-life circumstances, through repeated use by someone the characteristic smell of lavender at moderate concentrations may 'signal' the later sedative effects, and thus the smell might acquire, through classical conditioning, a calming effect by itself, for example, Schiffman and Siebert (1991) claim that an apricot fragrance paired with a relaxed state after progressive relaxation later 'triggers' the relaxed state by itself.

In passing, it may be noted that such dependence of odour response on so many other variables almost eliminates the possibility that traditional 'sterilised' types of laboratory-based experimentation will uncover the responses that are normal in the 'real-life' situation. As soon as you impose rigid control by the method of removing the usual contextual stimuli, you lose the determinants of the response. This leads to a consideration of the effects of lavender in applied situations, which, incidentally, correspond more to the real-life usage referred to in the literature and plays.

Applications of the psychological effects of lavender

The sedative effects of lavender identified above suggest that it might offer a less invasive, and more economic, means of improving sleep and reducing stress and anxiety than current drugs, and without their side-effects, for example, benzodiazepines may cause residual daytime sedation, performance decrements and anterograde amnesia (Vogel, 1992). Accordingly, clinical studies have been carried out to examine this possibility.

Based on Buchbauer *et al.*'s (1993) results on the sedative effects of lavender on mice, Bradshaw *et al.* (1998) found that lavender straw present during transportation halved the incidence of travel sickness symptoms from the 30 per cent measured in pigs bedded with wheat straw, although stress levels, as measured by salivary cortisol, were not changed. They suggest that this could either be due to the sedative effect of lavender or to the distracting effect of the novel odour.

Moving to humans, Hardy *et al.* (1995) conducted a trial on four insomniac psychogeriatric patients, under long-term drug treatment (i.e. with Temazepam, Promazine or Hemineverine). After 2 weeks monitoring of sleep, patients were taken off their drugs for 2 weeks. For a final

2 weeks an unobtrusive level of lavender oil was introduced using an odour diffuser for three one to one and half hour periods each night. As expected, patients slept less when the drugs were taken away. However, after introduction of the ambient lavender oil, sleep returned to the same level as under medication. This appeared to be a pure pharmacological effect; placebo and expectancy effects can be excluded as these patients had Alzheimer's disease. Patients were also reported to be less restless during sleep and to have better waking moods than under medication. Additionally, the lavender oil masked the unpleasant odours typically present in psychogeriatric wards.

In another clinical trial, Brooker et al. (1997) conducted a single-case experimental design across another four patients with senile dementia and behaviour problems. Each patient had between 8–12 half-hour trials over a 3-month period with four conditions randomised across time (ambient lavender, lavender oil and massage, massage, and no treatment). Behaviour was then recorded every minute for the hour after treatment. Although the clinical impressions were that all patients appeared to benefit from treatments, in this study significant improvements in behaviour were only observed for both lavender aroma and massage in one patient. Like Klemm et al. (1992), they comment that because of differences between individuals, the detailed study of individual responses to such interventions may be more suitable than group designs.

In contrast to these two small studies, Dunn et al. (1995) randomly allocated 122 patients to a massage, an aromatherapy massage with 1 per cent lavender oil or a rest period group, each receiving 1–3 treatment sessions of 15–30 min over a 5-day period. Physiological (BP, HR and resp.rate), psychological (mood and anxiety) and behavioural (motor and facial) measures were taken before and after each session. In the first session the lavender and massage group reported less anxiety than the control group, however, no other differences were observed on the other measures. They suggest that perhaps a higher concentration of lavender might have shown more effects.

Finally, Romine et al. (1999) used ambient lavender to investigate recovery after brisk walking for 2 min, with BP, mean arterial pressure, pulse pressure and HR taken pre- and post-exercise and then after a 10 min rest in a dimly lit room with ten subjects given no odour and ten subjects given lavender which was put in the air by an electrical pot-pourri device. Although all measures were all lowered in the lavender condition, no significant differences were found. They suggest that larger samples should be used to give the statistical power to clarify these findings. Dale and Cornwell (1994) compared the effects of lavender oil, synthetic lavender oil and distilled water in reducing perineal discomfort after childbirth when used in bath water over ten days. No significant differences between groups were found.

These studies give mostly negative results regarding the effects of lavender in real-life application. In addition to the confounding issues discussed in the laboratory studies, these trials illustrate the additional problems of conducting such trials, for example, standard measures may be insensitive or inappropriate since people with different conditions may react differently, also it may be difficult to control for expectational effects since therapists have a vested interest in showing that their interventions work. Perhaps these account for the lack of consistent effects in contrast to both the experimental evidence for sedative properties and the folklore for them given in the literature and plays.

Summary

Scientific research into the psychological effects of lavender is limited. However, there is a long history of it being regarded, and used, as a sedative or calming agent. The effects on cells and brain tissues also suggests both reduction in electrical activity and in anticonvulsant effects.

However, a review of research on the effects of lavender on EEG and psychological responses, both laboratory and clinically-based, reveals that responses to lavender may be determined, not only by these pharmacological sedative effects, but by individual, situational and expectational factors independent of the lavender odour itself. As a result, although experimental research indicates calming or sedative effects, these can be difficult to identify or pin down, especially in real-life applications.

References

Aoshima, H. and Hamamoto K. (1999) Potentiation of GABAa receptors expressed in *Xenopus* oocytes by perfume and phytoncid. Biosci. *Biotechnol. Biochem.*, 63, 743–8.

Baron, R.A. (1983) 'Sweet smell of success'? The impact of pleasant artificial scents on evaluations of job applicants. *J. Appl. Psychol.*, 68, 709–13.

Booth, D.A. (1979) Preference as motive. In Kroeze, J.H.A. (ed.), *Preference and Chemoreception*. IRL Press, Oxford.

Bradshaw, R.M., Marchant, J.N., Meredith, M.J. and Broom, D.M. (1998) Effects of lavender straw on stress and travel sickness in pigs. *J. Altern. Complement. Med.*, 4, 271–5.

Brooker, D.J.R., Snale, M., Johnson, E., Ward, D. and Payne, M. (1997) Single case evaluation of the effects of aromatherapy and massage on disturbed behaviour in sever dementia. *Brit. J. Clin. Psychol.*, 36, 287–96.

Buchbauer, G., Jirovetz, L., Czejka, M. and Nasel, C. (1993a) New results in aromatherapy research. *24th Intern. Symp. Essent. Oils*, Berlin, July 21–24, 1993, abstract, p.18.

Buchbauer, G., Jirovetz, L., Jager, W., Plank, C. and Dietrich, H. (1993) Fragrance compounds and essential oils with sedative effects upon inhalation. *J. Pharm. Sci.*, 82(6), 660–4.

Buchbauer, G., Jirovetz, L., Jager, W., Dietrich, H., Plank, C. and Karamat, E. (1991) Aromatherapy: evidence for sedative effects of the essential oil of lavender after inhalation. Zeitschrift fur naturforschung c- a, *J. Biosci.*, 46c, 1067–72.

Dale, A. and Cornwell, S. (1994) The role of lavender oil in relieving perineal discomfort following childbirth: a blind randomized clinical trial. *J. Adv. Nursing.*, 19(1), 89–96.

Degel, J. and Koster, E.P. (1999) Odors: implicit memory and performance effects. *Chem. Senses*, 24(3), 317–25.

Diego, M.A., Jones, N.A., Field, T., Hernandez-Reif, M., Schanberg, S., Kuhn, C., McAdam, V., Galamaga, R. and Galamaga, M. (1998) Aromatherapy positively affects mood, EEG patterns of alertness and math computations. *Int. J. Neurosci.*, 96(3–4), 217–24.

Dunn, C., Sleep, J. and Collett, D. (1995) Sensing an improvement: an experimemntal study to evaluate the use of aromatherapy, massage and periods of rest in an intensive care unit. *J. Adv. Nursing*, 21, 34–40.

Elisabetsky, E, Brum, L.F.S. and Souza, D.O. (1999) Anticonvulsant properties of linalool in glutamate-related seizure models. *Phytomed.*, 6(2), 107–13.

Elisabetsky, E., Marschner, J. and Souza, D.O. (1995) Effects of linalool on glutamatergic system in the rat cerebral cortex. *Neurochem. Res.*, 20, 461–5.

Hardy, M., Kirk-Smith, M. and Stretch, D. (1995) Replacement of drug treatment for insomnia by ambient odour. *The Lancet*, 346, 701.

Kirk-Smith, M.D. and Booth, D.A. (1987) Chemoreception in human behaviour: An analysis of the social effects of fragrances. *Chem. Senses*, 12(1), 159–66.

Kirk-Smith, M.D. and Booth, D.A. (1990) The effect of five odorants on mood and the judgement of others. In MacDonald, D.W., Muller-Schwartz, D. and Natynzcuk, S. (eds), *Chemical Signals in Vertebrates*. Oxford: Oxford University Press, pp. 48–54.

Klemm, W.R., Lutes, S.D., Hendrix, D.V. and Warrenburg, S. (1992) Topographical EEG maps of human responses to odors. *Chem. Senses*, 17, 347–61.

Knasko, S.C. (1992) Ambient odor's effect on creativity, mood, and perceived health. *Chem. Senses*, 17, 27–35.

Knasko, S.C., Gilbert, A.N. and Sabini, J. (1990) Emotional state, physical well-being and performance in the presence of feigned ambient odor. *J. Appl. Soc. Psychol.*, **20**, 1345–57.

Lis-Balchin, M. and Hart, S. (1999) Studies on the mode of action of the essential oil of lavender (*Lavandula angustifolia* Miller). *Phytother. Res.*, **13**, 540–2.

Lorig, T.S. and Roberts, M. (1990) Odor and cognitive alteration of the contingent negative alteration. *Chem. Senses*, **15**, 537–45.

Ludvigson, H.W. and Rottmann, T.R. (1989) Effects of ambient odors of lavender and cloves on cognition, memory, affect and mood. *Chem. Senses*, **14**, 525–36.

Mazzanti, G., Lu, M. and Salvatore, G. (1998) Spasmolytic action of the essential oil from *Hyssopus officinalis* L. var. *decumbens* and its major components. *Phytother. Res.*, **12** (Suppl.), S92–4.

Romine, I.J., Bush, A.M. and Geis, C.R. (1999) Lavender aromatherapy in recovery from exercise. *Percept. Motor Skills*, **88**(3), 756–8.

Rotton, J., Barry, T., Frey, J. and Soler, E. (1978) Air pollution and interpersonal attraction. *J. Appl. Soc. Psychol.*, **8**(1), 57–71.

Schiffman, S.S. and Siebert, J.M. (1991) New frontiers in fragrance use. *Cosmetics and Toiletries*, **106**(6), 39–45.

Teuscher, E., Melzig, M., Villmann, E. and Moritz, K.U. (1990) Untersuchungen zum wirkungsmechanismus atherischer Ole. *Zeitschrift fur Phytotherapie*, **11**, 87–92.

Torii, S., Fukuda, H., Kanemoto, H., Myananchi, S., Hamauzu, Y. and Kawasaki, M. (1988) Contingent negative variation (CNV) and the psychological effects of odour. In Van Toller, S. and Dodd, G. (eds), *Perfumery: The Psychology and Biology of Fragrance*. Chapman and Hall, London.

Vogel, G. (1992) Clinical uses and advantages of low doses of benzodiazepine hypnotics. *J. Clin. Psychiat.*, **53**(6), 19–22.

Yamada, K., Mimaki, Y. and Sashida, Y. (1994) Anticonvulsive effects of inhaling lavender oil vapour. *Biol. Pharm. Bull.*, **17**, 359–60.

15 Antimicrobial properties of lavender volatile oil

Stanley G. Deans

Introduction

The preservative properties of aromatic and medicinal plant volatile [essential] oils and extracts have been recognised since Biblical times, while attempts to characterise these properties in the laboratory date back to the 1900s (Martindale, 1910; Hoffman and Evans, 1911). Plant volatile oils are generally isolated from non-woody plant material by steam or hydrodistillation, and are variable mixtures of principally terpenoids, specifically monoterpenes [C_{10}] and sesquiterpenes [C_{15}] although diterpenes [C_{20}] may also be present, and a variety of low molecular weight aliphatic hydrocarbons [linear, ramified, saturated and unsaturated], acids, alcohols, aldehydes, acyclic esters or lactones and exceptionally nitrogen- and sulphur-containing compounds, coumarins and homologues of phenylpropanoids.

Terpenes

Terpenes are among the chemicals responsible for the culinary, medicinal and fragrant uses of aromatic and medicinal plants. Most terpenes are derived from the condensation of branched five-carbon isoprene units and are categorised according to the number of these units in the carbon skeleton (Dorman, 1999). Traditionally, plants and their extracts, including lavender [*Lavandula angustifolia* Mill.] have been used to extend the shelf-life of foods, beverages and pharmaceutical/cosmetic products through their antimicrobial and antioxidant properties (Baratta *et al.*, 1998a,b; Cai and Wu, 1996; Gallardo *et al.*, 1987; Janssen *et al.*, 1988; Jay and Rivers, 1984; Shelef, 1984; Péllisier *et al.*, 1994; Shapiro *et al.*, 1994; Ueda *et al.*, 1982; Youdim *et al.*, 1999).

Chemicals responsible for antimicrobial action

More recently, attempts have been made to identify the component[s] responsible for such bioactivities (Daferera *et al.*, 2000; Deans and Ritchie, 1987; Dorman and Deans, 2000; Deans *et al.*, 1994a,b; Jeanfils *et al.*, 1991; Lis-Balchin, 1997; Lis-Balchin and Deans, 1997; Lis-Balchin *et al.*, 1998; Vokou *et al.*, 1993). The antimicrobial properties of volatile oils and their constituents from a wide variety of plants have been assessed and reviewed (Carson *et al.*, 1995; 1996; Garg and Dengre, 1986; Inouye *et al.*, 1983; Jain and Kar, 1971; Jansen *et al.*, 1987; Larrondo *et al.*, 1995; Nenoff *et al.* 1996; Pattnaik *et al.*, 1995, 1996; Rios *et al.*, 1987; 1988; Sherif *et al.*, 1987). Investigations into the antimicrobial activities, mode of action and potential uses of plant volatile oils have regained momentum and there appears to be a revival in the use of traditional approaches to livestock welfare as well as food preservation. The activity of the oils would be expected to relate and reflect the respective composition of the volatile oils, the structural configuration of the constituent components and their functional groups along with potential synergistic interactions between components. A correlation of the antimicrobial activity of compounds under test in the study by Dorman and Deans (2000) and their relative percentage composition in the oils, with their chemical structure, functional groups and configuration suggested a number of observations on the structure/function relationship.

Structure versus function

Components with phenolic structures, such as carvacrol and thymol, were highly active against test bacteria, despite their low capacity to dissolve in water. The importance of the hydroxyl group in the phenolic ring was confirmed in terms of activity when carvacrol was compared to its methyl ester. The high activity of the phenolic components may be further explained in terms of the alkyl substitution into the phenolic nucleus, which is known to enhance the antimicrobial activity of phenols (Pelczar et al., 1988). The introduction of alkylation has been proposed to alter the distribution ratio between the aqueous and non-aqueous phases, including bacterial phases, by reducing the surface tension or altering the species selectivity. Alkyl-substituted phenolic compounds form phenoxyl radicals which interact with isomeric alkyl substituents (Pauli and Knobloch, 1987). This does not occur with etherified/esterified isomeric molecules, possibly explaining their relative lack of activity.

The presence of an acetate moiety in the structure appears to increase the activity of the parent compound as in the case of geraniol where geranyl acetate demonstrated an increase in activity. Aldehydes, notably formaldehyde and glutaraldehyde, are known to possess powerful antimicrobial activity. It has been suggested that an aldehyde group conjugated to a carbon=carbon double bond is a highly electronegative arrangement which may explain their activity (Moleyar and Narasimham, 1986), suggesting an increase in electronegativity increases the antibacterial activity (Kurita et al., 1979; 1981). Such electronegative compounds may interfere in biological processes involving electron transfer and react with vital nitrogen components such as proteins and nucleic acids, and therefore inhibit growth of the microorganisms. The aldehydes cis and trans citral displayed moderate activity against test bacteria while citronellal was less active. Alcohols are known to possess bactericidal rather than bacteriostatic activity against vegetative bacterial cells. The alcohol terpenoids studied did show some activity against test bacteria, acting as protein denaturing agents, solvents or dehydrating agents.

A number of oil components are ketones, wherein the presence of an oxygen function in the framework increases the antimicrobial properties of the terpenoids (Naigre et al., 1996). From this study, and by using the contact method, the bacteriostatic and fungistatic action of terpenoids was increased when carbonylated. Menthone was shown to have modest activity with *Clostridium sporogenes* and *Staphylococcus aureus* being the most significantly affected (Dorman and Deans, 2000). An increase in activity dependant upon the type of alkyl substituent incorporated into a non-phenolic ring structure appeared to occur in this study. An alkenyl substituent [1-methylethenyl] resulted in increased antibacterial activity, as seen in limonene [1-methyl-4-(1-methylethenyl)-cyclohexene], compared to an alkyl [1-methylethyl] substituent as in p-cymene [1-methyl-4-(1-methylethyl)-benzene]. The inclusion of a double bond increased the activity of limonene relative to p-cymene, which demonstrated no activity against the test bacteria. In addition, the susceptible organisms were principally Gram-negative, which suggests alkylation influences Gram reaction sensitivity of the bacteria. The importance of the antimicrobial activity of alkylated phenols in relation to phenol has been previously reported (Pelczar et al., 1988). Their data suggest that an allylic side chain seems to enhance the inhibitory effects of a component and chiefly against Gram-negative organisms.

Stereochemistry versus function

Furthermore, the stereochemistry has an influence on bioactivity. It was observed that α-isomers are inactive relative to α-isomers, for example, α-pinene; *cis*-isomers are inactive contrary to *trans*-isomers, for example, geraniol and nerol; compounds with methyl-isopropyl cyclohexane rings are most active; or unsaturation of the cyclohexane ring further increases the antibacterial activity, for example, terpinolene, terpineol and terpineolene (Hinou *et al.*, 1989).

Mode of action of terpenoids on membranes

Investigations into the effects of terpenoids upon isolated bacterial membranes suggest that their activity is a function of the lipophilic properties of the constituent terpenes (Knobloch *et al.*, 1986), the potency of their functional groups and their aqueous solubility (Knobloch *et al.*, 1988). Their site of action appeared to be at the phospholipid bilayer, caused by biochemical mechanisms catalysed by these phosholipid bilayers of the cell. These processes include the inhibition of electron transport, protein translocation, phosphorylation steps and other enzyme-dependant reactions (Knobloch *et al.*, 1986). Their activity in whole cells appears more complex (Knobloch *et al.*, 1988). Although a similar water soluble tendency is observed, specific statements on the action of single terpenoids *in vivo* have to be assessed singularly, taking into account not only the structure of the terpenoid, but also the chemical structure of the cell wall (Knobloch *et al.*, 1988). The plant extracts clearly demonstrate antibacterial properties, although the mechanistic processes are poorly understood.

Chemotherapeutic agents

Chemotherapeutic agents, used orally or systemically for the treatment of microbial infections of humans and animals, possess varying degrees of selective toxicity. Although the principle of selective toxicity is used in agriculture, pharmacology and diagnostic microbiology, its most dramatic application is the systemic chemotherapy of infectious diseases. The tested plant products appear to be effective against a wide spectrum of microorganisms, both pathogenic and non-pathogenic. Administered orally, these compounds may be able to control a wide range of microbes, but there is also the possibility that they may cause an imbalance in the gut microflora, allowing opportunist pathogenic bacteria, such as coliforms, to become established in the gastrointestinal tract with resultant deleterious effects. Further studies on therapeutic applications of volatile oils should be undertaken to investigate these issues, especially when consideration is made of the substantial number of analytical/bioactivity studies carried out on these natural products.

Antimicrobial activity of *Lavandula*

Antimicrobial activity against different bacterial species

There are relatively few reports on the antimicrobial properties of lavender, due mainly to its modest activity against bacteria and fungi. This is in contrast to the considerable antimicrobial status awarded to lavender by aromatherapists. In one of the larger studies in this field, Deans and Ritchie (1987) studied fifty volatile oils at four concentrations against a range of twenty-five bacterial genera. In the case of lavender, it was found to be most effective against the faecal indicator organism *Enterococcus faecalis*. With *Klebsiella pneumoniae*, in contrast, its presence resulted in enhancement of growth (Table 15.1a). In addition, some oil constituents were tested for antimicrobial action, and are tabulated in Table 15.1b.

The genus *Bacillus* has been shown to be susceptible to lavender volatile oil in a number of studies (Deans and Ritchie, 1987; Jeanfils *et al.*, 1991; Lis-Balchin *et al.*, 1998) with Jeanfils' group reporting that the antibacterial properties exceeded that of *Allium cepia* when tested against *Bacillus stearothermophilus* an important organism involved in food spoilage, being able to withstand elevated processing temperatures. However, the long-term stability over a period of 25 days showed the lavender extracts, obtained by both steam distillation and ethanol extraction, to be less active in general, compared with the corresponding onion extracts. These workers state, however, that the lavender extracts have a broader bactericide spectrum than the onion extracts. Moreover, two of the lavender extracts had a marked activity against members of the genus *Salmonella*.

Table 15.1a Antibacterial activity of lavender volatile oil against twenty-five test bacteria. Inhibition zone diameter in mm, including well diameter of 4 mm. Each value is the mean of three replicates. Taken from Deans and Ritchie, 1987

Organism	Source	Inhibition zone
Acinetobacter calcoacetica	NCIB 8250	10.0
Aeromonas hydrophila	NCTC 8049	8.0
Alcaligenes faecalis	NCIB 8156	4.0
Bacillus subtilis	NCIB 3610	12.5
Beneckea natriegens	ATCC 14048	4.0
Brevibacterium linens	NCIB 8456	8.5
Brocothrix thermosphacta	Sausage meat	5.5
Citrobacter freundii	NCIB 11490	8.5
Clostridium sporogenes	NCIB 10696	4.0
Enterobacter aerogenes	NCTC 10006	7.5
Enterococcus faecalis	NCTC 775	14.0
Erwinia carotovora	NCPPB 312	6.0
Escherichia coli	NCIB 8879	7.5
Flavobacterium suaveolens	NCIB 8992	4.0
Klebsiella pneumoniae	NCIB 418	*
Lactobacillus plantarum	NCDO 343	7.5
Leuconostoc cremoris	NCDO 543	4.0
Micrococcus luteus	NCIB 8165	9.5
Moraxella sp.	NCIB 10762	9.0
Proteus vulgaris	NCIB 4175	9.0
Pseudomonas aeruginosa	NCIB 950	4.0
Salmonella pullorum	NCTC 10704	7.0
Serratia marcescens	NCIB 1377	7.5
Staphylococcus aureus	NCIB 6571	8.0
Yersinia enterocolitica	NCTC 10460	9.0

Notes

NCIB – National Collection of Industrial Bacteria; NCTC – National Collection of Type Cultures; ATCC – American Type Culture Collection; NCPPB – National Collection of Plant Pathogenic Bacteria; NCDO – National Collection of Dairy Organisms.

* Enhancement of growth.

Table 15.1b Antibacterial activity of lavender oil constituents against twenty-five test bacteria. Inhibition zone diametre in mm, including well diametre of 4 mm. Each value is the mean of three replicates. Taken from Dorman and Deans, 2000. Bacterial sources as in Table 15.1a

Organism	A*	B	C	D
Acinetobacter calcoacetica	4.0	9.3	4.0	11.2
Aeromonas hydrophila	4.0	11.5	4.0	7.1
Alcaligenes faecalis	4.0	12.1	4.0	7.8
Bacillus subtilis	4.0	14.0	4.0	4.0
Beneckea natriegens	4.0	11.4	4.0	6.5
Brevibacterium linens	4.0	12.5	4.0	4.0
Brocothrix thermosphacta	4.0	8.1	4.0	5.9
Citrobacter freundii	7.8	27.5	6.0	5.9
Clostridium sporogenes	10.3	20.3	5.7	7.5
Enterobacter aerogenes	7.1	9.7	4.0	4.0
Enterococcus faecalis	4.0	16.7	9.2	7.8
Erwinia carotovora	7.4	12.3	8.7	4.0
Escherichia coli	11.2	13.8	8.9	7.8
Flavobacterium suaveolens	10.6	15.7	6.5	8.4
Klebsiella pneumoniae	7.0	12.6	8.1	7.9
Lactobacillus plantarum	4.0	25.3	4.0	4.0
Leuconostoc cremoris	4.0	4.0	4.0	4.0
Micrococcus luteus	4.0	13.4	7.6	6.3
Moraxella sp.	7.9	10.3	6.2	4.8
Proteus vulgaris	7.4	12.2	7.5	6.6
Pseudomonas aeruginosa	4.0	4.0	4.0	6.5
Salmonella pullorum	11.2	7.5	7.9	6.0
Serratia marcescens	6.5	8.8	4.0	5.4
Staphylococcus aureus	4.0	9.0	8.3	7.4
Yersinia enterocolitica	7.1	9.5	6.6	5.8

* Key to compounds: A – limonene; B – linalool; C – α-pinene; D – β-pinene.

Activity against Listeria monocytogenes

In a separate study, the impact of lavender volatile oil upon strains of *Listeria monocytogenes* was reported by Lis-Balchin and Deans (1997). Attempts were made to correlate oil composition with anti-*Listeria* activity. In some cases, two or more commercial samples of the same volatile oil were screened to determine whether the activity was consistent for that named oil and the results correlated with oil composition as determined by GC and GC-MS. There was a wide variation in the anti-*Listeria* activity of different lavenders, ranging from zero to 18 strains [out of 20] affected. The highest activity was found for the Bulgarian lavender which was stated to be 'extracted with supercritical CO_2': this sample had the highest linalyl acetate concentration [80 per cent] with 10 per cent lavandulyl acetate and 2.3 per cent linalool (Table 15.2). Spike lavender along with another Bulgarian lavender sample were lower in activity but had high linalool contents. Other lavenders and lavandin with a lower linalool content but a higher linalyl acetate content proved very ineffective against all strains of *L. monocytogenes*.

This study indicated considerable variation in the activity of the 'same' volatile oil against different strains of *L. monocytogenes*, this being in agreement with the findings of Aureli *et al.* (1992)

Table 15.2 Correlation between linalool and linalyl acetate content per cent of lavender volatile oils and antimicrobial activity. Taken from Lis-Balchin and Deans, 1997 and Lis-Balchin et al., 1998

Lavender oil	Linalool	Linalyl acetate	Anti-bacterial activity[1]	Anti-listeria activity[2]	Antifungal index[3]		
					Aspergillus niger	Aspergillus ochraceus	Fusarium culmorum
Lav 1	29.7	42.8	19	1	82	90	79
Bulgar lav	51.9	9.5	23	11	84	29	8
Bulgar lav CO_2	2.3	79.8	22	18	74	84	89
French lav 1	26.1	47.9	16	4	93	58	31
French lav 2	29.1	43.2	13	0	57	44	77
Lavandin	28.7	39.4	17	1	93	86	69
Spike lav	43.1	4.0	19	12	93	58	31

Notes
1 Number of activities recorded from total of twenty-five bacterial genera.
2 Number of activities recorded from total of twenty *Listeria monocytogenes* strains.
3 Activity recorded when calculated from formula $AI = [C - T]/C \times 100$, where C is the weight of mycelium in Control flasks and T is the weight of mycelium in the Test flasks following incubation in YES broth at 25°C for <10 days.

and Lis-Balchin et al. (1996a). The adulteration of volatile oils with synthetic components may give rise to a different proportion of enantiomers for a large number of components than in a pure botanical sample: this can greatly influence the bioactivity of the subsequent material (Lis-Balchin et al., 1996b; 1998).

Crude herb versus essential oil

An interesting study by Vokou et al. (1993) highlighted the beneficial effects of controlling plant pathogenic bacteria, *Erwinia carotovora* and others, in respect of suppressing potato sprout growth. Application of crude herb material, in place of extracted volatile oils, lead to similar results. The inhibition of sprouting was, however, reversible, thus allowing subsequent normal sprouting of the seed potato tubers. This phenomenon offers potential application on a commercial scale since the storage life of the crop is prolonged while treatment with lavender and other plants offers protection from microbial attack by certain microorganisms. This is particularly important for such regions as the Mediterranean where the problem of prolongation of tuber dormancy is exacerbated by elevated temperatures. Since lavender grows as a native in this region, its use should prove cost-effective.

Antifungal properties

In addition to antibacterial activities, lavender also possesses antifungal properties. The action of seven samples of lavender against three filamentous fungi is also shown in Table 15.2. Correlation between chemical composition and antimycotic activity is not always clear. The test organisms, *Aspergillus niger, A. ochraceus* and *Fusarium culmorum* reacted differently to the oils, with the plant pathogenic *F. culmorum* being less affected, with two exceptions, than the two *Aspergilli*. The spoilage organism *A. niger* was more inhibited in its growth, again with two exceptions, than the mycotoxigenic *A. ochraceus*.

Mode of antifungal action

Daferera et al. (2000) report that lavender oil was moderately inhibitory to radial growth and conidial germination of *Penicillium digitatum*, while conidial production was not affected by the oil at concentrations up to 1000 µg/mL. Lavender oil, along with others such as thyme and marjoram, were more effective in the inhibition of conidial germination than of radial growth.

Conidial germination is the first vital step in the sequence of events leading to the establishment of a germ-tube and subsequent hypha. The process begins with hydration followed by the action of lytic enzymes such as chitinase and α- and β-glucanases. This breaks down the thickened conidial cell wall to permit emergence of the germ-tube initial. Once this event takes place, there is a balance between the lytic and synthetic enzyme systems necessary for normal hyphal extension. An imbalance in either enzyme system leads to growth inhibition and/or death (McEwan, 1994). It would appear that these early ontogenic events are highly susceptible to disruption by the component[s] in lavender volatile oil, resulting in lack of successful conidial germination.

Acknowledgements

SAC received financial support from the Scottish Executive Rural Affairs Department.

References

Aureli, P., Costantini, A. and Zolea, S. (1992). Antimicrobial activity of some plant essential oils against *Listeria*. *Journal of Food Protection* 55, 344–8.

Baratta, M.T., Dorman, H.J.D., Deans, S.G., Biondi, D.M. and Ruberto, G. (1998a). Chemical composition, antibacterial and antioxidative activity of laurel, sage, rosemary, oregano and coriander essential oils. *Journal of Essential Oil Research* 10, 618–27.

Baratta, M.T., Dorman, H.J.D., Deans, S.G., Figueiredo, A.C., Barroso, J.G. and Ruberto, G. (1998b). Antimicrobial and antioxidant properties of some commercial essential oils. *Flavour and Fragrance Journal* 13, 235–44.

Cai, L. and Wu, C.D. (1996). Compounds from *Syzygium aromaticum* possessing growth inhibitory activity against oral pathogens. *Journal of Natural Products* 59, 987–90.

Carson, C.F., Cookson, B.D., Farrelly, H.D. and Riley, T.V. (1995). Susceptibility of methicillin-resistant *Staphylococcus aureus* to the essential oil of *Melaleuca alternifolia*. *Journal of Antimicrobial Chemotherapy* 35, 421–4.

Carson, C.F., Hammer, K.A. and Riley, T.V. (1996). *In vitro* activity of the essential oil of *Melaleuca alternifolia* against *Streptococcus* spp. *Journal of Antimicrobial Chemotherapy* 37, 1177–81.

Daferera, D.J., Ziogas, B.N. and Polissiou, M.G. (2000). GC-MS analysis of essential oils from some Greek aromatic plants and their fungitoxicity on *Penicillium digitatum*. *Journal of Agricultural and Food Chemistry* 48, 2576–81.

Deans, S.G. and Ritchie, G. (1987). Antibacterial properties of plant essential oils. *International Journal of Food Microbiology* 5, 165–80.

Deans, S.G., Kennedy, A.I., Gundidza, M.G., Mavi, S., Waterman, P.G. and Gray, A.I. (1994a). Antimicrobial activities of the volatile oil of *Heteromorpha trifoliata* [Wendl.] Eckl. & Zeyh. [Apiaceae]. *Flavour and Fragrance Journal* 9, 245–8.

Deans, S.G., Hiltunen, R., Wuryani, W., Noble, R.C. and Penzes, L.G. (1994b). Antimicrobial and antioxidant properties of *Syzygium aromaticum* [L.] Merr. & Perry: Impact upon bacteria, fungi and fatty acid levels in ageing mice. *Flavour and Fragrance Journal* 10, 323–8.

Dorman, H.J.D. (1999). Phytochemistry and bioactive properties of plant volatile oils: Antibacterial, antifungal and antioxidant activities. PhD thesis, Strathclyde Institute of Biomedical Sciences, University of Strathclyde, Glasgow, Scotland.

Dorman, H.J.D. and Deans, S.G. (2000). Antimicrobial agents from plants: Antibacterial activity of plant volatile oils. *Journal of Applied Microbiology* **88**, 308–16.

Gallardo, P.P.R., Salinas, R.J. and Villar, L.M.P. (1987). The antimicrobial activity of some spices on microorganisms of great interest to health. IV: Seeds, leaves and others. *Microbiologie aliments Nutrition* **5**, 77–82.

Garg, S.C. and Dengre, S.L. (1986). Antibacterial activity of essential oil of *Tagetes erecta* Linn. *Hindustan Antibiotic Bulletin* **28**, 27–9.

Hinou, J.B., Harvala, C.E. and Hinou, E.B. (1989). Antimicrobial activity screening of 32 common constituents of essential oils. *Pharmazie* **44**, H4.

Hoffman, C. and Evans, A.C. (1911). The use of spices as preservatives. *Journal of Indian Engineering and Chemistry* **3**, 835–8.

Inouye, S., Goi, H., Miyouchi, K., Muraki, S., Ogihara, M. and Iwanami, I. (1983). Inhibitory effect of volatile components of plants on the proliferation of bacteria. *Bokin Bobai* **11**, 609–15.

Jain, S.R. and Kar, A. (1971). The antibacterial activity of some essential oils and their combinations. *Planta Medica* **20**, 118–23.

Janssen, M.A., Scheffer, J.J.C. and Baerheim-Svendsen, A. (1987). Antimicrobial activities of essential oils: A 1976–1986 literature review on possible applications. *Pharmaceutische Weekblad [Scientific Edition]* **9**, 193–7.

Janssen, M.A., Scheffer, J.J.C., Parhan Van Atten, A.W. and Baerheim-Svendsen, A. (1988). Screening of some essential oils for their activities on dermatophytes. *Pharmaceutische Weekblad [Scientific Edition]* **10**, 277–80.

Jay, J.M. and Rivers, G.M. (1984). Antimicrobial activity of some food flavouring compounds. *Journal of Food Safety* **6**, 129–39.

Jeanfils, J., Burlion, N. and Andrien, F. (1991). Antimicrobial activities of essential oils from different plant species. *Landbouwtijdschrift-Revue de l'Agriculture* **44**, 1013–19.

Knobloch, K., Weigand, N., Weis, H.M. and Vigenschow, H. (1986). *Progress in Essential Oil Research* [E.J. Brunke (ed.)], pp. 429. Walter de Gruyther, Berlin.

Knobloch, K., Pauli, A., Iberl, N., Weis, H.M. and Weigand, N. (1988). Modes of action of essential oil components on whole cells of bacteria and fungi in plate tests. In *Bioflavour* [P. Schreier (ed.)], pp. 287–99. Walter de Gruyther, Berlin.

Kurita, N., Miyaji, M., Kurane, R., Takahara, Y. and Ichimura, K. (1979). Antifungal activity of molecular orbital energies of aldehyde compounds from oils of higher plants. *Agriculture and Biological Chemistry* **43**, 2365–71.

Kurita, N., Miyaji, M., Kurane, R., Takahara, Y. and Ichimura, K. (1981). Antifungal activity of components of essential oils. *Agriculture and Biological Chemistry* **45**, 945–52.

Larrondo, J.V., Agut, M. and Calvo-Torras, M.A. (1995). Antimicrobial activity of essences from Labiatae. *Microbios* **82**, 171–2.

Lis-Balchin, M. (1997). Essential oils and aromatherapy: their modern role in healing. *Journal of the Royal Society of Health* **117**, 324–9.

Lis-Balchin, M. and Deans, S.G. (1997). Bioactivity of selected plant essential oils against *Listeria monocytogenes*. *Journal of Applied Microbiology* **82**, 759–62.

Lis-Balchin, M., Hart, S., Deans, S.G. and Eaglesham, E. (1996a). Comparison of the pharmacological and antimicrobial action of commercial plant essential oils. *Journal of Herbs, Spices and Medicinal Plants* **4**, 69–86.

Lis-Balchin, M., Ochocka, R.J., Deans, S.G. and Hart, S. (1996b). Bioactivity of the enantiomers of limonene. *Medical Science Research* **24**, 309–10.

Lis-Balchin, M., Deans, S.G. and Eaglesham, E. (1998). Relationship between bioactivity and chemical composition of commercial essential oils. *Flavour and Fragrance Journal* **13**, 98–104.

Martindale, W.H. (1910). Essential oils in relation to their antiseptic powers as determined by their carbolic coefficients. *Perfumery and Essential Oil Research* **1**, 266–96.

McEwan, M. (1994). The antifungal effects of plant essential oils and their production by transformed shoot culture. PhD thesis, Strathclyde Institute of Biomedical Sciences, University of Strathclyde, Glasgow, Scotland.

Moleyar, V. and Narasimham, P. (1986). Antifungal activity of some essential oil components. *Food Microbiology* 3, 331–6.

Naigre, R., Kalck, P., Roques, C., Roux, I. and Michel, G. (1996). Comparison of antimicrobial properties of monoterpenes and their carbonylated products. *Planta Medica* 62, 275–7.

Nenoff, P., Haustein, U.F. and Brandt, W. (1996). Antifungal activity of the essential oil of *Melaleuca alternifolia* [tea tree oil] against pathogenic fungi *in vitro*. *Skin Pharmacology* 9, 388–94.

Pattnaik, S., Subramanyam, V.R., Kole, C.R. and Sahoo, S. (1995). Antibacterial activity of essential oils from *Cymbopogon*: Inter- and intra-specific differences. *Microbios* 84, 239–45.

Pattnaik, S., Subramanyam, V.R. and Kole, C.R. (1996). Antibacterial and antifungal activity of ten essential oils *in vitro*. *Microbios* 86, 237–46.

Pelczar, M.J., Chan, E.C.S. and Krieg, N.R. (1988). Control of microorganisms by physical agents. In *Microbiology*, pp. 469–509. McGraw-Hill International, New York.

Pélissier, Y., Marion, C., Casadebaig, J., Milhau, M., Djenéba, K., Loukou, N.Y. and Bessière, J.M. (1994). A chemical, bacteriological, toxicological and clinical study of the essential oil of *Lippia multiflora* [Verbenaceae]. *Journal of Essential Oil Research* 6, 623–30.

Pauli, A. and Knobloch, K. (1987). Inhibitory effects of essential oil components on growth of food-contaminating fungi. *Zeitschrift für Lebensmittel Untersuchung und Forschung* 185, 10–13.

Rios, J.L., Recio, M.C. and Villar, A. (1987). Antimicrobial activity of selected plants employed in the Spanish Mediterranean area. *Journal of Ethnopharmacology* 21, 139–52.

Rios, J.L., Recio, M.C. and Villar, A. (1988). Screening methods for natural products with antibacterial activity: a review of the literature. *Journal of Ethnopharmacology* 23, 127–49.

Shapiro, S., Meier, A. and Guggenheim, B. (1994). The antimicrobial activity of essential oils and essential oil components towards oral bacteria. *Oral Microbiology and Immunology* 9, 202–4.

Shelef, L.A. (1984). Antimicrobial effects of spices. *Journal of Food Safety* 6, 29–44.

Sherif, A., Hall, R.G. and El-Amamy, M. (1987). Drugs, insecticides and other agents from *Artemisia*. *Medical Hypotheses* 23, 187–93.

Ueda, S., Yamashita, H., Nakajima, M. and Kumabara, Y. (1982). Inhibition of microorganisms by spice extracts and flavouring compounds. *Nippon Shokuhin Kogyo Gakkaishi* 29, 111–6.

Vokou, D., Vareltzidou, S. and Katinakis, P. (1993). Effects of aromatic plants on potato storage: sprout suppression and antimicrobial activity. *Agriculture Ecosystems & Environment* 47, 223–25.

Youdim, K.A., Dorman, H.J.D. and Deans, S.G. (1999). The antioxidant effectiveness of thyme oil, α-tocopherol and ascorbyl palmitate on evening primrose oil oxidation. *Journal of Essential Oil Research* 11, 643–8.

16 Lavender oil and its use in aromatherapy

Maria Lis-Balchin

Introduction

Aromatherapy is broadly defined as 'treatment with odours' (Buchbauer, 1992). However, in England it involves the application of a very diluted essential oil (EO) or mixture of EOs (1–2 per cent) in a carrier oil like almond oil, which is massaged into the skin; either on hands, feet, head or the total body (Ryman, 1991; Worwood, 1991; Westwood, 1991; Price, 1993). Aromatherapy can also mean the addition of drops of EO to the bath or a basin of hot water, or the volatilization of the EO(s) using various burners. It usually involves counselling about diet, exercise, life-style etc. by the aromatherapist, who may have absolutely no qualifications to offer such advice. In France, Germany and other parts of Europe, aromatherapy takes on a different meaning as it involves the internal usage of EOs as medicines, and is practised by medically-qualified doctors (Lis-Balchin, 1997). This includes oral, rectal and vaginal introduction of EOs into the body, as well as massaging in of more concentrated EOs into the body. This 'clinical aromatherapy' has also spread to England and the United States, where totally unqualified people practise internal usage of various EOs after making their diagnosis of the client's medical condition. Other herbs, novel medicinal plant extracts, hydrolats, herbal oils, phytols, infusions etc. are often included in the aromatherapy treatment, although the herbal knowledge of the aromatherapists may be lacking (Lis-Balchin, 1999). Many of the attributes of lavender oil were mistakenly taken from herbals, for example, Culpeper (1653), who used alcoholic extracts or teas, not distilled EOs and was also interested in astrology, hence every plant had an assigned planet. The definition and application of aromatherapy can therefore range from a mixture of the paranormal aspects of plant essences and their energetics to the medical and scientific aspects of the EOs as chemicals.

Aromatherapy basics

Aromatherapy applications include (Table 16.1):

1. A diffuser, which can be powered by electricity, giving out a fine mist of the EO.
2. A burner, with water added to the fragrance to prevent burning of the EO: about 1–4 drops of EO is added to 10 ml water. The burner can be warmed by candles or electricity. The latter would be safer in a hospital/ children's room/ and even a bedroom.
3. Ceramic or metal rings placed on an electric light bulb with a drop or two of EO. This results in a rapid burn-out of the oil and also lasts for a very short time due to the rapid volatalization of the EO in the heat.
4. A warm bath with drops of EO added. This results in the slow volatalization of the EO, and not in absorption of the EO through the skin as stated in aromatherapy books, as the EO

Table 16.1 Clinical studies on aromatherapy

Source	Study	EOs used
Cornwell and Dale (1994)	Peritoneal repair after childbirth	Lavender in bath
Burns and Blaney (1994)	Aromatherapy in childbirth	Numerous oils and methods
Dunn et al. (1995)	Aromatherapy in intensive care unit	Lavender foot massage
Buckle (1993)	Active lavender?	Lavender, used two types

does not mix with water. Pouring in an EO mixed with milk serves no useful purpose as the EO will still not mix with water; the pre-mixing of the EO in a carrier oil, as for massage, result in a nasty oily scum around the bath.

5 A bowl of hot water with drops of EO, usually used for soaking feet or used as a bidet. Again the EO will not mix with the water. This is useful for respiratory conditions and colds, where the EO can be breathed in when the head is over the container and a towel placed over the head. This is an old way of treatment and has been used successfully with Vicks, Obas oil, *Eucalyptus* oils for numerous years.

6 Compresses using EO drops on a wet cloth, either hot or cold, to relieve inflammation, treat wounds etc. Again, the EO is not able to mix with the water and can be concentrated in one or two areas.

7 Massage of body, hands, feet, back or all over using 2–4 drops of EO (single EO or mixture) diluted in 10 ml carrier oil (fixed, oily), for example, almond oil or jojoba, grapeseed, wheatgerm oils etc.

Oral intake

Another method involves the 'medicinal' properties of the EO after oral intake. This is not to be condoned, although practised by a number of aromatherapists, unless effected under a medically qualified aromatherapist.

EO drops are 'mixed' in a tumbler of hot water or presented on a sugar cube or 'mixed' with a teaspoonful of honey and taken internally. This is not true aromatherapy, as almost all the rest of the methods are based on EO volatalization and therefore largely the effect of the EO on the central nervous system (CNS) via the nose and thence the Limbic System which can cause a secondary effect on other parts of the body.

There are some conventional medically-approved and tested peppermint oil capsules (e.g. Colpermin) on the market, used for treating irritable bowel syndrome, IBS, but this could not be called 'aromatherapy' as the aroma is encapsulated, so why would any internal usage have such a status?

Direct effects on the skin can also occur, for example, antiseptic action of the EO or counterirritant effect which can cause reddening of the skin and an increased blood flow to the area and could presumably ease pain and swelling. The latter could only be effected by a few EOs, for example, thyme, clove, oregano. Many EOs can be effective antiseptics, but this may not be the outcome when used at a 1–2 per cent dilution in a carrier oil. The use of EOs in the treatment of wounds or scars is again outside the terminology of 'aromatherapy' and under the umbrella of conventional usage as medicines, be it as antiseptics or healing factors. The use of lavender oil, in a complex mixture of EOs, plant extracts and other components has been utilized by a plastic surgeon (Dr Giller) in Israel, but who has as yet not carried out proper controlled studies, but this, like the use of lavender oil for burns (Gattefose, 1937), is surely nothing to do with aromatherapy.

Massage

The massage applied is usually gentle effleurage with some petrissage (kneading) with and without some shiatsu, lymph drainage in some cases and is sometimes more vigorous, according to the aromatherapist's skills (Price, 1993). The massage should be relaxing, but also able to increase the circulation of the blood and lymphatic system 'in order to release toxins' (of dubious scientific validity), break down tension in muscles and tone weak muscles.

EO blending

The actual blending of EOs is considered an art form, but is basically simple: to 10 ml of carrier oil, in a brown bottle, 1–4 drops of the same or different EO are added and the bottle stopperred. The contents are then gently mixed without shaking the bottle too much and creating air bubbles. Some aromatherapists swear that different people can create a different mixture simply by using their own energetics (probably heat provided by the hands), but scientifically, it may simply be a question of slight changes in the volume of EO applied by different people. The actual size of drops depends largely on the droppers provided with each individual bottle and can vary by over 500 in volume/weight (personal observations).

Lavender oil and its supposed functions

The following quotes will illustrate the tendency to mix the 'esoteric' with the 'medicinal' value of lavender and also show that misinterpretations of past herbal uses of plants have added to the general muddle:

'*Lavandula angustifolia* has a "yang" quality and "its ruling planet is Mercury" (Tisserand, 1985).' That author lists its properties as: 'analgesic, anticonvulsive, antidepressant, antiseptic, antispasmodic, antitoxic, carminative, cholagogue, choleretic, cicatrisant, cordial, cytophylactic, deodorant, diuretic, emmanogogue, hypotensive, nervine, sedative, splenic, sudorific, tonic, vermifuge and vulnerary'. Its uses are for: 'abcess, acne, alopecia areata, asthma, blenorrhoea, blepharitis, boils, bronchitis, carbuncles, catarrh, chlorosis, colic, conjunctivitis, convulsions, cystitis, depression, dermatitis, diarrhoea, diphtheria, dyspepsia, earache, eczema, epilepsy, fainting, fistula, flatulence, gonorrhoea, halitosis, headache, hypertension, hysteria, influenza, insomnia, laryngitis, leucorrhoea, migraine, nausea, nervous tension, neurasthenia, oliguria, palpitations, paralysis, pediculosis, psoriasis, rheumatism, scabies, scrofula, gall stones, sunstroke, throat infections, tuberculosis, typhoid fever, ulcers of cornea and leg, vomiting, whooping cough and wounds'.

'Lavender helps balance the mind and emotions' (Westwood, 1991) and has 'immunity' as its key word; it is indicated for over-analytical, anxiety, fear of failure, hyperactivity, hysteria, imbalance, immune system, impatience, insecurity, insomnia, irritation, irrationality, mood swings, overwork, panic, paranoia, possessiveness, greed for power, feeling pressurized, lack of relaxation, stage fright, tension, poor time-management, lack of tolerance and tranquillity, workaholic, worry. The physical conditions indicating its use are: baldness, immune system, sore throats, stiffness, dermatitis, eczema, itchy or scarred skin and general first aid.

Scientific verification for EO therapeutic effects

It remains to be seen whether aromatherapy has any actual medicinal benefits (Vickers, 1996; Nelson, 1997) other than stress-alleviating, through massage, and whether these are attributable only to massage with the true EOs, especially as there is a wide difference in the actual percentage chemical composition of EOs obtained from different geographical sources and also different samples from plants grown in various countries where differences in hybridization has occurred and even the same plants grown under different climatic conditions show differences (Lis-Balchin, 1997). The commercial EOs are also often admixed, adulterated with other EOs from different plants and mixed with synthetics, which all add to differences in bioactivity (Lis-Balchin et al., 1998).

Of the known effects of EOs, there are few examples of direct 'aromatherapy', that is, true effects due to 'aroma' rather than due to the chemical nature of the EO. Most of the work on true 'aromatherapy' has been undertaken by perfumery companies, who used fragrances rather than pure EOs.

Scientific proof of EO efficacy: aroma science

Known bioactivities of EOs

These include: antimicrobial effects, effects on the CNS, pharmacological effects and various other miscellaneous effects.

Antimicrobial activity

EOs have been used externally to eradicate fungal or bacterial infections since the Black Death (and before); doctors would wrap scarves soaked in EOs like camphor round their necks and over their mouths when visiting patients (Valnet, 1982). This did not prevent the death *per se*, as the doctors rarely went near the patients, and only poked them with a long stick from a distance, to ascertain whether they were alive or dead. In the Second World War, wounded soldiers had their wounds treated with EOs until penicillin and other antibiotics became available (Valnet, 1982). Many EOs have considerable antimicrobial activity (Maruzella and Henry, 1958; Maruzella and Sicurella, 1960; Youzef and Tawil, 1980; Moleyar and Narasimham, 1987; Lis-Balchin and Deans, 1997; Lis-Balchin et al., 1998). Plant EOs like thyme and oregano, are extremely potent antimicrobials *in vitro* and can have a considerable effect on a wide range of different bacteria (Lis-Balchin, 1995; Lis-Balchin et al., 1998). Lavender is not strongly antimicrobial by comparison, and there is a substantial difference between the bioactivity in different lavender oils from different sources (French lavender, spike lavender etc.) and also the same oil from a different supplier (Lis-Balchin et al., 1998).

The antimicrobial activity of some of the EOs is often regarded by aromatherapists as proof of the aromatherapeutic usage. However, the actual mode of application of such EOs is far removed from the proper definition of aromatherapy, which is treatment with odours.

Effect on CNS

(1) *Effect on mood* – Many fragrances have been shown to have an effect on mood and in general, pleasant odours generate happy memories, more positive feelings and a general sense of

well-being (Warren and Warrenburg, 1993). Much of this type of research has been conducted by perfumery companies to boost sales (Jellinek, 1956) and some EOs have also been applied in hospitals in the United States to create a more happy and positive atmosphere and also in offices and factories in Japan to enhance productivity. Even in the times of Culpeper (1653), lavender was stated to have effects on the minds as well as the body: there is, however, some discrepancy about the species involved, as spike lavender was used medicinally and the *L. angustifolia* was used in perfumery, but could also have been used as a the mind-bending agent!

(2) *Psychological and physiological effects* – Many essential oil (EO) vapours have been shown to depress Contingent Negative Variation (CNV) brain waves in human volunteers (Table 16.2) and these EOs are considered to be sedative. Others increase CNV and are considered stimulant. The effects of inhaling different EOs on the CNV can be compared to the effect on mouse motility and the direct effect of the EO on smooth muscle *in vitro* (Table 16.2). Although there is a great difference in the application of the oils and the measurement of their effect, there is surprisingly, a frequent agreement. Lavender (of unspecified genus) was found to have a sedative effect on man (judging by CNV studies) (Kubota *et al.*, 1992; Torii *et al.*, 1988; Manley, 1993) and also had a 'positive' effect on mood, EEG patterns and Math computations (Diego *et al.*, 1998) and also caused reduced motility in mice (Buchbauer, 1991; 1993; Jager *et al.*, 1992; Kover *et al.*, 1987; Ammon, 1989). However, Karamat *et al.* (1992) found that lavender had a stimulant effect on decision times in human experiments. The effect of three different 'lavenders' on the smooth muscle of the guinea-pig *in vitro* (Lis-Balchin *et al.*, 1997c) showed mainly a sedative effect, but the true *L. angustifolia*, be it a commercial EO, always gave an initial contraction followed by a relaxation. There was also a good agreement between the effect of individual EOs (and mixes of three or more) on smooth muscle *in vitro* and the predicted effect on clients, as judged by aromatherapists (Lis-Balchin and Hart, 1997a).

Table 16.2 Sedative and stimulant EOs

EO	Sedative	Stimulant
Lavender	1, 2, 4, 5, 8, 11	5, 6
Jasmine	5	1, 2, 4, 9
Peppermint	5	1, 2, 3, 9
Rose	5, 7	2, 3
Rosemary	1, 4, 5	8, 11
Sandalwood	1, 2, 3, 4, 5	
Valerian	1, 3, 5	3, 8

Sources
1 Kubota *et al.* (1992): CNV studies in man.
2 Torii *et al.* (1988): CNV studies in man.
3 Manley (1993): CNV studies in man.
4 Buchbauer (1991; 1993); Jager *et al.* (1992): motility of mice.
5 Lis-Balchin (1995); Lis-Balchin *et al.* (1996): smooth muscle *in vitro*.
6 Karamat *et al.* (1992) decision times.
7 Kikuchi *et al.* (1991) heart rate in man.
8 Kovar *et al.* (1987); Ammon (1989) in mice.
9 Ludvigson and Rottman (1989) in man.
10 Badia (1991) in man.
11 Diego *et al.* (1998) in man.

Lorig and Roberts (1990) showed that at low concentrations, the expectancy of the odour produced changes in CNV rather than the odour itself. Eighteen subjects were presented with equal intensities of jasmine, lavender and galbanum in high concentration, and three mixes which were labelled low levels of jasmine, galbanum or lavender but were in fact all the same. The CNV effects were studied against no odour. The CNV changes for lavender were lowest, followed by galbanum and jasmine: the three mixtures produced the same results, that is, for the expected odours rather than for the same mixture in each case. The subjects assessed the primary odours according to intensity and pleasantness, but few correlations between these and CNV changes were found. Most subjects hated lavender.

King (1988) studied anxiety-reducing effects of fragrances using electromyography. The effect of ambient odour compared to no odour in sessions one week apart on creativity, mood and perceived health was studied in fifteen men and fifteen women (Knasko, 1992). The odours varied: lemon, lavender and dimethyl sulfide (DMS, an unpleasant odour). With lemon, fewer health symptoms were reported compared with the non-odour day. Subjects in the DMS group were less happy than those in the lavender group on both odour and non-odour days. The order of giving the odours affected the outcome, which was not very helpful in the study. However, there were no differences in arousal or creativity.

Inhalation studies in a group of forty staff and students, of rosemary oil versus lavender oil using EEG and simple maths computations, showed that lavender increased β-power, suggesting drowsiness and the subjects stated that they were more relaxed, while rosemary instigated decreased frontal α- and β-power, suggesting increased alertness, and the subjects were faster and more accurate in the maths Diego et al. (1998). These results seem to show that odour has an effect on performance *per se*, but Knasko et al. (1990) used no odour, but lied to their subjects that odour would be given: they also got an improvement in carrying out tasks, that is, mind over matter!

An interesting single study involving an anosmic showed that, as in eight neurologically normal people, there were changes in cerebral blood flow (CBF) on inhaling 1,8-cineole, thus showing a positive brain effect to the oil despite the inability to smell it (Nasel et al., 1994). The results showed that in each case there was a global increase in CBF, without preference for a specific area.

There is some evidence that certain EOs can lower blood pressure, if it is elevated, for example, nutmeg (Warren and Warrenburg, 1993), but this has not been investigated with lavender oil.

Lack of evidence for direct action on tissues following skin absorption

There is virtually *no* scientific evidence, as yet, regarding the direct action of EOs, applied through massaging of the skin, on specific internal organs. This is despite some evidence that certain EO components can be absorbed (Table 16.3) when massaged into the skin or by inhalation and that some have a beneficial effect on ureteric stones when taken orally (Engelstein et al., 1992). Jager et al. (1992) showed that linalool and linalyl acetate was detectable in the blood after massage with lavender oil; the lavender oil could have been inhaled and not absorbed through the skin entirely. Buchbauer et al. (1992) found benzyl alcohol, benzaldehyde and 2-phenylethanol I the blood after inhalation of limeblossom and passiflora as well as the components themselves. Kovar et al. (1987) showed the presence of 1,8-cineole in the blood after rosemary oil was inhaled or given orally. Buchbauer et al. (1993b) also showed the presence of benzaldehyde and carvone in the brain after inhalation of the two components individually.

Table 16.3 Evidence for transfer of components of EO into blood/brain when applied to skin, orally or by inhalation

Source	EO administered	Components found in blood/brain
Jager et al. (1992)	Lavender (massage)	Linalool, linalyl acetate (blood)
Buchbauer et al. (1992)	Limeblossom and benzyl alcohol, benzaldehyde passiflora and 2-phenylethanol (inhalation)	Benzyl alcohol, benzaldehyde, 2-phenylethanol (blood)
Kovar et al. (1987)	Rosemary (oral and inhalation)	1,8-cineole (blood)
Buchbauer et al. (1993b)	Benzaldehyde, carvone (inhalation)	Benzaldehyde, carvone (brain)
Fuchs et al. (1997)	Carvone (massage)	Carvone (blood)

Furthermore, although many EOs are very active on many different animal tissues *in vitro* (Lis-Balchin and Hart, 1997a), we have no idea as yet whether their activity in minute amounts (as used in aromatherapy massage) can benefit the patient through direct action on target organs or tissues (Vickers, 1996) rather than through the odour pathway leading into the mid-brain's 'Limbic System' and thence through the normal sympathetic and parasympathetic pathways. There is also no proof that synergism occurs when mixtures of EOs are used (Lis-Balchin et al., 1997d).

EOs are used by some doctors in France and Germany as orally or rectally-introduced medicines; these involve large doses, sometimes in excess of 45 ml per day (Franchomme and Penoel, 1990) rather than the almost homeopathic doses used by aromatherapists in their massage oil mixes in the United Kingdom and the United States, which rarely exceed 0.5 ml per massage. Most of the EOs are of course volatalized during the massage.

Miscellaneous effects of EOs

Many plant EOs have been shown to have an antioxidant effect and this property has also been shown to effect the lipoprotein membrane composition of the cell walls, giving a reversal of the ageing profile of saturated fatty acids compared to the unsaturated fatty acid composition in young animals (Dorman et al., 1995). This has raised the banner of the benefits of EOs on longevity and youthful appearance of the skin. Unfortunately, lavender is not an antioxidant EO!

Clinical studies

Intensive care

In an intensive care unit (ICU), 122 patients were randomly allocated (of which 111 completed the trial) to receive massage, aromatherapy massage with lavender oil or a period of rest (Dunn et al., 1995). Physiological, psychological and behavioural indices were measured before and after the procedures. This 14-month trial was conducted with patients in ICU for about 5 days and receiving 1–3 therapy sessions in 24 h. Only half received all three therapy sessions, less than a quarter received two and the rest just one session. Eleven patients died or were removed from the unit before completion. Massage was either on the back, outside of limbs or scalp, that is, whatever was available, for any time between 15–30 min. Lavender (*L. vera*) at 1 per cent was used. Verbalization was kept at a minimum, that is, no counselling. Physiological assessments included: blood pressure, heart rate and respiratory rate. The behavioural assessments used were specially designed for poor verbal communication including positive and negative responses in observable behaviour, for example, motor activity, somatic changes, facial expressions, these scores were repeated three times on a 4-point scale. The neurological status of unconscious patients was recorded using the Glasgow coma scale. Patient assessment on conscious patients was on a 4-point scale and measured their anxiety level, mood and ability to cope with the situation. The patient age range was an incredible 2–92 years! The results showed no difference between massage with lavender and the other groups except in a single parametre, namely the anxiety level. The results were not unexpected, bearing in mind the differences in the conditions of massage and other factors.

Dementia with agitation

A single case study evaluation of the use of aromatherapy (using lavender oil) with and without massage, and massage alone on four patients with severe dementia and disturbed behaviour was assessed using ten treatments of each aromatherapy type against ten no-treatment periods (Brooker et al., 1997). In the study, 8–12 treatments were randomly given to each patient over a 3-month period. Of the four patients, two actually became more agitated following their treatment sessions. Only one patient seemed to have benefited. The numbers were too small to make concrete significant assessments, but there was, however, no statistically significant benefit in combining aroma with massage yet again.

Perineal repair

Perineal repair following childbirth was assessed by Cornwell and Dale (1995), using 6 drops of lavender oil included in the bath water, for 10 days, and was compared against 'synthetic' lavender oil and a placebo (distilled water containing a GRAS additive, which had an undisclosed odour). No significant differences between groups were found for discomfort, but lower scores in discomfort means for days 3 and 5 were recorded for the lavender group (the meaning of this was however unclear, as this was insignificant).

High dose chemotherapy

A study at the London Clinic into the use of EOs for the treatment of chemotherapy-induced side-effects in a group of patients undergoing high dose chemotherapy, with stem cell rescue for breast cancer (Gravett et al., 1995) was reported at an aromatherapy conference but no scientific publications followed. Patients were not randomly allocated, but a division was made between the patients treated by two consultants, no-double blinding was used either, therefore there was possible bias from patients and nurse observers. One of the problems treated with aromatherapy was skin rashes. These can be due to some direct skin toxicity by chemotherapy drugs, sensitivity reactions to this and other drugs given. Dermatologists usually prescribed emollients with mild steroids if this was severe. The 'aromatherapy formulation' was substituted and used as follows: Lavender 10 drops, German chamomile 5 drops, Tea tree 5 drops, Stenophylla 5 drops, Calendula CO_2 added to make a 0.2 per cent solution. This was mixed in 50 ml sweet almond oil. The Calendula was not really Calendula but Tagetes (French Marigold), which is a known sensitizer. Reasons for the use of these protocols included: 'The extract has a gooey consistency and is an effective anti-inflammatory agent. The phytols are a series of solvents that have been developed in collaboration with ICI and are not CFCs but boil at around −30°C. Phytols are used to produce taxol which forms part of the cancer treatment. (These statements are entirely erroneous – present author.)

No significant differences were found in any of the experiments with patients and it is surprising that the hospital allowed the study through, due to the seriousness of the patients' condition and lack of scientific verification for the aromatherapy treatment.

Aromatherapy in childbirth

Studies of the use of aromatherapy in childbirth (Burns and Blaney, 1994) were not very conclusive, mainly because a large number of different EOs were used at different times and in different ways. There was also a bias towards the use of a few oils, one of which was lavender. Some of the women, and also some midwives, seemed to think that it speeded up the childbirth, but there were no controls.

'Double-blind' study using two different types of lavender oil

A so-called 'double-blind' study using two different types of lavender oil, one of which was 'real', but no analyses of the oils were given (Buckle, 1993) proved that there was no significant difference in the state of the patients using either oil as the author completely disregarded the use of controls and the assumption was made that (a) there would be a beneficial effect through the use of lavender and that (b) lavender exerted an effect through its absorption by the skin rather than through its inhalation (as patients wore oxygen masks, which were, however, inadequate in preventing inhalation). It seemed rather unbelievable that the author of the study then concluded that due to the small (insignificant) changes in the effect between the two EOs, it would rule out massage, touch or placebo as being sole contributory factors and that the 'changes' were due to the different EOs.

Epilepsy

An interesting application of EOs by a neuropsychiatrist (Betts et al., 1995) showed that out of thirty epileptic patients treated with conventional therapy with the added use of EOs, nine became free from seizures for some considerable time and another twenty patients were greatly improved. It was interesting to also note that the majority of the patients chose ylang ylang oil from a whole range of different odours. This study is very often quoted as an example of 'alternative' medicine, but it is in fact only an example of the complementary application of conventional and alternative disciplines working together.

Cognition, memory, affect and mood

The effect of ambient odours of lavender and cloves on cognition, memory, affect and mood of seventy-two volunteers was studied in two sessions separated by a week (Ludvigson and

Rottman, 1989). The group was divided into three each time. It is not really clear whether the same odour was given to the same group on the two occasions, or whether it was possible purely by chance. Lavender adversely influenced arithmetic reasoning in the first session; the subject's affective reactions to the experiment were, however, favourable. Cloves on the other hand decreased the willingness to return. The second session's results were very complex, but it became obvious that those who had had an odour once did not want to return. However, there were no simple effects of odour apparent. There was no effect on memory in this study, which was very surprising.

Alopecia areata

A randomized double-blind study on patients with *Alopecia areata* showed that a combination of lavender oil with thyme, cedarwood, rosemary in jojoba and grapeseed oil, massaged into the scalp daily, for 7 months resulted in an improvement in the condition (Hay et al., 1998).

Childhood atopic eczema

A study of massage therapy, with or without EOs as a complementary therapy for the treatment of childhood atopic eczema showed that initially there was no significant difference between the treatments, but after a period of rest, there seemed to be an apparent detrimental effect of the EO massage on the eczema, possibly due to sensitization (Anderson et al., 2000).

Problems arising in aromatherapy studies

Scant attention has been given to the actual chemical composition of the EOs used and even the exact botanical origin type of the oil was ignored, for example, there are many main commercial types of 'lavender oil', all differing in both genus and composition! There are a large number of chemotypes of *Lavandula* EOs, and of these, often vague chemotypes are preferentially used by some aromatherapists (Lis-Balchin, 1999). The danger here is that little is known about them, the composition is very variable and no toxicity tests have been carried out on most of them. Only commercially-used EOs have been tested for toxicity, as these are mainly used in the food and cosmetics industries; however large-scale admixing, deterpenation, dilution and adulteration also occurs giving rise to a multitude of different lavender oils.

Anecdotal evidence

Many studies in Italy and France, have offered very little in the way of scientific evidence on the efficacy of EOs on patients and were reported in a rather unscientific way. There may well have been some success in the treatment of depressed patients (Rovesti and Colombo, 1973), but there is little data supplied as to the precise diagnosis of the patients, their symptoms, which symptoms were relieved, the number of patients involved and the statistical significance or otherwise of the results. Under such circumstances, the evidence is at best anecdotal. Most of the recent work has never been published in peer-reviewed journals but has been quoted as gospel, for example, Franchomme and Penoel (1990) and only single case studies are presented on the efficacy of treatment, which were used mostly internally.

There is also long-standing evidence for the benefits of inhaling certain EOs to relieve coughs, congestion etc. in the respiratory tract using mixtures of *Eucalyptus globulus*, pineneedle and camphor (Martindale, 1992) which could also alleviate sleeplessness and save on diapezams.

Lavender oil has been used in one trial relatively successfully, however, only four geriatrics were involved in the study (Hardy et al., 1995).

Toxicity of lavender

More clinical and toxicological research is needed, in order to extend the use of aromatherapy. From the toxicological aspect, there is the danger of causing dermatitis in sensitive people (Rudzki et al., 1976): lavender oil is not implicated greatly, but there was a report of occupational allergy to a lavender shampoo used by a female hairdresser (Brandao, 1986). The hairdresser had allergy problems on her hands due to a variety of products, but reacted more strongly to a lavender shampoo and lavender oil itself. Menard (1961) reported a similar case, but this time the hairdresser was allergic to the eau de Cologne containing lavender, rather than to the lavender alone. Patch tests have shown a few allergies due to photosensitization and also pigmentation was reported (Brandao, 1986; Nakayama et al., 1976).

There is also the danger of airborne contact allergic dermatitis through overuse of EOs and their storage (Schaller and Corting, 1995), which produced a severe response in a man who had been active with EOs, and proved long-lasting due to the sequestering of the odorants in the house even after the removal of all the bottles.

There may also be danger in the overuse of EOs during pregnancy and childbirth. Studies during childbirth in particular should take into account the baby's health, as there is always the danger of over-sedation of the infant and the subsequent lack of breathing reflex (personal communication). Clinical studies should include EOs which are not generally employed in order to assess the efficacy of frequently used EOs and eliminate aromatherapist's bias, as well as indicate whether the effect is actually due to any EO in particular. Studies using case notes of past clients could also be of help in assessing the efficacy of usage of certain EOs for different clinical conditions.

Future clinical application of aromatherapy

What could be achieved by using aromatherapy as an adjunct to clinical medicine especially in hospitals and general practice? So far there have been many 'successes' in various areas, notably hospices. There are no miracle cures, but an alleviation of suffering and possibly pain, mainly through touch, relaxation due to gentle massage and the presence of someone who cares and listens to the patient. This is probably also the case in geriatric wards, in general wards, in the treatment of severely physically and mentally-challenged children and adults.

There is a need for this kind of healing contact, and aromatherapy with its added power of odour, fits this niche. Future studies may reveal the individual benefits of different EOs for different ailments, but, in practice this may not be of the greatest importance as aromatherapy (especially when combined with massage) offers relief from stress and this in itself is of the greatest benefit for most people.

Nurses and other healthcare professionals have expressed the wish to learn and train in the use of aromatherapy, in favour of all the other alternative therapies (Trevelyan, 1996). The medical profession is also turning towards this branch of alternative medicine, as it seems to be useful in the treatment of patients whose symptoms are largely based on stress, and who do not respond to conventional medicine. The future of aromatherapy may, however, be in doubt, if there is no scientific verification of its efficacy forthcoming and if there is a clamp-down on the EO industry (which includes aromatherapists selling or just using EOs) as to the safety and standard of the

EOs. EOs are hazardous in the wrong hands and in the United Kingdom there are no restrictions on who can practise with them, regardless of their qualifications and knowledge of their potential dangers, despite some regulatory bodies (like the Aromatherapy Organisation Council, AOC and the individual organisations to which aromatherapists can belong, for example, International Federation of Aromatherapist, IFA); in the rest of Europe, there are more stringent regulations, with EOs being sold mainly by pharmacists, and aromatherapy practised by medically qualified practitioners or herbalists, and these are probably to be implemented in the United Kingdom in the near future. The regulations round the world will then doubtlessly be amended.

References

Ammon, H.P.T. (1989) Phytotherapeutika in der Kneipp-therapie. *Therapiewoche*, 39, 117–27.
Anderson, C., Lis-Balchin, M. and Kirk-Smith, M. (2000) Evaluation of EO therapy on childhood atopic eczema. *Phytother. Res.*, 14, 452–6.
Badia, P. (1991) Olfaction sensitivity in sleep: the effects of fragrances on the quality of sleep. *Perf. Flav.*, 16, 33–4.
Betts, T., Fox, C., Rooth, K. and MacCullum, R. (1995) An olfactory countermeasure treatment for epileptic seizures using a conditioned arousal response to specific aromatherapy oils. *Epilepsia*, 36, Suppl. 1. S130.
Brandao, F.M. (1986) Occupational allergy to lavender. *Contact Dermatitis*, 15, 249–50.
Brooker, D.J., Snape, M., Johnson, E., Ward, D. and Payne, M. (1997) Single case evaluation of the effects of aromatherapy and massage on disturbed behaviour in severe dementia. *Br. J. Clin. Psychol.*, 36, 287–96.
Buckle, J. (1993) Aromatherapy: does it matter which lavender EO is used? *Nurs. Times*, 89, 32–5.
Buchbauer, G. (1992) Biological effects of fragrances and EOs. *Perf. Flav.*, 18, 19–24.
Buchbauer, G. and Jirovetz, L. (1994) Aromatherapy – use of fragrances and EOs as medicaments. *Flav. Fragr. J.*, 9, 217–22.
Buchbauer, G., Jirovetz, L., Jager, W., Dietrich, H. Plank, C. and Karamat, E. (1991) Aromatherapy: evidence for the sedative effects of the EOs of lavender after inhalation. *Z. Naturforsch.*, 46, 1067–72.
Buchbauer, G., Jirovetz, L., Jager, W., Plank, C. and Dietrich, H. (1993a) Fragrance compounds and essential oils with sedative effects upon inhalation. *J. Pharm. Sci.*, 82(6), 660–664.
Buchbauer, G., Jirovetz, l., Czejka, M., Nasel, C.H. and Dietrich, H. (1993b) New results in aromatherapy research. *24th Int. Symp. Essential Oils*, Berlin.
Burns, E. and Blaney, C. (1994) Using aromatherapy in childbirth. *Nurs. Times*, 90, 54–8.
Cornwall, S. and Dale. A. (1995) Lavender oil and perineal repair. *Modern Midwife*, Mar. 5, 31–33.
Culpeper, N. (1653) *The English Physitian Enlarged*, George Sawbridge, London.
Diego, M.A., Jones, N.A., Field, T., Hernandez-Reif, M., Schanberg, S., Huhn, C., McAdam, V., Galamaga, R. and Galamaga, M. (1998) Aromatherapy positively affects mood, EEG patterns of alertness and Math computations. *Intern. J. Neuroscience*, 96, 217–24.
Dorman, H.J.D., Youdim, K.A., Deans, S.G. and Lis-Balchin, M. (1995) Antioxidant-rich plant volatile oils: *in vitro* assessment of activity. *26th Int. Symp. Essential Oils*, Hamburg, Germany, Sept. 10–13.
Dunn, C., Sleep, J. and Collett, D. (1995) Sensing an improvement: an experimental study to evaluate the use of aromatherapy, massage and periods of rest in an intensive care unit. *J. Adv. Nursing*, 21, 34–40.
Engelstein, D., Kahan, E. and Servadio, C. (1992) Rowatinex for the treatment of uterolithiasis. *J. d' Urologie*, 98, 98–100.
Franchomme, P. and Penoel, D. (1990) *Aromatherapie Exactement*. Roger Jollois, Paris.
Fuchs, N., Jager, W., Lenhardt, A., Bohm, L., Buchbauer, I. and Buchbauer, G. (1997) Systemic absorption of topically applied carvone: influence of massage technique. *J. Soc. Cosmet. Chem.*, 48, 277–82.

Gattefosse, R.M. (1937) *Aromatherapy* (translated 1993), C.W. Daniel and Co. Ltd., Saffron Walden.
Gravett, P.J., Finn, M. and Hallasey, S. (1995) An investigation of the use of EOs for the treatment of chemotherapy-induced side-effects in a group of patients undergoing high dose chemotherapy, with stem cell rescue for breast cancer. Paper given at *AROMA '95*, Warwick University.
Hardy, M., Kirk-Smith, M.D. and Stretch, D.D. (1995) Replacement of chronic drug treatment for insomnia by ambient odour, *Lancet*, 346, 701.
Hay, I.C., Jamieson, M. and Ormerod, A.D. (1998) Randomized trial of aromatherapy. Successful treatment for Alopecia areata. *Arch. Dermatol.*, 134, 1349–52.
Jager, W., Buchbauer, G., Jirovetz, L. and Fritzer, M. (1992) Percutaneous absorption of lavender oil from a massage oil. *J. Soc. Cosmet. Chem.*, 43, 49–54.
Jellinek, P. (1956) *Die Psychologischen Grndlagen der Parfumerie*, Alfred Hutig Verlag, Heidelberg.
Karamat, E., Ilmberger, J., Buchbauer, G. *et al.* (1992) Excitatory and sedative effects of essential oils on human reaction time performance. *Chem. Senses*, 17, 847.
King, J.R. (1988) Anxiety reduction using fragrances. In: *Perfumery: The Psychology and Biology of Fragrance* (S. Toller and G.H. Dodds (eds)), Chapman & Hall, New York.
Kovar, K.A., Gropper, B., Friess, D. and Ammon, H.P.T. (1987) Blood levels of 1,8-cineole and locomotor activity of mice after inhalation and oral administration of rosemary oil. *Planta Med.*, 53, 315–18.
Knasko, S.C. (1992) Ambient odours effect on creativity, mood and perceived health. *Chem. Senses*, 17, 27–35.
Knasko, S.C., Gilbeert, A.N. and Sabini, J. (1990) Emotional state, physical well-being and performance in the presence of feigned ambient odour. *J. Appl. Soc. Psychol.*, 20, 1345–7.
Kubota, M., Ikemoto, T., Komaki, R. and Inui, M. (1992) Odor and emotion-effects of essential oils on contingent negative variation. *Proc. 12th Int. Congress on Flavours, Fragrances and Essential Oils*, Vienna, Austria, Oct. 4–8. pp. 456–61.
Lawless, J. (1992) *The Encyclopaedia of Essential Oils*. Element, Dorset.
Lis-Balchin, M. (1995) *The Chemistry and Bioactivity of Essential Oils*. Amberwood Publishing Ltd., Surrey.
Lis-Balchin, M. (1997) Essential oils and 'aromatherapy': their modern role in healing. *J. Roy. Soc. Health*, 117, 324–9.
Lis-Balchin, M. (1999) Possible health and safety problems in the use of novel plant essential oils and extracts in aromatherapy. *J. Roy. Soc. Promotion Health*, 119, 240–3.
Lis-Balchin, M. and Deans, S.G. (1997) Bioactivity of selected plant essential oils against *Listeria monocytogenes*. *J. Appl. Microbiol.*, 82, 759–62.
Lis-Balchin, M. and Hart, S. (1997a) Correlation of the chemical profiles of essential oils mixes with their relaxant or stimulant properties in man and smooth muscle preparations *in vitro*. *Proc. 27th Int. Symp. Ess. Oils*, Vienna, Austria, 8–11 Sept. 1996, Ch. Franz, A. Mathé and G. Buchbauer (eds), Allured Pub. Corp., Carol Stream, Ill. pp. 24–8.
Lis-Balchin, M. and Hart, S. (1997b) The effect of essential oils on the uterus compared to that on other muscles *in vitro*. *Proc. 27th Int. Symp. Ess. Oils*, Vienna, Austria, 8–11 Sept. 1996. Ch. Franz, A. Mathe and G. Buchbauer (eds), Allured Pub. Corp., Carol Stream, Ill. pp. 29–32.
Lis-Balchin, M., Hart, S., Deans, S.G. and Eaglesham, E. (1997c) Comparison of the pharmacological and antimicrobial action of commercial plant essential oils. *J. Herbs, Spices, Med. Plants*, 4, 69–86.
Lis-Balchin, M., Deans, S.G. and Hart, S. (1997d) A study of the changes in the bioactivity of essential oils used singly and as mixtures in aromatherapy. *J. Alt. Complement. Med.* 3, 249–55.
Lis-Balchin, M., Deans, S.G. and Eaglesham, E. (1998) Relationship between the bioactivity and chemical composition of commercial plant essential oils. *Flav. Fragr. J.*, 13, 98–104.
Lorig, T.S. and Roberts, M. (1990) Odour and cognitive alteration of the contingent negative variation. *Chem. Senses*, 15, 537–45.
Ludvigson, H.W. and Rottman, Th.R. (1989) Effects of ambient odours of lavender and cloves on cognition, memory, affect and mood. *Chem. Senses*, 14, 525–36.
Manley, C.H. (1993) Psychophysiological effect of odor. *Crit. Rev. Food Sci. Nutr.*, 33, 57–62.
Martindale (1992) *The Extra Pharmacopoeia*, 26th edn, The Pharmaceutical Press, London.
Maruzella, J.C. and Henry, A. (1958) The *in vitro* antibacterial activity of essential oils and oil combinations. *J. Am. Pharmaceut. Assoc.*, 47, 294–6.

Maruzella, J.C. and Sicurella, N.A. (1960) Antibacterial activity of essential oil vapours. *J. Am. Pharmaceut. Assoc.*, 49, 629–94.

Menard, E. (1961) Les dermatoses profesionelles. *Concours Medicale*, 83, 4308–11.

Moleyar, V. and Narasimham, P. (1987) Antibacterial activity of essential oil components. *Int. J. Food Microbiol.*, 16, 337–42.

Nakayama, H., Harada, R. and Toda, M. (1976) Pigmented cosmetic dermatitis. *Int. J. Derm.*, 15, 673–5.

Nasel, C., Nasel, B., Samec, P., Schindler, E. and Buchbauer, G. (1994) Functional imaging of effects of fragrances on the human brain after prolonged inhalation. *Chem. Senses*, 19, 359–64.

Nelson, N.J. (1997) Scents or nonsense: aromatherapy's benefits still subject to debate. *J. National Cancer Institute*, 89, 1334–6.

Price, S. (1993) *Aromatherapy Workbook*. Thorsons, London.

Rovesti, P. and Colombo, E. (1973) Aromatherapy and aerosols. *Soap. Perfumery and Cosmetics*, 46, 475–7.

Rudzki, E., Grzywa, Z. and Bruo, W.S. (1976) Sensitivity to 35 essential oils. *Contact Derm.*, 2, 196–200.

Ryman, D. (1991) *Aromatherapy*. Piatkus, London.

Schaller, M. and Korting, H.C. (1995) Allergic airborne contact dermatitis from essential oils used in aromatherapy. *Clin. Exp. Dermatol.*, 20, 143–5.

Tisserand, R. (1985) *The Art of Aromatherapy*, Revised edn, C.W. Daniel Co. Ltd, Saffron Walden.

Torii, S., Fukuda, H., Kanemoto, H., Miyanchio, R., Hamauzu, Y. and Kawasaki, M. (1988) Contingent negative variation and the psychological effects of odor. In: *Perfumery: The Psychology and Biology of Fragrance*. S. Toller and G.H. Dodds (eds), Chapman & Hall, New York.

Trevelyan, J. (1996) A true complement. *Nurs. Times*, 92, 42–3.

Valnet, J. (1982) *The Practice of Aromatherapy*. C.W. Daniels Co. Ltd., Saffron Walden.

Vickers, A. (1996) *Massage and Aromatherapy. A Guide for Health Professionals*. Chapman & Hall, London.

Warren, C. and Warrenburg, S. (1993) Mood benefits of fragrance, *International Flavors and Fragrances*, 18, March/April.

Westwood, C. (1991) *Aromatherapy. A Guide for Home Use*. Amberwood Publishing Ltd., Dorset.

Worwood, V.A. (1991) *The Fragrant Pharmacy*. Bantam Books, London.

Yousef, R.T. and Tawil, G.G. (1980) Antimicrobial activity of volatile oils. *Pharmazie*, 35, 698–701.

17 Perfumery uses of lavender and lavandin oils

Rhona Wells and Maria Lis-Balchin

Arctander (1960) lists several different lavender and lavandin oils as well as lavandin and lavender concretes and absolutes.

Lavandin absolute

The lavandin absolute from concrete is produced by extracting with absolute alcohol, the alcoholic extracts are then chilled and then evaporated continuously under reduced vacuum towards the end of the distillation. The lavandin absolute is a viscous dark green liquid of herbaceous odour, resembling the flowering lavandin. It is sweeter than the essential oil.

It is used for fougeres, new-mown-hay types, herbaceous, floral fragrances, forest-notes and refreshing colognes. It blends well with patchouli, clove oil, bergamot and lime and rounds off rough ionones.

Lavandin absolute can also be produced from distillation, as some of the components are reasonably water-soluble and in a 20 ton still, a vast amount of lavandin can be extracted with benzene or petroleum ether and thence re-extracted with alcohol.

Lavandin oil

Lavandin oil was first produced in the late 1920s, but has since escalated well above that of true lavender. Lavandin is a hybrid, produced by crossing true lavender (*Lavandula angustifolia* Miller) with spike lavender *L. latifolia* (L.). There are many such hybrids growing all over Europe and other parts of the world and they yield a much higher return per acre due to the larger plants. The oil is pale yellow to almost colourless and has a strongly herbaceous odour with a distinctive top-note which is fresh camphene-cineole-like and disappears very rapidly from a blotter (Arctander, 1960).

Lavandin oil is used in large quantities for a fresh note in perfumes. It is well adapted for detergent products and needs no strong fixatives, but in soaps there is a necessity for good fixation.

Lavandin oil blends well with natural and synthetic products in perfumery including: clove oil, eugenol, bay leaf, cinnamon leaf etc. It also blends well with: amyl salicylate, citronella, cypress, decyl alcohol, geranium oil, pine needle, thyme and oregano oils and patchouli. Fixation is accomplished usually by sesquiterpene fractions from labdanum, nitromusks, coumarin and oakmoss (Arctander, 1960).

Lavender oil

Lavender oil (*L. angustifolia* Miller or *L. officinalis*) is steam distilled from the freshly cut flowering tops of lavender. Most of the lavender plants were originally grown and distilled in the higher areas of Mediterranean France (600–1500 m), but nowadays a substantial amount comes from China. Exact figures for the production of the oil is difficult to obtain due to the immense amount of adulteration, mixing, cutting, and addition of synthetics or simply synthetic lavender oil itself.

The true oil is almost colourless and has a sweet, floral herbaceous refreshing odour with a pleasant, balsamic-wood undertone. It has a fruity-sweet top-note which is very transient, and the whole oil has low tenacity. It is used in colognes, lavender-waters, fougeres, chypres, ambres, floral and non-floral perfumes. It blends well with bergamot and other citrous oils, clove, flouve, liatris, oakmoss, patchouli, rosemary, sage, pine etc. It also blends with amyl salicylate, citronellol, geraniol and their esters, musks and numerous other aromachemicals (Arctander, 1960).

Lavender absolute

This is produced by solvent extraction followed by absolute extraction from *L. angustifolia* Miller or *L. officinalis* (see above).

Lavender oils and extracts used in perfumery

Historical use of lavender as perfumes

The Romans steeped the leaves and stems of lavender in their bath water, and thus apparently lavender became associated with cleanliness (Trueman, 1975). It was then introduced as a cottage plant in Britain.

Still rooms

Lavender water (Table 17.1) was used extensively till the sixteenth century when lavender was distilled in the country house 'Still-rooms' of most rich households. Ladies of leisure would entertain friends in these still rooms to watch the distillations in progress and enjoy the aromas. It was one of the most important duties of the mistress of the household to be in charge of the stills (Irvine, 1995). In those days, there was a tendency for the distilled 'waters' to be made from a wide mixture of different herbs and spices and the addition of sweet wine gave the resulting waters a very high alcoholic boost; it seems likely that many of the 'waters' were in fact alcoholic beverages rather than just perfumes.

Table 17.1 Modern lavender water

Compound	Parts
Lavender oil, French	45
Bergamot oil, FCF	25
Lemon oil, Sicilian	6
Neroli oil, reconstituted	4
Musk ketone	3
Sweet orange oil	3
Geranium oil, reunion	4
Benzoin resinoid	4

Source: Curtis and Williams, 1994.

Such scents were used liberally in medicines, as household deodorants and on the person; floors were strewn with scented herbs, which included lavender, meadowsweet and rushes. Perfume 'cakes' made of a variety of herbs and spices, were set to smoulder on the embers to sweeten the air. Fumigations were carried out daily in some households to ward off airborne illnesses.

English lavender water and perfume

The lavender waters used in many English country houses in the sixteenth century, were later re-created by Atkinson, Jardley and Caron and have been ever since then a very popular male fragrance, although it was originally created for women.

Atkinson's 'English Lavender' (1910), was in fact the first eau de toilette for men, and was followed by Yardley's 'Old English Lavender' (1913) (Table 17.2), which was an essential accessory to the tweed suit! Then came Caron's 'Pour un Homme' (1934).

Yardley had prospered for several hundred years, but at the turn of the century, the firm went under, as newer more trendy perfumes have taken over the market. The original Yardley's lavender was made using lavender oil with the addition of neroli, attar of roses, French rose absolute and musk. The firm, Yardley's, which originated as bucklers in 1770, sold lavender grown in Norfolk which was macerated in bear grease for dressing men's hair by 1817 (Irvine, 1996). By the end of the nineteenth century it was the largest manufacturer of lavender products.

Victorian hey-day for lavender perfumes

The highest popularity for lavender perfumes was in Victorian times, when pure perfumes were preferred, after the muskiness and lasciviousness of the perfumes in the sixteenth, seventeenth and eighteenth century and the pure 'essences' of lavender alone, as well as those of rose and violet were found to be clean and fresh and wholesome. In days gone by, the price of perfumes was very high and only the rich could afford them; then, with the production of cheap synthetic

Table 17.2 Old English lavender

Component	Parts
Lavender oil (Mitcham)	60
Lavender oil French (40–42%)	25
Bergamot oil	3
Labdanum tincture	5
Oakmoss tincture	4
Geraniol	2
Lemon oil	1.5
Benzyl acetate	1
Musk ketone	1.2
Neroli oil	0.5
Vanillin	0.3
Coumarin	0.25
Indole (5% dilution)	0.25
Alcohol, 95%	2296.0
Distilled water	800.0
Total	3200.0

Table 17.3 Eau de cologne (1834)

Ingredient	kg
Lavender oil	1.2
Bergamot oil	6.2
Neroli oil	0.8
Lemon oil	3.1
Clove oil	1.6
Rosemary oil	0.8
Alcohol 90°	to 100 litres

Note: Encyclopaedia Roret Recipe.

perfumery ingredients, lavender-type perfumes became very cheap and available to all. Lavender, which had long been associated with fresh, clean linen was also used in soaps and other cosmetic products and is still currently used in fabric softeners and other household products, as a 'clean laundry' smell is still appreciated as being evocative of heady lavender fields.

Eau de cologne

One of the most famous uses of lavender is in the original eau de colognes. It was in the middle of the seventeenth century that an Italian barber, Gian Paolo de Feminis left his native country to settle in Cologne. He was a perfumer and merchant as well as a barber and created a toilet water, 'Acqua della Regina' otherwise known as 'Eau Admirable', which he first marketed for medicinal purposes and although Feminis said that the recipe had been passed on to him by a monk, it was more likely to have been developed by English military doctors in India to combat dysentery (Irvine, 1996). Later his grandson Giovanni Maria Farina restyled the product and at first advertised it as an antidote for all poisons and an outstanding prophylactic against the plague, but later sold it for its odour as 'Kolnische Wasser' or 'Eau de Cologne' (Wells and Billot, 1988). Later, the house of Roger de Gallet acquired the sole interest to the cologne. Later the '4711 cologne' was created (1879) (Barille and Laroze, 1995). Richard Wagner adored it and ordered it by the quart, three at a time. Napoleon Bonaparte had twelve Winchesters delivered regularly during his various campaigns for freshening up. This was following in the traditions of Louis XV, who adored perfumes and Madame Pompadour, who spent more on perfume than on anything else.

Eau de cologne has many compositions, but the most famous contained oils of bergamot, lemon, neroli, clove, rosemary and lavender (Table 17.3). Classic eau de cologne is a fresh, harmonious blend of citrus oils, with lavender as a middle note. Other formulations included more citrus oils like sweet orange, petitgrain, mandarin, grapefruit, lime, orange flower, melissa (balm) and clary sage. Other ingredients included traces of: thyme, wormwood, calamus, nutmeg, hyssop, caraway, aniseed, cinnamon and benzoin; the carnation base of clove was always one of the most important ingredients. The addition of linalool, linalyl acetate and ethyl linalyl acetate tend, like lavender itself, to add depth and richness. Rose notes also help modify the odour.

Use of lavender oil in famous perfumes

At the beginning of the century, lavender was apparently first incorporated into men's fragrances in Italy. It is also used as a pivotal cornerstone of fougère accords, a typically masculine scent

Table 17.4 Fougère-type perfume

Compound	Parts
Lavender oil, French	14
Bergamot oil, FCF	8
Coumarin	12
Rose base	5
Jasmine base	4
Oakmoss absolute	6
Patchouli oil, light	2
Vetivert oil, Bourbon	10
Geranium oil, Bourbon	2
Iso-amyl salicylate	3

Source: Curtis and Williams, 1996.

family (Table 17.4). Nowadays, lavender is still used as a top note in men's eau de toilettes, to add freshness to the blend (Pavia, 1995).

Geranium is frequently used in conjunction with lavender in the true men's lavender scents such as Moustache (Rochas) or Pino Silvestre, and also the classical fougère blends, where the top notes are primarily lavender-linked and the heart lends itself well to the dry floral aspects of geranium. It also adds floral aspects to green fragrances such as Grey Flannel and Monsieur Lanvin.

Jicky (Guerlain) introduced in 1889, has the familiar cologne freshness of citrus fruits and herbs. Lemon lavender and bergamot are blended with mint, verbena, sweet marjoram and soft coumarin, which becomes the spicy heartnote together with synthetic linalool interwoven with sandalwood and rosewood and then one gets the aphrodisia of musk, ambergris and civet for sensuality. Heart notes (middle notes) contain geranium. It was one of the very first 'synthetic' perfumes although it only contained a very small amount of synthetics, and yet provided a completely new odour.

Opium (Yves Saint Laurent) produced in 1976, contains lavender, linalool, linalyl acetate with bergamot, orange, lemon, mandarin as top note and a typical mellis accord built from a combination of benzyl salicylate 12 per cent; eugenol with clove and pimento; patchouli, 8 per cent; hydroxycitronellal, 10 per cent; vertofix and coumarin, 2.5 per cent. Next comes vanillin and ethyl vanillin and then a combination of balsamic and resinous notes including benzoin, styrax, opoponax and tolu to give the oriental effect.

Lavender and aromatherapy

Lavender has a fresh clean, slightly herbal note, and has also enjoyed considerable revival of fashion with the popularity of aromatherapy and aromachology, where the relaxing effect of lavender is used in consumer products. Lavender is the most common essential oil used by aromatherapists due to its pleasant odour, relaxant effect on most people and above all its low price.

References

Arctander, S. (1960) *Perfume and Flavor Materials of Natural Origin*, Elizabeth, NJ, USA.
Barille, E. and Laroze, C. (1995) *The Book of Perfume*, Flammarion, Paris.

Curtis, T. and Williams, D.G. (1994) *Introduction to Perfumery*, Part III, Ellis Horwood, London.
Irvine, S. (1995) *Perfume*, Aurum Book, Haldane Mason Ltd., London.
Trueman, J. (1975) *The Romantic Story of Scent*, Aldus/Jupiter.
Wells, F.V. and Billot, M. (1988) *Perfumery Technology*, 2nd edn, Ellis Harwood Ltd., Chichester.

Other general references for perfumery

Edwards, M. (1997) *Perfume Legends*, HM Editions, Paris.
Lamparsky, D. ed. (1991) *Perfumes, Art, Science and Technology*, Elsevier Science, New York.
Lefkowith, C.M. (1994) *The Art of Perfume*, Thames and Hudson, London.
Morris, E.T. (1984) *Fragrance. The Story of Perfume from Cleopatra to Chanel*, Charles Scribner's Sons, New York.
Pavia, F. (1995) *The World of Perfume*, Knickerbocker Press, New York.
Piesse, S. (1890) *Histoire des Parfums*, J.B. Baillière et fils, Paris.
Poucher, W.A. (1994) *The Production, Manufacture and Application of Perfumes*, Chapman & Hall, New York.
Haarman and Reimer (1989) *The Book of Perfume*, 5 Vols., R. Gloss & Co., Germany.

18 Miscellaneous uses of lavender and lavender oil

Use in hair products, food flavouring, tissanes, herbal pillows and medicinal products

Maria Lis-Balchin

Introduction

Lavender and its various cultivars and their essential oils (EOs) have been used for centuries in various products, both as a perfume and for their medicinal value. Nowadays, lavender is used mainly as the EO in manufactured products like cosmetics, perfumes, soaps etc, but there has been a general reversion to the use of the dried lavender plant itself in the homespun industry of making products like herbal pillows and lavender bags.

Old recipes for lavender preparations in the home

Recipe for lavender water (prepared at home)

Into a quart bottle put in 1 oz lavender oil, one drop of musk and one and a half pints of spirit of wine; shake; allow to settle; shaken again after a few days and poured into perfume bottles. It could also be made with rose-water and orange flower water. Brandy could be used with the lavender oil and it was to be diluted with water (Grieve, 1937).

Lavender vinegar

Lavender vinegar was a refreshing toilet water made by mixing 6 parts rosewater, 1 part of lavender spirits and 2 parts of Orleans vinegar (Grieve, 1937). Fresh flowers could also be directly steeped in the vinegar. Alternative recipes included: dried rose leaves, lavender flowers and jasmine flowers mixed in vinegar, then rosewater would be added and allowed to rest for 10 days.

Old medicinal concoctions using lavender

Lavender drops

This was a compound tincture of lavender oil, useful as a colouring and flavouring for medicines but also effective for fainting.

Red lavender

This consisted of lavender mixed with rosemary and cinnamon bark, nutmeg and sandalwood and macerated in spirit of wine for several days used in a dose of teaspoonful to some water for indigestion (Grieve, 1937). The British pharmacopoeia officially recognised red lavender for 200 years. In the eighteenth century it was known as palsy drops and red hartshorn. The first formula was complicated and used thirty ingredients in a distillation: fresh lavender, sage, rosemary, betony. Cowslips, lily of the valley, with French brandy; cinnamon, nutmeg, mace, cardamom were digested with all the rest for 24 h and then musk, ambergris, saffron, red roses and red sanders wood was tied up in a bag and suspended in to perfume and colour it.

Red lavender is sometimes mentioned in the literature, for example in 'Man and Wife' by Wilkie Collins (1870).

Modern uses

BPC products

Compound Lavender Tincture BPC 1949
Dose: 2–4 ml
Lavender Spirit BPC 1934
Dose: 0.3–1.2 ml

Uses of lavender medicinally

Bertram (1995) suggests the following uses for *Lavandula angustifolia*: nervous headache, neuralgia, rheumatism, depression, sluggish circulation, chillblains, insomnia, windy colic, physical and mental exhaustion, neurasthenia, sense of panic, fainting (1–3 drops in honey). He continues: for toothache, sprains, sinusitis, bladder infection, to relieve stresss, calm and relax and also migraine. For transient high blood pressure, combine with lime blossom (1 : 3), The uses are nearly identical to those suggested both by Culpeper (1653) and Gerard (1597), both of whom were referring to *L. stoechas*, not this species!

Different lavender preparations

Use dried flowers of lavender, 0.5–2 g, 3 × daily (Bertram, 1995).
Tea: 1 teaspoon per cup of boiling water, infuse 15 min. Dose: 0.33 of the cup.
Home liniment: handful of flowers, fresh (50 g) in 500 ml alcohol (e.g. vodka) and macerate for 8 days in cool, shady place, shaking daily. Filter and massage into affected area.
Tincture BHP (1983): 1 : 5 parts 60% alcohol. Dose: 2–4 ml.
Lavender bath: 30 g fresh flowers and tips to 500 ml water, boiled, strained, added to bath as tonic.

Natural food flavours

Lavandin oil, spike lavender oil and lavender oil, absolute and even concrete are used as natural food flavours. Reported uses in the food industry (Fenaroli, 1998) include: baked goods, frozen dairy, soft candy, gelatin, pudding, non-alcoholic and alcoholic beverages (Table 18.1).

Cooking with lavender at home

Several suggestions are made in the Norfolk lavender booklet (1999), including: herring or trout stuffed with thin slices of lemon and lavender sprigs; duck or chicken stuffed with lavender and white grapes; roast lamb or pork with lavender and redcurrants.

Table 18.1 Lavandin, spike lavender and lavender uses as natural food flavours reported uses in the food industry (Fenaroli, 1998) include

Food	Lavender absolute (ppm)	Lavender oil	Lavandin	Spike lavender
Baked goods	23.00	11.37	16.78	43.50
Frozen dairy	18.00	9.21	12.83	35.61
Soft candy	19.42	10.37	14.13	35.15
Gelatin, pudding	15.71	7.67	10.24	35.00
Non-alcoholic beverages	6.97	4.21	5.56	12.02
Alcoholic beverages	8.00	4.31	5.56	12.15

Lois Vickers (1991) suggests a number of fascinating recipes for cooking with lavender, garnishing foods and use as crystallised flowers. *L. angustifolia* cultivars like 'Hidcote' of 'Munstead' are suggested for crystallisation using the whisked white of an egg made in its own volume of water for painting the flower head, then coating them with icing sugar and baking for at least 2 h in a cool oven. Lavender can be used in salads mixed with nasturtium, marigold and borage flowers, mixed in with avocado slices and chicory leaves! Lavender marmalade is suggested using the usual recipe of Seville oranges (1 kg), one lemon, 2 l of water, 2 kg of sugar and the dried lavender flowers in a muslin bag (35 g). Lavender yoghurt ice and lavender meringues are suggested as deserts. Perfumed lavender sugar, lavender syrups and liqueurs also whet the appetite for lavender-holics!

Tissanes or teas

Tissanes or teas are sold in great quantities in Europe, and have now become relaxant drinks in the United Kingdom, even being sold in sachets in supermarkets. They often include cinnamon and fennel. Unsubstantiated claims exist that a herbal tea of 1 tsp. dried lavender flowers to 600 ml of boiling water, infused for 5 min and drunk with honey cures cystitis, vaginitis and leucorrhoea (probably due to the antiseptic effects of honey); it is also good for convalescents and for treating oily skins, as a rinse for oily dark hair and used to relieve headaches and as a tonic against faintness, spasms, colic or vertigo (Rose, 1982).

Scented candles

These can be easily made using grated candle wax in a saucepan, heating gently and then carefully adding the fragrance and lastly putting in the wick.

Pot-pourris

Pot-pourris were very fashionable in the sixteenth century and are still used today, but in a different format. Originally, the scented petals and leaves of rose, lavender, lemon verbena, lemon balm, thyme, mint, calendula, partly dried, were mixed with lemon and orange peel and put into an earthenware jar and layered together with salt for several weeks until it 'caked' (Flanders, 1995). The cake was then taken out and broken up and perfumed with spices, like cloves, cinnamon and nutmeg and then mixed with fixative oils or orris powder or gum benzoin, frankincense or myrrh and again left to 'cure' for many weeks. The pot-pourri did not look attractive though it smelt wonderful, and was kept in a china jar, often with pierced lids to allow the smell to waft out. Alternatively, the lid was removed on occasion, when the jar was placed in front of an open fire and the aroma would escape into the room.

Nowadays, fragments of dry scented wood are used together with dried lavender etc. and the contents are re-livened with drops of EOs of lavender etc. when needed. The contents can be altered for different seasons. The pot-pourri can be made to look more attractive by the addition of whole cloves, petals, dried leaves etc.

Veterinary products

Spike lavender is included in some veterinary shampoos and other products as an insect repellent, especially for fleas (Potter, 1988).

Household cleaning

Elizabethans used to scrub floors with bunches of lavender and the oil was used to polish furniture (Ryman, 1991). The fragrance is still the most used of all in any cleaners including detergents, but true lavender is nowadays replaced by synthetics. Home-made furniture polish can be made using 240 g grated beeswax, 60 g grated household soap, 600 ml genuine turpentine and the same volume of water, heating first the beeswax and turpentine in a bain-marie to melt it, then adding the soap dissolved separately in water and then adding lavender oil (Flanders, 1995).

Perfumes, beauty products and scents for the home

Lavender is added to lotions, bath products, after-shaves, soaps, colognes, toilet waters, perfumes, etc. Soaps can be scented with lavender oil by grating the unscented soap into a saucepan and then adding drops of fragrance and gently heating and stirring until the mixture solidifies again (Flanders, 1995). Alternatively, soap balls can be made in the cold by mixing the finely grated soap with some water and the fragrance in a mortar with a pestle and then making the mixture into balls by hand.

Herbal pillows

Herbal pillows have been used as an aid to sleeping for hundreds of years, but became old-fashioned and were only recently re-discovered in England due to the advent of 'aromatherapy' and the re-discovery of the sedative properties of lavender (Buchbauer et al., 1991).

Lavender bags

These were used in centuries past and were even more popular in the Victorian era. The bags were made simply from pieces of cotton, round or square, which were filled with dried lavender flowers and tied up with ribbon. These bags were used for perfuming contents of draws, wardrobes and rooms in general and to keep insects at bay. Sometimes, larger pieces of material were sewn together and filled with lavender and used as drawer liners. Lavender would sometimes be mixed with peppermint, thyme, rosemary and other dried herbs and spices and used as a more effective insect repellent. These and other shaped containers made of material, containing dried lavender are making a come-back.

New patents

Therapeutic bath salts

A mixture of lavender oil, ylang ylang oil, rosewood oil and patchouli oil with epsom salts, LiCl and Copper gluconate have been patented by McLean (1999) as a muscle relaxant.

Therapeutic skin creams

A composition for the control of dermatomycoses and dermatophytoses with *Tinea pedis*, included lavender and *Equisetum* (Koniger, 1997).

Another preparation for minor skin irritations contains lavender, tea tree and eucalyptus in a carrier oil (Elliott, 1997) and is said to reduce irritation, promote healing, resist insects and take advantage of the science of aromatherapy.

Wound treatment

A base of Centella asiatica with lavender oil, thyme oil, rosemary oil and Aesculus tincture, with possible addition of Medicago or Carlina and cedar oil (Llopart, 1982).

Treatment of skin and scalp

Lavandula, Mentha. Allium and *Salvia* are included in an infusion in vinegar of *Geum* (Iris, 1989).

Fly and mosquito repellent

Lavender absolute, benzoin, dimethylbenzylcarbinyl acetate, with jasmine absolute, racemic borneol, d-limonene and/or hydrolinalool is patented by Warren (1997).

Lastly, lavender water is poured onto gravesites in Malaysia in libation ceremonies.

References

Bertram, T. (1995) *Encyclopaedia of Herbal Medicine*, 1st edn, Grace Publishers, Dorset.
Buchbauer, G., Jirovetz, L., Jager, W., Dietrich, H., Plank, C., and Karamat, E. (1991) Aromatherapy: evidence for sedative effects of the essential oil of lavender. *Z Naturforschung*, 46c, 1067–72.
Collins, Wilkie (1870) *Man and Wife*, A novel F. S. Ellis.
Culpeper, N. (1653) *The English Physitian Enlarged*, George Sawbridge, London.
Fenaroli, G. (1998) *Fenaroli's Handbook of Flavor Ingredients,* Vol. 1, 3rd edn, CRC Press, Boca Raton.
Flanders, A. (1995) *Aromatics*, Mitchell Beazley, London.
Gerard, J. (1597) *The Herball or General History of Plants*, John North, London.
Grieve, M. (1937) *A Modern Herbal*, Reprinted 1992. Tiger Books International, London.
Iris (1989) IRIS S.a.s. di Salvioli Michele e C. (Soliera, IT) US Patent 4,855,131.
Koniger, H. (1997) US Patent 5,641,481.
Llopart, J.-P. (1982) US Patent 4,318,906.
Lois Vickers (1991) *The Scented Lavender Book*, Edbury Press.

McLean, L. (1999) Therapeutic bath salts for use in relaxation, inflammation and pain therapy. Issued Sept. 28. US patent 5,958,462.
Potter (1988) *Potter's New Cyclopaedia of Botanical Drugs and Preparations*, Revised edn, Williamson, E.M. and Evans, F.J. (eds), C.W. Daniel Co. Ltd., Saffron Walden.
Rose, J. (1982) *Herbal Body Book*, GD/Perigee Book, New York.
Ryman, D. (1991) *Aromatherapy*, Piatkus, London.
Warren, C. B., Marin, A. B., and Butler, J. F. (1997) Use of unsaturated aldehyde and alkanol derivatives for their mosquito repellency properties. US Patent No. 5,665,781.

19 New research into *Lavandula* species, hybrids and cultivars

Maria Lis-Balchin

Introduction

'All or almost all has been said about lavender' – a quote taken from M. Abrial in 1937 when he presented a novel lavandin (Vinot and Bouscary, 1971). This is not the case, as there are enormous gaps in the scientific knowledge about *Lavandula*. This is despite numerous books and papers on the subject, many of which just paraphrase the others. Much of what is written about lavenders stems from past, well-known hearsay evidence, without any scientific verification or has been made up by the authors themselves. An example of the latter is taken from Kourik (1998) who states that the ancient Egyptians used lavender in their mummification processes and constructed stills to extract its essential oil. There is however, no mention of lavender in any Egyptological text, although other herbs, spices and plants are very much in evidence (Germer, 1985; Manniche, 1989; Wilkinson, 1998).

Husbandry

Increasing yield with application of foliar sulphur and phosphorus as soil fertilizers

The foliar application of S as 1–2 per cent sulphuric acid and or P as 1–2 per cent phosphoric acid on *Lavandula angustifolia* in Egyptian plantations, showed an increase both in height and fresh and dry weights; the best yield of essential oil was after foliar application of 1 per cent S. It was found that there was also an effect on uptake of other minerals from the soil using foliar S and/or P (Hussein *et al.*, 1996).

The best yield of lavender was obtained with urea at 40 kg/feddan (1 feddan = 0.42 hectares), compared to using ammonium sulphate, ammonium nitrate or ammonium chloride; the best yield of essential oil was obtained at half the urea dose, that is, 20 kg/feddan (El-Sherbeny *et al.*, 1997).

Aromatic lavender as rotation crops

Five-month old *L. vera (angustifolia)* plants which were colonized with a selected isolate of *Glomus intraradices*) were used in a field which had been fumigated and unused for 5 years (Camprubi and Calvet, 1996). Half the field was again fumigated with metham sodium: the two fields areas were then treated with 'without lavender and 6 months after planting the plants were cut down and seeds of Troyer citrange and Cleopatra mandarin were sown; the pre-treated aomatic areas resulted in significantly higher numbers of arbuscular mycorrhizal propagules and rootstock seedlings also grew better. The net effect was to introduce mycorrhizal fungi into agricultural soil using this technique with lavender.

Germination of lavender seeds

L. angustifolia seeds were treated with GA3 at 50–800 ppm, which stimulated germination under laboratory conditions (Macchia *et al.*, 1996), the best concentration was found at 200 ppm. At this concentration, seeds could then be stored for up to 3 months and retained 65 per cent of their viability.

Authentification of lavender essential oil

Due to the large-scale adulteration of commercial lavender oils, new ways of authentification have had to be developed. The tests which can be usefully employed have been studied by Charles (2000) and include:

Refractive index (RI)
Optical rotation (OP)
Specific gravity (SG)
Gas chromatography (GC)
   Using different columns: DB-5, DB-WAX
   Chiral columns RT-β-DEXse
with Mass spectroscopy (MS)
or with FID
Radioactive (^{14}C)
Stable isotope ratio (δ^{13}C and δD)

The chiral analyses are especially good in the case of synthetic linalool and linalyl acetate adulteration and this is easily confirmed by the radioactive analyses. Most of the initial, ISO testing is now obsolete (RI, OP and SG), with ordinary GC of limited benefit unless gross adulteration has occurred with unusual components or solvents.

Pharmacological studies

Effect on lipid peroxidation in vivo

Lavender oil and a variety of other essential oils were tested for effects on the lipid-peroxidation-antioxidase defence system and lipid metabolism in 150 patients with chronic bronchitis (Siurin, 1997). There was a lowering of plasma levels of dienic conjugates and ketones, activation of catalase in red blood cells which the author stated was characteristic of an antioxidant effect on exposure to essential oils of rosemary, basil, fir and eucalyptus. However, lavender had a normalizing effect on the level of total lipids and the ratio of cholesterol to its alpha fraction.

Effects on lipid peroxidation in vitro

Aqueous methanolic extracts of *L. angustifolia*; rosmarinic acid, caffeic acid, luteolin 7-O-glucoside, methyl carnosoate investigated in two different biological systems of lipid peroxidation (LPO). All the extracts and individual components were active. More effect was found on the enzyme-dependent system than the enzyme-independent system. This suggests that there is a direct effect on the enzyme, acting as an inhibitor (Hohmann *et al.*, 1999).

Local anaesthetic effect

Anaesthetic activity of *L. angustifolia* essential oil was evaluated *in vivo* in the rabbit conjunctival reflex test and *in vitro* using the rat phrenic nerve-hemidiaphragm preparation (Ghelardini *et al.*, 1999) against two citrus fruit essential oils (*Citrus reticulata* 'Blanco' and *Citrus limon*), with no medical uses. *L. angustifolia* and its components, linalool and linalyl acetate were able to drastically reduce contractions in the rat phrenic nerve diaphragm and in the eye, there was a dose-dependent increase in the number of stimuli necessary to provoke the reflex, apparently confirming *in vivo* the local anaesthetic activity observed *in vitro*. The activity in the rat diaphragm however, does not seem to be indicative to the present author of an anaesthetic action.

Attenuation of blood pressure increase in handgrips

Nagai *et al.* (2000) studied the effect of pleasant odour inhalation using lavender, rose and jasmine on diastolic blood pressure during exercise and found that it was decreased by 24 per cent. In contrast, the blood pressure increase during static handgrips, where a power of 30–40 per cent of maximum was maintained was not affected by the odours. The authors concluded that as the blood pressure increase during the static handgrip is a lower brainstem reflex, the results indicated that the inhalation of these odours suppresses the muscle sympathetic vasoconstrictor activity and attenuates the blood pressure increase by affecting the CNS higher than the midbrain.

Anti-inflammatory effect of L. latifolia

L. latifolia is used in Paraquay for treating bronchitis, asthma, rheumatism and has also been applied topically for swellings. The aerial parts of *L. latifolia* (650 g) were extracted with hot 70 per cent ethanol and then concentrated to give a 70 per cent extract (164 g). The extract was then partitioned between water and chloroform to give a chloroform-soluble and water-soluble fraction and a precipitate (Shimizu *et al.*, 1990). The chloroform fraction was then further partitioned to yield coumarin, 7-methoxy-coumarin, trans-phytol and caryophyllene oxide. Coumarin in the chloroform fraction as well as the fraction itself was shown to have a weak effect on carageenin-induced paw edema in rats on topical application. The authors also reported that caryophyllene oxide showed an inhibitory effect on histamine-induced contractions in guinea-pig ileum, but did no further tests with acetylcholine or atropine therefore these are not very meaningful results.

'Hypoglycaemic' effect of Lavandula species, but not in diabetes

Gamez *et al.* (1987, 1988) studied the effects of water-soluble extracts of *L. dentata*, *L. latifolia*, *L. stoechas* and *L. multifida* on hypoglycaemic rats. Tests in normal rats, that is, normaglycaemic showed that both the 10 per cent and 20 per cent infusion and 10 per cent suspension administered orally decreased the blood glucose, that is, had a hypoglycaemic effect. *L. stoechas* was more active as an infusion while the other species as a suspension. *L. dentata* and *L. stoechas* both showed an hypoglycaemic effect on glucose-induced hyperglycaemic rats. However, there was no hypoglycaemic effect on alloxan-induced diabetic rats. This indicates that the hypoglycaemic effect can only occur if the pancreas is intact and is therefore absolutely useless in diabetes.

Perillyl alcohol and its effect on cancers

Perillyl alcohol is found in small amounts in lavender, peppermint, spearmint, cherries and celery seeds. Many experiments using perillyl alcohol in animals have shown a positive effect on regression of various tumours including: pancreatic, mammary, and liver tumours (Belanger, 1998). It is also a chemopreventative agent for colon, skin and lung cancer. It is also a chemotherapeutic agent for treating neuroblastoma and prostate and colon cancer. Preliminary trials in human cancers have not been successful as remission has not occurred and there are gastrointestinal and other side effects.

Psychological effects

Effects on flight controllers

In order to optimize the environment in which flight controllers work, several essential oils and and aromas were used including lavender, anise and brandy mint. After 20 days of exposure the health condition of the flight controllers improved (Leshchinskaia et al., 1983). They felt less tired at the end of the day. When no photocides were used, the REG-wave amplitude decreased and also the tonic tension of cerebral vessels increased indicative of fatigue; when the aromas were employed the changes were in the opposite direction.

Basic emotions induced by lavender

The autonomic nervous system (ANS) responses were studied with regard to emotional factors induced by odorants which included lavender, ethyl acetoacetate, camphor, acetic acid and butyric acid (Vernet-Maury, 1999). Fifteen subjects inhaled these five odorants at different times and their various ANS parametres were monitored. These included: skin potential and resistance, skin blood flow, and temperature, instantaneous respiratory frequency and instantaneous heart rate. The subjects also evaluated their feelings after inhalation, which ranged from pleasant to unpleasant on an 11 point hedonic scale. Analysis of variance showed that lavender elicited happiness as did ethyl acetoacetate, camphor elicited either happiness, surprise or sadness, based on past association with the odour. Butyric and acetic brought unhappiness with negative emotions like anger and disgust expressed. More than sixty subjects in three similar experiments, showed similar autonomic responses which were correlated with basic emotions; therefore a multiparametric autonomic analysis could identify a type of basic emotion and its intensity.

EEG responses to different odours

Four odours were studied on the EEG of four men and two women at rest while sitting (Lee et al., 1994). The odours were: lavender, citrus and floral and were released for 10 min from a duct to fill the room completely. Subjective estimation, at the lower concentration, indicated that citrus was the most comfortable, but this was not significant. The rate of alpha wave (Oz) in the period of giving the citrus was significantly higher than for lavender. The rate of beta wave (Oz) during floral exposure was significantly higher than that of lavender. At the higher concentration, the regression coefficient of the power of the spectra of frequency-fluctuation of alpha wave for lavender was significantly higher than for the other two odours and at the lower concentration it was higher during the dosing than at other periods. This measurement is therefore suggested to be useful for evaluating psychophysiological responses.

Effects on EEG activity and sensory evaluation

Thirteen female subjects were assessed under four odour conditions using lavender, chamomile, sandalwood and eugenol applied respectively to each person (Masago et al., 2000). Both EEG and sensory evaluation by the subjects were evaluated. The factor of a 'comfortable feeling' was highest for lavender and descended in order of eugenol, chamomile and sandalwood. The alpha-1 (8–10 Hz) of EEG at the parietal and posterior temporal regions decreased significantly after the onset of inhalation of lavender. Some smaller changes occurred with eugenol and chamomile but not with sandalwood. Therefore the authors concluded that alpha-1 activity decreased significantly only when the subject felt comfortable. There seems therefore to be a correlation between alpha-1 activity and subjective evaluation.

Insecticidal and antimitotic effects

Naturally-occurring insecticidal monoterpenoids studied using recombinant homomeric insect GABA receptors

The actions of naturally-occurring insecticidal monoterpenoids was studied using recombinant homomeric insect GABA receptors from *Drosophila melanogaster* (Priestley, 1999). Some sesquiterpenoids, known as picrodendrins, for example, picrotoxin are antagonistic at native GABA receptors and at heterologously expressed, recombinant *Drosophila melanogaster* RDL_{ac} homomeric GABA receptors and are also thought to have potential neurological effect. In view of this, linalool, terpinen-4-ol and thymol were investigated. The RDL_{ac} – encoded CRNA was injected into the cytoplasm of *Xenopus laevis* oocytes and the responses of the expressed homomeric RDL_{ac} receptors were detected by voltage – clamp electrophysiology. The monoterpenoids were pre-applied continuosly for 2 min prior to co-application with GABA for 30 sec. The amplitude of the currents were increased in the presence of thymol but not the other two monoterpenoids. Thymol is therefore considered to be a novel allosteric modulator for insect GABA receptors.

Acaricidal effect of lavender

L. angustifolia essential oil and its component linalool were effective against *Psoroptes cuniculi*, (mites from rabbits) when presented in the air and linalool was isolated from the dead mites in the ether extract when *L. angustifolia* oil was used, indicating this was an active component. (Perrucci *et al.*, 1996)

Essential oils as pediculicides

Lavender, juniper, eucalyptus, geranium, lemon and rosemary were used as mixtures in a pilot study to determine possible novel, safe pediculicides (Weston *et al.*, 1997). *Pediculus humanus capitis*, the head louse was not found to be suitable for *in vitro* studies and therefore *Pediculus humanus, the clothing mite*, was studied. The oils and some of their components were tested *in vitro*, impregnated onto filter paper against the human lice. Some components like α-pinene, camphene and terpineole were very effective. The clinical study involved twenty children, treated with the essential oil mix in a cream/alcohol vehicle at 5 per cent and was very effective. A mild allergic response was, however, obtained in one case, and found to be to rosemary, which was then removed from subsequent mixtures. This was a very small study and more data would be required for acceptance of essential oils as lice killers.

Effects of lavender against house dust mites

Lavender oil was tested against tea tree oil and lemon oil *in vitro* against the house dust mite, *Dermatophagoides pteronyssinus* (Priestley *et al.*, 1998). The various mites cause diseases like scabies and various veterinary infestations. *D. pteronyssinus* causes asthma. The insecticidal assay was set up using filter paper impregnated with the test substances which made contact with the mites studied in a closed system. The immobility was assessed after 30 min and the mortality after 2 h. Lavender was intermediate between tea tree, with the highest potential and lemon oil with the lowest.

Miscellaneous research

Calyx-bore floral fragrances in lavender

In contrast to most families of Angiosperms, the Lamiaceae produce floral scent on the calyx and not the corolla (Mattern and Vogel, 1994). Leaves, calyces and corollas had homologous glandular trichomes (peltate glandular hairs and short and long capitate glandular hairs). Anatomical differences were very few between species and only a few species exhibited these differences. The glandular trichomes of the calyx were more dense and more exposed than those of leaves in most species.

Pheromones in lepidoptera

The abdominal hair brushes of the male hawk moth, *Acherontia atropos (Lepidoptera)*, were found to contain twenty-four volatiles, including linalool, linalyl propionate and caryophyllene (Bestmann *et al.*, 1993). These volatiles were not thought to be derived from the food consumed by these insects as indicated by studies on different diets fed to the insects. It was most likely that they were synthesised *de novo*.

Tissue-culture-based screening

Tissue-culture-based screening for selection of high biomass and phenolic producing clonal lines of lavender using *Pseudomonas* species and azetidine-2-carboxylate was reported by Al-Amier *et al.* (1999)

Allelopathy

L. angustifolia was planted for different times in an experimental garden in Russia and the effect on the number of micro-organisms, their group composition and N_2-fixing ability was studied (Sidorenko *et al.*, 1995). There were allelopathically active substances in the rhizospheres of 3-year-old plants which promoted growth and numbers and therefore of micro-organisms which assimilated organic and also mineral N_2-containing compounds. The number of organisms was less depressed than with 1-year-old plants, but potential N_2-fixing soil activity was lower. The numbers of micro-organisms were higher in inter-rows than in the rhizosphere.

Dietary supplements for depression, anxiety and sleep disorders

Cauffield and Forbes (1999) surveyed a wide range of dietary supplements including Kava Kava, St John's Wort, melatonin, valerian, chamomile as well as lavender against depression, anxiety and sleep disorders and contrasted some of these against low-dose antidepressants like amitriptyline. Lavender was included in the list of alternatives for sleep disorders and was considered low risk.

Patch testing with lavender oil

Patch testing was carried out using lavender over a 9-year period in Japan using Finn Chambers and Scanpor tape in a 2-day closed patch test (Sugiura et al., 2000). The lavender oil was used at 20 per cent in petrolatum on the upper back of each patient suspected of suffering from cosmetic contact dermatitis. The comparison between the frequency of positive patch testing to lavender compared to that of other essential oils was determined each year. During the period 1990–98, the positive rate for lavender was 3.7 per cent (0–13.9 per cent). There was a dramatic increase in 1997, which coincided with the importation of the aromatherapy trend for using lavender oil. This involved using dried lavender flowers in pillows, drawers, cabinets and rooms as well. In 5 out of 11 cases of a positive patch test to lavender in 1997, dried flowers were also used by the patients. In 1998 this increased to 8 out of 15 cases. The authors concluded that the positive patch test increases were therefore due to the use of dried lavender rather than fragrances in cosmetic products. The present author finds that rather a simplistic conclusion as there was virtually no change in the number of patients using dried lavender products in 1997–98 and new aroma-chemicals, which appeared in greater abundance in recent years were not taken into account.

References

Al-Amier, H., Mansour, B.M.M., Toaima, N., Korus, R.A. and Shetty, K. (1999) Tissue-culture-based screening for selection of high biomass and phenolic producing clonal lines of lavender using *Pseudomonas* and azetidine-2-carboxylate. *J. Agr. Food Chem.*, 47, 2937–43.

Bellanger, J.T. (1998) Perillyl alcohol: application in oncology. *Altern. Med. Rev.*, 3, 448–57.

Bestmann, H.J., Erler, J., Vostrowsky, O. and Wasserthal, L.T. (1993) Pheromones.92. Odorous substances from the abdominal hair brushes of the male sphingid moth *Acherontia atropos* L. (*Lepidoptera, Sphingidae*). *Zeit. Natursch. C- J. Biosci.*, 48, 510–4.

Camprubi, A. and Calvet, C. (1996) A field innoculation system for citrus nurseries using pre-cropping with mycorrhizal aromatic plants. *Fruits (Paris)*, 51, 133–7.

Cauffield, J.S. and Forbes, H.J. (1999) Dietary supplements used in the treatment of depression, anxiety and sleep disorders. *Lippincott. Prim. Care Pract.*, 3, 290–304.

Charles, D.J. (2000) Authenticity Assessment of Lavender Essential Oil, 31st International Essential Oil Symposium (ISEO), Hamburg, Sept. 11–13.

El-Sherbeny, S.E., El-Saeid, H.M. and Hussein, M.S. (1997) Response of lavender (*Lavandula angustifolia*) plants to different nitrogen sources, *Egypt. J. Hort.*, 24(1), 7–17.

Gamez, M.J., Jimenez, J., Risco, S. and Zarzuelo, A. (1987) Hypoglycaemic activity in various species of the genus *Lavandula*. Part 1. *Lavandula stoechas* L. and *Lavandula multifida* L. *Pharmazie*, 42, 706–7.

Gamez, M.J., Zarzuelo, A., Risco, S., Utrilla, P. and Jimenez, J. (1988) Hypoglycaemic activity in various species of the genus *Lavandula*. Part 2. *Lavandula dentate* and *Lavandula latifolia*. *Pharmazie*, 43, 441–2.

Germer, R. (1985) *Flora des pharaonische Agypten*, Mainz.

Ghelardini, C., Galeotti, N., Salvatore, G. and Mazzanti, G. (1999) Local anaesthetic activity of the essential oil of *Lavandula angustifolia*. *Planta Med.*, 65, 700–3.

Hohmann, J., Zupko, I., Redei, D., Csanyi, M., Falkay, G., Mathe, I and Janicsak, G. (1999) Protective effects of the aerial parts of *Salvia officinalis, Melissa officinalis* and *Lavandula angustifolia* and their constituents against enzyme-dependent and enzyme-independent lipid peroxidation. *Planta Med.*, 65, 576–8.

Hussein, M.S., El-Saeid, H.M. and El-Sherbeny, S.E. (1996) Yield and quality of lavender in relation to foliar application of sulphur and phosphorus. *Egypt. J. Hort.*, 23, 167–78.

Kourik, R. (1998) *The Lavender Garden*, Chronicle Books, San Fransisco.

Lee, C.F., Katsuura, T., Shibata, S., Ueno, Y., Ohta, T., Hagimoto, S., Sumita, K., Okada, A., Harada, H. and Kikuchi, Y. (1994) Responses of electroencephalogram to different odors, *Ann. Physiol. Anthropol.*, 13, 281–91.

Leshchinskaia, I.S., Makarchuk, N.M., Lebeda, A.F., Krivenko, V.V. and Sgibnev, A.K. (1983) Effect of phytoncides on the dynamics of the cerebral circulation in flight controllers during their occupational activity. *Kosm. Biol. Aviakosm. Med.*, 17, 80–83.

Macchia, M., Moscheni, E. and Angelini, L.G. (1996) Germination of lavender seeds: aspects linked to dormancy and methods of increasing vigour. *Atti convegno internazionale: Cltivazione e miglioramento di plante officinali*, Trento, Italy, 2–3 June, 1994, 391–7.

Manniche, L. (1989) *An Ancient Egyptian Herbal*. British Museum Publications Ltd. London.

Mattern, G. and Vogel, S. (1994) Anatomical investigations: comparison of calyx and leaf glands, *Beit. Biol. Planz.*, 68, 125–56.

Musago, R., Matsuda, T., Kikuchi, Y., Miyazaki, Y., Iwanaga, K., Harada, H. and Katsuura, T. (2000) Effects of inhalation of essential oils on EEG activity and sensory evaluation. *J. Physiol. Anthropol. Appl. Human Sci.*, 19, (1), 35–42.

Nagai, M., Wada, M., Usui, N., Tanaka, A. and Hasebe, Y. (2000) Pleasant odors attenuate the blood pressure increase during rhythmic handgrips in humans. *Neuroscience Lett.*, 289, 227–29.

Perrucci, S., Macchioni, G., Cioni, P.C., Flamini, G., Morelli, I. and Taccini, F. (1996) The activity of volatile compounds from *Lavandula angustifolia* against *Psoroptes cuniculi*. *Phytother. Res.*, 10, 5–8.

Priestley, C.M., Burgess, I. and Williamson, E.M. (1998) Effects of essential oils on house dust mites. *J. Pharm. Pharmacol.*, 50, Suppl., 193.

Priestley, C.M., Sattelle, D.B. and Williamson, E.M. (1999) An investigation into the actions of naturally-occurring insecticidal monoterpenoids using recombinant homomeric insect GABA receptors, *J. Pharm. Pharmacol.*, 51, Suppl., 101.

Shimizu, M., Shogawa, H., Matsuzawa, T., Yonezawa, S., Hayashi, T., Arisawa, M., Suzuki, S., Yoshizaki, M., Morita, N., Ferro, E., Basualdo, I. and Berganza, L.H. (1990) Anti-inflammatory constituents of topically applied crude drugs. IV. Anti-inflammatory effect of paraguayan crude drug 'Alhucema' (*Lavandula latifolia* Vill.), *Chem. Pharm. Bull.*, 38, 2281–4.

Sidorenko, O.D., Gorbuniova, E.O. and Voronina, E.P. (1995) Allelopathic action of lavender on soil micro-organisms. *Biol. Bull. Russian Acad. Sci.*, 22, 101–3.

Sugiura, M., Hayakawa, R., Kato, Y., Sugiura, K. and Hashimoto, R. (2000) Results of patch testing with lavender oil in Japan. *Contact Dermatitis.*, 43, 157–160.

Siurin, S.A. (1997) Effects of essential oil on lipid peroxidation and lipid metabolism in patients with chronic bronchitis. *Klin. Med.* (Mosk), 75(10), 43–5.

Vernet-Maury, E., Alaoui-Ismaili, O., Dittmar, A., Delhomme, G. and Chanel, J. (1999) Basic emotions induced by odorants: a new approach based on autonomic pattern results. *J. Auton. Nerv. Syst.*, 75, 176–83.

Vinot, M. and Bouscary, A. (1971) Studies on lavender (VI) the hybrids, *Recherches*, 18, 29–44.

Weston, S.E., Burgess, I. and Williamson, E.M. (1997) Evaluation of essential oils and some of their component terpenoids as pediculicides for the treatment of human lice. *J. Pharm. Pharmacol.*, 49, Suppl. 4, 120.

Wilkinson, A. (1998) *The Garden in Ancient Egypt*. The Rubicon Press, London.

20 Further research into *Lavandula* species

Cell cultures of *L. vera* and rosmarinic acid production

Mladenka Paunova Ilieva-Stoilova, Atanas Ivanov Pavlov and Elena Georgieva Kovatcheva-Apostolova

Introduction

Lavandula spp. are widely cultivated aromatic crops in Mediterranean areas, because lavender essential oil constitutes a very important natural product of a high industrial interest (Ivanov and Nikolov, 1988; Charlword *et al.*, 1986). Although these species can be vegetatively propagated, the poor rooting ability of the stem cutting, as well as the lack of selected clones restrain its economical exploitation (Calvo and Segura, 1989). For these reasons, an interest has been developed recently in producing these valuable compounds using a biotechnological approach. The tissue culture technology offers the ways for overcoming these limitations by providing methods for both rapid multiplication and improvement of the species. The first report on the *in vitro* multiplication of the *Lavandula* spp. from axillary buds was published in 1980 (Quazi, 1980). Later, a complex investigation on the *Lavandula latifolia Medicus* (a species which is one of the most widely cultivated in the Mediterranean areas) was performed by Calvo *et al.* (1988; 1988a; 1989; 1989a) and Jordan *et al.* (1990). They found that the hypocotyl explants of *L. latifolia Medicus* were with the potential for mass propagation (Calvo *et al.*, 1988a). On this basis, they succeeded in establishing the culture requirements for promoting morphogenesis from hypocotyl and from cotyledon and root explants as well (Calvo *et al.*, 1988a; Calvo and Segura, 1989) and from cultured leaves of *L. latifolia* (Calvo and Segura, 1989a). They also extended this research to procedures and factors, promoting plant regeneration from callus and isolated cells of *L. latifolia Medicus* (Calvo *et al.*, 1988a; Jordan *et al.*, 1990). The knowledge gained from these studies may enable mass production of *L. latifolia* plants. Also, plants regenerated from single cells can be an important source of genetic diversity and some cloned variations which could be used for selecting strains that produce large amounts of essential oils (Jordan *et al.*, 1990).

Lavender essential oil is an important raw material for perfumery and cosmetic industries and that is why its production by lavender plant cell cultures has been investigated as well, but progress is slow. Small amounts of monoterpenoids, constituents of lavender essential oil, have been detected in shoots, regenerated from callus cultures of *L. augustifolia* (Webb *et al.*, 1984). On the other hand, undifferentiated suspension cultures were used for biotransformation of monoterpenoids by Lappin *et al.* (1986; 1987). They established a rapidly growing suspension culture of *L. augustifolia* and had examined its ability to biotransform a number of monoterpenoids and structurally related compounds. They found that the cell culture reduces the monoterpenoid aldehydes (added to the culture medium) to the corresponding alcohols. The latter, once formed, disappeared from the culture medium, either incorporated into higher terpenoids or into other secondary metabolites. In either case, such metabolic activity would be a serious barrier to the utilization of lavender cell cultures for commercial biotransformation of monoterpenoids.

Cultured photomixotrophic L. vera cells grown under light have been screened as a potential biotin producer (Watanabe et al., 1982; Watanabe and Jamada, 1982). Selection of high producer cell lines was carried out by irradiation of gamma ray and addition of pimelic acid to the culture medium. The biotin content reached about 0.5 $\mu g\,g^{-1}$ fresh weight (Watanabe et al., 1982; Watanabe and Jamada, 1982; 1982a; Tanaka, 1989). Biotin-producing green L. vera callus was preserved successfully in liquid nitrogen and it was found that the callus recovered after the freeze preservation retaining its biosynthetic capability for biotin (Watanabe et al., 1983; 1985a).

Watanabe et al. (1985) reported that when these cells are cultivated in a medium supplemented with S-containing compounds (such as L-cystine or DL-homocystine, L-cysteine or LD-homocysteine) a blue pigment is accumulated. Recently, Tsuro and Inorie (1996) also reported the production of this pigment in leaf-derived callus of L. vera DC. Banthorpe et al. (1985) investigated terpenoid metabolism in callus cultures of L. augustifolia Mill. subsp. augustifolia (≡L. officinalis Chaix≡L. vera DC.), and found that these cultures also appear to secrete an intensive blue pigment into the supporting medium under a wide variety of conditions. The blue pigmentation of the agar medium increases with the increase of sucrose concentration, by accidental bacterial or fungal infection of cultures or with the supply of an amino acid in the medium. They established that the blue coloration actually resulted from the partial secretion of yellow pigments, produced by the callus: these secreted products complex with Fe^{2+}, present in the MS nutrient medium (Dixon, 1985). The structure of the yellow pigment was established as an ester of 2–4(3,4-dihydroxyphenyl)ethemyl alcohol and 3-(3,4-dihydroxypheyl)-2-propenoic acid (Banthorpe et al., 1985).

Nakajima et al. (1986) immobilized cultured cells of L. vera using synthetic resin pre-polymers and followed the time course of their pigment production and how it was influenced by the carbon source (sucrose) (Nakajima et al., 1989).

Recently, Lopes-Arnaldos et al. (1997) studied the antioxidant activity of extracts from lavender (L. x intermedia) cell cultures. They established that the antioxidant activity of the extracts was due to the rosmarinic acid (RA), which is their main phenolic constituent.

Cell culture of *Lavandula vera* MM

Callus culture

It is known that the callogenesis depends on the type and concentration of auxins and cytokinins added to the general used high-salt media (Dixon, 1985). A scheme for obtaining callus culture from lavender was developed on the basis of Lismaier and Skoog (LS) culture medium with 3 per cent sucrose (Dixon, 1985) (Table 20.1). The L. vera callus culture (marked as L. vera MM) was

Table 20.1 Callogenesis of L. vera on LS media, supplemented with different concentrations of the growth regulators

Growth regulators, mg/L	Variants																
	1	2	3	4	5	6	7	8	9	10	11	12	13	14	15	16	17
2.4 D	1.00	2.00	3.00	4.00	5.00	6.00	—	—	—	—	—	—	—	—	—	—	0.20
Kinetin	0.05	0.05	0.05	0.05	0.05	0.05	0.05	0.05	0.05	1.00	1.00	0.05	0.20	0.50	1.00	1.50	—
NAA	—	—	—	—	—	—	1.00	2.00	3.00	4.00	5.00	5.00	5.00	5.00	5.00	5.00	—

obtained from stem explants of the oleaginous variety 'Druzba'. They were cultured at 25–28°C on LS, supplemented with 0.05 mg/L kinetin, 1 mg/L α-naphthylacetic acid (NAA) (variant 7 – Table 20.1). The resulting callus was subcultured on all the variants of the culture media showed in Table 20.1. The best growth of the callus culture was established on LS culture medium supplemented with 0.2 mg/L 2,4-dichlorophenoxyacetic acid (2,4 D) (variant 17 – Table 20.1) and it was subcultured for more than six years on the same medium at 25–28°C in the dark at 2–3 week intervals. The *L. vera* MM callus culture was homogenous and friable with a white to beige color.

Suspension culture

The suspension culture *L. vera* MM was obtained by continuous adaptation of callus clumps under the conditions of repeated cultivation in LS liquid medium. The cultivation was performed in 500 mL flasks with 1/5 net volume. The flasks were agitated on a shaker (11 rad/s) in the dark at 26–28°C. As a result of this adaptation, a homogeneous bluish-white cell suspension of *L. vera* MM with steady growth characteristics was obtained. At the beginning of the growth cycle, the cell aggregates consisted of 4–6 cells. During the linear phase of growth of the cell suspension, the shape of the plant cells changed from oval to ovoid and the number of constituent cells in an aggregate increased to 10–12. Vacuolation of the plant cells is also worth noting.

Identification of the main phenolic acids biosynthesized by L. vera MM cell suspension

The investigations of *L. vera* MM cell suspension showed that the methanol extract of cell biomass has antimicrobial and antioxidant activities due to the biosynthesis of RA and other phenolic compounds (Ilieva *et al.*, 1994; 1995). For their identification, methanolic extracts from fresh cells of *L. vera* MM grown in LS medium for 9 days were used (Kovatcheva *et al.*, 1996). The UV

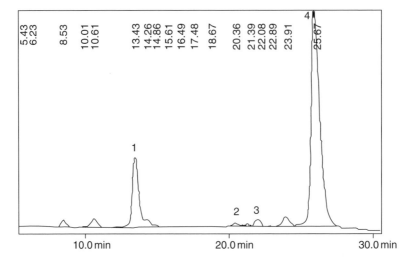

Figure 20.1 HPLC of the methanolic extract of cell biomass from *L. vera* MM. Caffeic acid (1), p-coumaric acid (2), ferulic acid (3), and RA (4).

Table 20.2 Content of phenolic acids in cell biomass of *L. vera* MM

Compound	Content, mg/g dry weight
Rosmarinic acid	5.15 ± 0.73
Caffeic acid	0.37 ± 0.02
p-Coumaric acid	Trace
Ferulic acid	Trace

Data present the mean values and standard deviation.

spectrum showed maxima between 280 and 340 nm, indicative of the presence of phenolic acids. Rosmarinic, caffeic, p-coumaric and ferulic acids were identified in the extracts by TLC and HPLC (Figure 20.1). The main component was RA (Table 20.2). It was isolated and purified from ethyl acetate extracts of cell biomass as a yellowish powder and its identity was confirmed by means of HPLC with photodiode array detector, ^{1}H NMR, ^{13}C NMR and mass spectroscopy. The spectra obtained are identical with the spectrum of pure RA (Roth, Natural Product, Germany), as well as with the data obtained by other authors (Fukui et al., 1984).

The basic physiological characteristics of *L. vera* MM cell culture

The time courses of the growth of *L. vera* MM cell culture and biosynthesis of RA were followed in parallel with the utilization of the main nutrient components (Figure 20.2) (Ilieva and Pavlov, 1997).

The cell suspension of *L. vera* MM grew intensively (maximum accumulated biomass – 13 g/L DB on the 8th day), hence there was no clearly outlined lag phase in its cell growth cycle (Figure 20.2a). The biosynthesis of RA took place between the 4th and the 8th day, being especially intensive between the 6th and the 8th day, when the maximum amount of RA accumulated in the cells was determined (about 68 mg/L) (Figure 20.2a). It is worth noting that RA was synthesized for a relatively short period (4 days) which is indicative of the great potential of *L. vera* MM for the biosynthesis of RA.

The process of fast hydrolysis of sucrose to glucose and fructose, which was completed on the 4th day of the cultivation was clearly outlined (Figure 20.2b). Intensive consumption of the monosaccharides, was observed simultaneously with the inversion of sucrose. The cell culture of *L. vera* MM tended to have a slight preference for glucose, which was completely exhausted on the 4th day.

Consumption of fructose was slower and determinable amounts of it were found even on the 8th day of the cultivation. It should be noted, however, that the period of intensive biosynthesis of RA by the *L. vera* MM plant culture began with the uptake of glucose by the medium (Figure 20.2b).

The time course of nitrogen uptake by the *L. vera* MM cell suspension (Figure 20.2c) showed that intensive consumption of both nitrogen forms (NH_4 and NO_3) was observed at the beginning of the cultivation. By the 6th day of cultivation, the ammonium ions were exhausted and the nitrate ion concentration in the culture medium reached 0.35–0.40 g/L, remaining constant until the end of the growth cycle. The nitrogen source is important both for the growth of the cell culture *L. vera* MM and for the biosynthesis of RA (Figure 20.2a,c). Ammonium ions

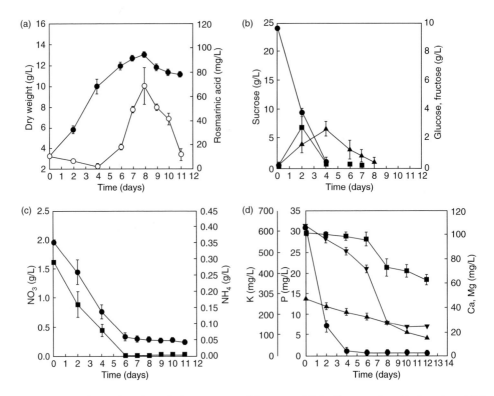

Figure 20.2 The basic physiological characteristics of the *L. vera* MM cell suspension. (a) Growth and RA production; (—●—) dry weight, (—○—) RA. (b) Carbohydrate utilization; (●) Suc, (—■—) Glu, (—▲—) Fru. (c) Uptake of the nitrogen sources; (—●—) NO_3, (—■—) NH_4. (d) P, Ca, Mg, and K uptake; (—●—) P, (—■—) Ca, (—▲—) Mg, (—▼—) K.

are especially necessary in the initial stage of the culture growth and it is not until they have been consumed to a certain extent that nitrate reductases are activated, reducing nitrate ions to ammonium ions and thus making possible their consumption (Hirasuna et al., 1991). On the other hand, it is manifest that the beginning of intensive biosynthesis of RA by *L. vera* MM plant cell culture was connected with the ammonium limitation in the culture and the low level of nitrate ions in the medium (Figure 20.2c).

The time courses of consumption of phosphorus, potassium, magnesium, and calcium are shown in Figure 20.2d. Phosphorus was taken up most intensively and was almost completely consumed by the 4th day of the cultivation. Intensive biosynthesis of RA by *L. vera* MM cell culture commenced after phosphorus had almost completely been consumed by the medium, so in this case phosphorus plays a certain regulatory role in the process of biosynthesis of RA.

Regarding the uptake of calcium, potassium and magnesium from the nutrient medium (Figure 20.2d), it may be noted that no direct relationship between their uptake and the process of biosynthesis of RA by *L. vera* MM was detected.

In conclusion, following from the time course of the utilization of the major nutrient components during cultivation (Figure 20.2), the intensive biosynthesis of RA began after the complete consumption of glucose, ammonium and phosphate ions, and after a relatively low level of nitrate ions in the medium had been reached. Therefore, we must underline that the biosynthesis of RA by *L. vera* MM is related to the concentration of sucrose, nitrogen and phosphate in the cultivation medium. This fact has been fundamental for our next investigations connected with the influence of the nutrient elements on the growth and the RA production by *L. vera* MM.

The influence of macro nutrient elements on the growth and RA production of *L. vera* MM cell culture

Effect of sucrose concentration

The biosynthesis of RA by any of its plant cell producers is considerably affected by the amount of sucrose added to the nutrient medium, but the sucrose concentration in the nutrient medium needed for obtaining the maximum yield of RA is different for the different plant cell producers (Zenk *et al.*, 1977; Ulbrich *et al.*, 1985; Hyppolyte *et al.*, 1992). The results obtained from investigations of the effect of sucrose concentration in the medium on the biosynthesis of RA by the *L. vera* MM cell culture (Ilieva and Pavlov, 1997) showed that 7 per cent sucrose was most effective (Figure 20.3b). The intensive biosynthesis of RA was apparent for a relatively short period (day 6 to day 12 of cultivation, that is, during the linear and early stationary phase) (Figure 20.3). The latter is a favorable prerequisite for an effective biotechnological method for its production. The high yield in this case, as well as in the other cases of increased sucrose concentration is to some extent due to the increased amount of cell biomass (in this case twice as much compared to the control) and the amount of RA, which is an intracellular component, is increased. Of more importance is the fact, however, that the added 7 per cent sucrose is converted to RA more effectively compared with the control or with the other cases under study. In this case, the RA measured a three-fold higher yield regarding the sucrose concentration in the medium (9 mg/g sucrose versus 3 mg/g sucrose for the control cultivation) (Figure 20.4). At this point, it should be noted that the cultivation of *L. vera* MM at sucrose concentrations over 7 per cent was not as good. In the case of 8 per cent sucrose (Figure 20.3a), as a result of the substantial amount of synthesized biomass (34 g/L DB at day 12), an increase in the viscosity of the medium ensued and the mass-exchange properties of the cultivation system deteriorated. This made the effective consumption of nutrient components impossible and cultivation very difficult.

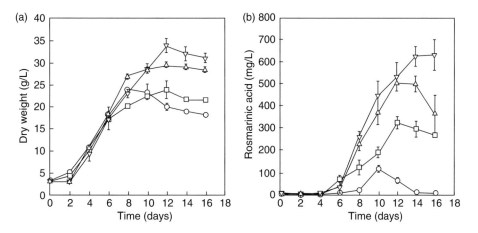

Figure 20.3 Time course of growth (a) and RA biosynthesis (b) by *L. vera* MM cell suspension culture, cultivated in media with different amounts of sucrose. (—○—) 5% sucrose; (—□—) 6% sucrose; (—△—) 7% sucrose; (—▽—) 8% sucrose.

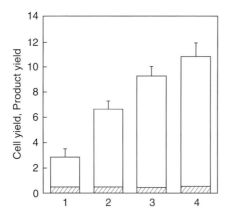

Figure 20.4 Effect of sucrose concentration in the nutrient medium on yields of the cell biomass (g DB/g sucrose) and RA (mg RA/g sucrose) by *L. vera* MM cultivated in media with: 3% sucrose (control-1); 6% sucrose (2); 7% sucrose (3), and 8% sucrose (4). (▭) Product yield; (▨) cell yield.

Effect of nitrogen concentration

The influence of the nitrogen source and concentration on secondary metabolism has not been thoroughly clarified though its influence on the yields of secondary metabolites with phenolic structure for some cell cultures has been reported (Do and Corumer, 1991; Hirasuna *et al.*, 1991; Srinivasan and Ryn, 1993) including the production of RA by *Anchusa officinalis* (De-Eknamkul and Ellis, 1985). Usually, the start of the biosynthesis of secondary metabolites is related to comparatively low levels of nitrogen in the medium.

The nitrogen source, introduced to the nutrient medium as ammonium and nitrate ions, is important for both growth and biosynthesis of RA in the cell culture of *L. vera* MM (Ilieva and Pavlov, 1999). On the other hand, in the case of *L. vera* MM plant cell culture the beginning of intensive biosynthesis of RA was correlated with the limitation of nitrate and ammonium ions in the medium. Consequently, if a smaller amount of ammonium ions is sufficient to ensure the onset of growth, it may be expected that the lag-period of biosynthesis of RA can be reduced without affecting the amount of biomass synthesized. The fuller and faster the consumption of nitrate ions, the sooner the level is reached for onset of RA biosynthesis.

The results obtained from cultivation of *L. vera* MM in nutrient media containing different initial concentrations of ammonium ions proved that their reduction really insured the intensive growth of the culture and intensive biosynthesis of RA (Figure 20.5). The reduction of ammonium ions to 0.25 of their amount in the control ensured intensive growth of the culture and a dry biomass of 16 g/L on day 6 was achieved (Figure 20.5a). The intensive biosynthesis of RA began earlier, that is, on day 2 instead of day 4 as in the standard LS medium (Figure 20.2a) and 230 mg RA/L were accumulated on day 6 (Figure 20.5b). During cultivation in a medium with 50 per cent of the ammonium ions, growth was more intensive from the very beginning and 18 g DB/L were synthesized by day 6 (Figure 20.5b). In this case, RA accumulation started on day 4, but was much more intensive and reached 250 mg/L by day 8. During cultivation in nutrient medium with a doubled concentration of ammonium ions, culture growth was slightly suppressed (Figure 20.5a), and the biosynthesis of RA was completely restrained (Figure 20.5b).

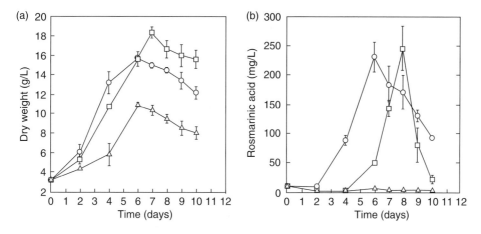

Figure 20.5 Time course of growth (a) and RA biosynthesis (b) by *L. vera* MM cell suspension culture cultivated in LS nutrient medium with different concentration of ammonium ions. (–○–) $0.25 \times NH_4$; (–□–) $0.5 \times NH_4$; (–△–) $2 \times NH_4$.

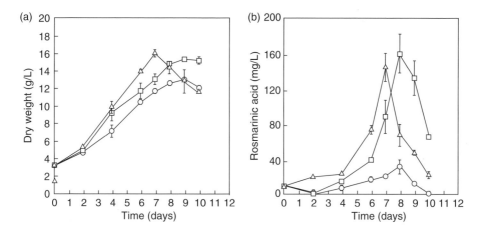

Figure 20.6 Time course of growth (a) and RA biosynthesis (b) by *L. vera* MM cell suspension culture cultivated in LS nutrient medium with different concentration of nitrate ions. (–○–) $0.8 \times NO_3$; (–□–) $1.2 \times NO_3$; (–△–) $1.5 \times NO_3$.

Reduction of the concentration of nitrate ions to 0.8 of the standard concentration in the medium did not affect the culture growth (Figure 20.6a), but the amount of RA decreased by about 50 per cent (Figure 20.6b). The 1.2-fold nitrate ion concentration led to a slight increase in the yield of cell biomass (Figure 20.6a) and the amount of RA synthesized was about 170 mg/L (Figure 20.6b), that is, 2.4 times as much as in the control (Figure 20.2a). The 1.5-fold concentration of nitrate ions in the nutrient medium (Figure 20.6) did not lead to a substantial increase in the cell yield and the maximum amount of RA synthesized was 150 mg/L (2.1 times as much as in the control).

Effect of phosphate concentration

Phosphorus is usually introduced into nutrient media in the form of phosphate. It is quickly consumed by plant cell suspensions and is stored in the vacuoles chiefly in terms of orthophosphates (Su, 1995). Phosphate is needed for the biosynthesis of nucleic acids, phospholipids and nucleotides, as well as for the phosphorylation of sucrose (Ashihara and Tokoro, 1985). The relationship between the concentration of phosphorus in the nutrient medium and the biosynthesis of RA varies for the different cell cultures (De-Ekmankul and Ellis, 1985; Getlowski and Petersen, 1993). It is known, however, that the biosynthesis of RA from all its cell producers is closely correlated with cell growth (De-Ekmankul and Ellis, 1985; Hyppolyte et al., 1992; Getlowski and Petersen, 1993). In such cases it is generally believed that the effect of phosphorus may come from its role in regulating the cell growth and the cell energetics (Su, 1995). That is why we have investigated the growth and biosynthesis of RA by *L. vera* MM cell suspension cultivated in nutrient media with different concentrations of phosphate (Ilieva and Pavlov, 1996). The results obtained are presented in Figure 20.7.

Reducing the concentration of phosphate to 1/2 of the concentration in the standard LS nutrient medium delayed the growth of *L. vera* MM (Figure 20.7a), although the amount of synthesized biomass from *L. vera* MM cell suspension was not much lower compared to the amount of biomass synthesized during control cultivation (Figure 20.2a). The lower concentration of phosphate also led to a lower rate of biosynthesis of RA (c. 38–40 mg/L versus 68 mg/L for control cultivation). The two-fold increase in the concentration of phosphate had a favorable effect on the growth of *L. vera* MM (17 g/L) (Figure 20.7), notwithstanding its delay at the beginning compared to the control cultivation. Doubling the concentration of phosphate led to a considerable increase in the amount of biosynthesized RA which is 140 mg/L at day 9 (Figure 20.7b), that is, twice as much as in the control (Figure 20.2a). The subsequent three-fold increase in the concentration of phosphorus in the nutrient medium had a favourable effect on the process of biosynthesis of cell biomass, but considerably suppressed the biosynthesis of RA (Figure 20.7).

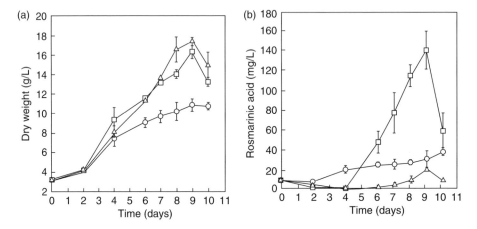

Figure 20.7 Time course of growth (a) and RA biosynthesis (b) by *L. vera* MM cell suspension culture cultivated in LS nutrient medium with different amounts of KH_2PO_4. (–○–) 85 mg/L – KH_2PO_4; (–□–) 340 mg/L – KH_2PO_4; (–△–) 510 mg/L – KH_2PO_4.

Optimization of the nutrient medium for RA production by L. vera MM

The results from the studies on the influence of macro nutrient elements on the growth and the RA production by *L. vera* MM, indicate that: the 8 per cent sucrose, 50 per cent of the standard concentration of ammonium ions, 1.2-fold of the standard nitrate concentrations and two-fold increase in the concentration of phosphate in the nutrient medium were found to have the most favourable effect both on the process of growth and biosynthesis of RA by the *L. vera* MM. Limitation the sucrose concentration was defined in the process of research with regard to some peculiarities of the cultivation of larger amounts of plant cell cultures, which is important for the following large-scale process. It was established that the increase in the sucrose concentration over 60 g/L was not expedient since that would lead to a high concentration of the biomass and consequently to impairment of the mass-exchange characteristics (Ilieva and Pavlov, 1997). This in turn led to a low efficiency of control and management of the biosynthesis process. So, in optimizing the LS nutrient medium, a full factor experiment ($n = 3$) was used at a constant concentration of sucrose in the nutrient medium – 60 g/L.

Polynomial regression models were elaborated to give a quantitative description of the processes of biosynthesis of RA (Y_1) and cell biomass (Y_2) as a result of the variation of the concentration of: $NH_4^+ - 0.09\,g/L \leq X_1 \leq 0.23\,g/L$; $NO_3^- - 2.44\,g/L \leq X_2 \leq 3.02\,g/L$ and $KH_2PO_4 - 0.170 \leq X_3 \leq 0.425\,g/L$.

$$\hat{Y}_1 = 739.3 - 218.4x_1 + 272.5x_2 - 165.1x_3 - 39.4x_1x_2 + 164.9x_1x_3 - 68.8x_2x_3,\ R_m = 0.9901$$
$$\hat{Y}_2 = 29.10 - 2.00x_1 + 1.08x_2 - 0.52x_3 - 0.69x_1x_2 + 0.96x_1x_3 - 0.87x_2x_3,\ R_m = 0.9643$$

Optimization procedures according to the modified simplex method allowed us to establish the optimal conditions for the biosynthesis of RA by *L. vera* MM – $X_1^* = 0.09\,g/L$; $X_2^* = 3.02\,g/L$ and $X_3^* = 0.170\,g/L$, where $Y_1^*\max = 1786.74\,mg/L$ (twenty-seven times higher compared with the cultivation in the standard LS medium) (Pavlov *et al.*, 2000). As a result, a modified nutrient medium (Table 20.3) was established for the cultivation of *L. vera* MM to achieve a maximum yield of RA.

Table 20.3 Modified nutrient medium for the production of RA from *L. vera* MM

Components	Concentration, g/L in	
	LS nutrient medium	Modified nutrient medium
Sucrose	30.00	60.00
NH_4NO_3	1.65	0.412
KNO_3	1.90	5.37
KH_2PO_4	0.17	0.17
$MgSO_4 \cdot 7H_2O$	0.37	0.37
$CaCl_2 \cdot 2H_2O$	0.44	0.44
$MnSO_4 \cdot 4H_2O$	0.0223	0.0223
KI	0.83×10^{-3}	0.83×10^{-3}
H_3BO_3	6.2×10^{-3}	6.2×10^{-3}
$ZnSO_4 \cdot 7H_2O$	8.6×10^{-3}	8.6×10^{-3}
$CuSO_4 \cdot 5H_2O$	0.025×10^{-3}	0.025×10^{-3}
$Na_2MoO_4 \cdot 2H_2O$	0.25×10^{-3}	0.25×10^{-3}
$CoCl_2 \cdot 6H_2O$	0.025×10^{-3}	0.025×10^{-3}
$FeSO_4 \cdot 7H_2O$	0.02786	0.02786
Na_2EDTA	0.03726	0.03726
Mesoinosite	0.1	0.1
Thiamin – HCl	0.4×10^{-3}	0.4×10^{-3}

Conclusion

The *L. vera* MM cell culture we obtained synthesized RA, which is well known to be one of the best natural antioxidants (Loliger, 1983). RA also has an antimicrobial, antiviral and anti-inflammatory effect (Parnham and Kesselring, 1985; Ilieva *et al.*, 1994; Lopez-Arnaldos *et al.*, 1997), which makes it a valuable product with potential for commercial applications. That gave us grounds to conduct a thorough investigation on the cell culture in order to determine its capacity for biosynthesis of RA. The results are definitely of technological significance and are a good basis for the scale-up of the process of biosynthesis of RA from plant cell suspension *L. vera* MM.

References

Ashihara, H. and Tokoro, T. (1985) Metabolic rate of inorganic phosphate absorbed by suspension cultured cells of *Catharanthus roseus. J. Plant Physiol.*, 118, 227–35.

Banthorpe, D.V., Bilyard, H.J., and Watson, D.G. (1985) Pigment formation by callus of *Lavandula angustifolia. Phytochemistry*, 24, 2677–80.

Calvo, M.C. and Segura, J. (1988) *In vitro* morphogenesis from explants of *Lavandula latifolia* and *Lavandula stoechas* seedlings. *Scientics Hort.*, 36, 131–7.

Calvo, M.C. and Segura, J. (1989) *In vitro* propargation of Lavender. *Hort. Sci.*, 24(2), 375–6.

Calvo, M.C. and Segura, J. (1989a) Plant regeneration from cultured leaves of *Lavandula latifolia* Medicus: Influence of growth regulators and illumination conditions. *Plant Cell Tissue and Organ Culture*, 19, 33–42.

Calvo, M.C., Jordan, A., and Segura, J. (1988a) Plant regeneration from isolated cells of *Lavandula latifolia* Medicus. *In Vitro Cellular and Developmental Biol.*, 24(9), 943–6.

Charlword, B.V., Hegarty, P.K., and Charlword, K.A. (1986) The synthesis and biotransformation of monoterpenes by plant cells in culture. In: PH. Morris, A.H. Scragg, A. Stafford, and M.W. Fowler (eds), *Secondary Metabolism in Plant Cell Cultures*, Cambridge: Cambridge University Press, pp. 15–34.

De-Eknamkul, W. and Ellis, B.E. (1985) Effects of macronutrients on growth and rosmarinic acid formation in the cell suspension cultures of *Anchusa officinalis. Plant Cell Reports*, 4, 46–9.

Dixon, R.A. (1985) Isolation and maintenance of callus and cell suspension cultures. In: R.A. Dixon (ed.), *Plant Cell Culture – A Practical Approach*, pp. 1–20. Oxford: IRL.

Do, CH.B. and Cormier, F. (1991) Effect of high ammonium concentrations on growth and anthocyanin formation of grape (*Vitis vinifera* L.) cell suspension cultured in a production medium. *Plant Cell Tissue and Organ Culture*, 27, 169–74.

Fukui, H., Yazaki, K., and Tabata, M. (1984) Two phenolic acids from *Lithospermum erythrorhizon* cell suspension cultures. *Phytochemistry*, 23, 2398–9.

Gertlowski, C. and Petersen, M. (1993) Influence of the carbon source on growth and rosmarinic acid production in suspension cultures of *Coleus blumei. Plant Cell Tissue and Organ Culture*, 34, 183–90.

Hirasuna, T.J., Shuler, M.L., Lackney, V.K., and Spaiswick, R.M. (1991) Enhanced anthocyanin production in grape cell cultures. *Plant Sci.*, 78, 107–20.

Hyppolyte, L., Marin, B., Baccen, J.C., and Jonard, R. (1992) Growth and rosmarinic acid production in cell suspension cultures of *Salvia officinalis* L. *Plant Cell Reports*, 11, 109–12.

Ilieva, M. and Pavlov, A. (1996) Rosmarinic acid by *Lavandula vera MM*, cell suspension: Phosphorus effect. *Biotechnol. Lett.*, 18(8), 913–16.

Ilieva, M. and Pavlov, A. (1997) Rosmarinic acid production by *Lavandula vera MM* cell suspension culture. *Appl. Microbiol. Biotechnol.*, 47, 683–8.

Ilieva, M. and Pavlov, A. (1999) Rosmarinic acid production by *Lavandula vera MM* cell suspension culture: nitrogen effect. *World Journal of Microbiology and Biotechnology*, 15(6), 671–5.

Ilieva, M., Kozhuharova, L., Pavlov, A., and Kovatcheva, E. (1994) Antimicrobial activity of phenolic compounds from cell culture of *L. vera MM*. *Proceedings of the International Euro Food Tox IV Conference-Bioactive*

Substances in Food of Plant Origin, II. Centre for Agrotechnology and Veterinary Sciences, Polish Academy of Sciences, Olztyn, Poland. pp. 172–7.

Ilieva, M., Pavlov, A., Kovatcheva, E., and Mihneva, M. (1995) Growth and phenolics production of cell suspension culture of Lavandula vera MM, Biotechnol. Biotechnol Eq., 9(4), 27–9.

Ivanov, A.I. and Nikolov, S.D. (1988) Pharmacognosy, pp. 295. Sofia: Meditsina i Physkultura.

Jordan, A.M., Calvo, M.C., and Segura, J. (1990) Morphogenesis of callus and single-cell cultures of Lavandula latifolia Medicus – Journal of Horticultural Science, 65(1), 49–53.

Kovatcheva, E., Pavlov, A., Koleva, I., Ilieva, M., and Mihneva, M. (1996) Rosmarinic acid from Lavandula vera MM cell culture. Phytochemistry, 43(6), 1243–4.

Lappin, G.J., Stride, J.D., and Tampian, J. (1987) Biotransformation of monoterpenoids by suspension cultures of Lavandula angustifolia. Phytochemistry, 26(4), 995–7.

Lappin, G.J., Tampion, J., and Stride, J.D. (1986) The biotransformation of monoterpenoides by plant tissue culture. In PH. Morris, A.H. Scragg, A. Stafford, and M.W. Fowler (eds), Secondary Metabolism in Plant Cell Cultures. Cambridge: Cambridge University Press, pp. 113–6.

Loliger, J. (1983) Natural antioxidants. In: Rancidity in Foods. J.C. Allen and Hamilton R.J. (eds), p. 100. London: Appl. Sci.

Lopez-Arnaldos, T., Zapata, J.M., Calderon, A.A., and Ros Barcelo, A. (1997) Antioxidant activity of lavandin (Lavandula x intermedia) cell cultures in relation to their rosmarinic acid content, ACS SYMPOSIUM SERIES 660, Spices-Flavor Chemistry and Antioxidant Properties, pp. 206–18.

Jordan, A.M., Calvo, M.C., and Segura, J. (1990) Morphogenesis of callus and single-cell cultures of Lavandula latifolia Medicus. J. Horti. Sci., 65(1), 49–53.

Nakajima, H., Sonomato, K., Morikana, H., Sato, F., Ichimura, K., Yamada, Y., and Tanaka, A. (1986) Entrapment of Lavandula vera cells with synthetic resin pre-polymers and its application to pigment production. Appl. Microbiol. Biotechnol., 24, 266–70.

Nakajima, H., Sonomato, K., Sato, F., Ichimura, K., Yamada, Y., and Tanaka, A. (1989) Influence of carbon source on pigment production by immobilized cultured cells of Lavandula vera. J Fermentation Bioengi., 68(5), 330–3.

Parnham, M.J. and Kesselring, K. (1985) Rosmarinic acid. Drugs of the Future, 10, 756–7.

Pavlov, A., Ilieva, M., and Panchev, I. (2000) Nutrient medium optimization for rosmarinic acid production by Lavandula vera MM cell suspension. Biotechnol. Prog., 16, 668–70.

Quazi, M.H. (1980) In vitro multiplication of Lavandula spp. Amer. Bot., 45, 361–2.

Srinivasan, V. and Ryn, D.D. (1993) Improvement of shikonin productivity in L. eryrhrorison cell culture by alternating carbon and nitrogen feeding strategy. Biotechnol. Bioengi., 42, 793–9.

Su, W.W. (1995) Bioprocessing technology for plant cell suspension cultures. Appl. Biochem. Biotechnol., 50, 189–229.

Tanaka, A. (1989) Cultured plant cells as the catalysts for bioreactor. In: Fiechter, Okada, Tanner (eds), Bioproducts and Bioprocesses, pp. 3–12, Berlin: Springer-Verlag.

Tsuro, M. and Inorie, M. (1996) Production of blue pigment in leaf-derived callus of Lavender (Lavandula vera DC). Breeding Sci., 46, 361–6.

Watanabe, K. and Yamada, Y. (1982) Selection of variants with high levels of biotin from cultured green Lavandula vera cells irradiated with gamma rays. Plant & Cell Physiol., 23(8), 1456–8.

Watanabe, K. and Yamada, Y. (1982a) Selection of high biotin producing strains of cultured Lavandula vera cells. Proceedings of the 5th International Congress on Plant Tissues Cell Culture (Plant Tissue culture), pp. 357–8.

Watanabe, K., Jano, S.I., and Yamada, Y. (1982) The selection of cultured plant cell lines, producing high levels of biotin. Phytochemistry, 21(3), 513–6.

Watanabe, K., Mitsuda, H., and Yamada, Y. (1983) Retention of Lavandula vera callus after freeze-preservation. Plant & Cell Physiol., 24(1), 119–22.

Watanabe, K., Yamada, Y., Ueno, S., and Mitsuda, H. (1985a) Change of freezing resistance and retention of metabolic and differentiation potentials in cultured green Lavandula vera cells which survived repeated freeze-than procedures. Agric. Biol. Chem., 49(6), 1727–31.

Watanabe, K., Sato, F., Furuta, M., and Yamada, Y. (1985) Induction of pigment production by S-containing components in cultured *Lavandula vera* cells. *Agric. Biol. Chem.*, 49(2), 533–4.

Webb, J.K., Banthorpe, D.V., and Watsen, D.G. (1984) Monoterpene synthesis in shoots regenerated from callus cultures. *Phytochemistry*, 23(4), 903–4.

Ulbrich, B., Wiensner, W., and Arens, H. (1985) Large scale production of rosmarinic acid from plant cell cultures of Coleus blumei Benth. In: K.H. Neumann, W. Barz, E. Reinhard (eds), *Primary and Secondary Metabolism of Plant Cell Cultures*, pp. 293–303. Berlin: Springer-Verlag.

Zenk, M.N., EL-Shigi, H., and Ulbrich, B. (1977) Production of rosmarinic acid by cell suspension cultures of *Coleus blumei*. *Naturwissenschaften*, 64, 585–6.

21 *Lavandula* x *heterophylla* and *L.* x *allardii*

A puzzling complex

Susyn Andrews

Two lavender hybrids, *L.* x *heterophylla* and *L.* x *allardii*, which have long been mentioned in the literature and are often thought to be related, if not one and the same. A historical review of the two hybrids will hopefully put the situation into perspective.

Time chart of *L.* x *heterophylla* poiret

1802   *L. heterophylla* was first mentioned by Viviani, an Italian botanist at Genoa, who obtained this plant from the Pavia Botanic Garden in northern Italy.

1804   Desfontaines listed a *L. heterophylla* among the plants growing at the School of Botany at the Jardin des Plantes in Paris.

1806   Persoon (Pers.) described *L. heterophylla* as the plant mentioned by Viviani.

1808   *L. heterophylla* Pers. is listed in the Appendix of a catalogue of plants at the Jardin des Plantes.

1813   Poiret, who wrote up the account of *Lavandula* in the botanical encyclopaedia, described *L. heterophylla* as a hybrid which was cultivated in the Jardin des Plantes. He was obviously referring to the plant that Desfontaines listed. Poiret also mentioned another species (*heterophylla*) which was unknown to him, and this was growing in the Dinegro Garden in Genoa. This was probably Viviani's plant.

Balbis (1813), a physician and botanist in Turin, northern Italy, listed *L. heterophylla* Pers. among the list of plants cultivated at the Botanic Garden there. Again it must be the Viviani plant.

1826   Gingins de la Sarraz (1793–1863), a Swiss historian and botanist, in his Histoire Naturelle des Lavandes, divided *L. heterophylla* into two varieties. Variety 'a' is the plant grown in the Jardin des Plantes in Paris, which was said to be a vigorous hybrid between *L. dentata* and *L. vera*. This had an interrupted spike and greenish glabrous leaves. Variety 'b' is the lavender found in the gardens of northern Italy, and said to be a less vigorous hybrid between *L. dentata* and *L. spica*. This had a contracted spike and broad tomentose leaves.

1937   Chaytor, in her treatment of the taxonomy of *Lavandula* agreed with de la Sarraz's (1826) findings. She inferred that he had suggested that his variety 'a' had the parentage *L. dentata* × *L. angustifolia*, while variety 'b' was *L. dentata* × *L. latifolia*.

Collections of variety 'b' have been seen by Andrews at the British Museum of Natural History in London and at Kew Gardens. These are all from Viviani's plant. The grey, tomentose leaves are densely crowded at the base of the branch and the spike is contracted and dense.

Collections of variety 'a' remained impossible to locate until a visit to the Paris Herbarium in January 1993 revealed an unusual collection comprising two sheets from the herbarium of

Monsieur A.N. Desvaux (1784–1850) who was a former director of Angers Botanic Gardens. The label (written by Desfontaines) clearly states *L. x heterophylla* Desf. and also gives the synonym *L. hybrida* Willd.

When I looked at the microfiches of the Willdenow (1765–1812) herbarium, which is held at the Botanical Museum of Berlin-Dahlem, I found that No. 10758 was labelled *L. hybrida* Wiild. Unfortunately, the rest of the writing is obscure and so I will have to see the original specimen sometime. The two herbal sheets of Desvaux are now on loan to Kew Gardens.

One of the sheets of Herb. A.N. Desvaux var. 'a' shows some smaller branches similar to the Viviani plant (var. 'b'), but there are also some with longer inflorescences as well and the leaves are also contracted at the base. The other sheet shows another part of the same plant with the long thin leaves typical of *L. x allardii*.

Time chart of *L. x allardii* Hy

1890 A hybrid lavender appeared in the garden of Maulevrie near Angers in France. Both *L. dentata* and *L. latifolia* were growing in this garden.

1895 Monsieur F.C., Hy (1853–1918), a French clergyman and cryptogamist as well as a professor at the Catholic University of Angers described this new hybrid. Hy based his description on his own collections of the plant, which he gathered at Maulevrie in July and September 1894. He agreed that *L. dentata* and *L. latifolia* were the parents of this hybrid.

1896 Hy collected more material in August 1896.

1937 Chaytor (1937) described *L. x allardii* as differing from *L. x heterophylla* in its greater height, in that the leaves are not crowded together at the base of the branches, in the thin, often entire leaves and long internodes, and in the interrupted more slender spikes. However, she thought that Hy's specimens agreed more closely with the original description of *L. x heterophylla* and variety 'a', while the parentage of *L. allardii* as given by Hy agreed with variety 'b'.

1992 Herbarium specimens of Hy 422 and Hy s,n, have been seen by the author (Andrews) in the Paris Herbarium and in the British Museum of Natural History, while Hy 4015 was only seen in Paris and more recently in the Herbarium of the Conservatoire et Jardin Botanique in Geneva. All these specimens are now on loan to Kew for further study.

Preliminary conclusions

Based on bracteoles

When one is classifying lavenders, the bracteoles are an important and stable character. *L. dentata* does not have bracteoles, while those of *L. angustifolia* can hardly be seen by the naked eye. Those of *L. latifolia* are long and linear and the bracteoles of both *L. x allardii* and *L.* x *heterophylla* match those of *L. latifolia*.

Thus it would be possible that *L. x allardii* and *L. x heterophylla* have the same parents, namely *L. dentata* × *L. latifolia*. If this is correct, then the epithet *heterophylla* will have to take precedence over *allardii*, as the former is an earlier name.

Based on further samples of L. x heterophylla

Apart from the type specimens of *L. x heterophylla* and *L. x allardii* mentioned in the charts, I have not seen any recent herbarium material of the former, while material of the latter has been available from the Royal Horticultural Society (RHS) at Wisley, South Africa and Kenya. The RHS material came from a Mrs Walter K. Howard of Wayne, Pennsylvania, United States as *L. heterophylla* in 1982 and was planted against the walls of the laboratories in Wisley, but did not survive. DeBaggio (1991) talked about *L. heterophylla* in the United States as a 1.5 × 1 m plant with lusty grey-green foliage and a pleasant, assertive fragrance and further details could prove most enlightening. The question is when and from where did this come into the United States? It would be interesting to see more specimens from there. I have yet to see living material of typical *L. x allardii* or *L. x heterophylla*.

However, it is worth distinguishing two clones of *L. allardii*. Clone A is found in South Africa, while clone B is widespread in Australia. Both are more vigorous than the typical *L. x allardii* of Hy and I have seen living plants in the National Collections at Norfolk Lavender and Jersey Lavender. David Christie, the owner of the latter collection obtained his material of Clone A from relatives in South Africa and he passed cuttings onto Henry Head of Norfolk Lavender.

'African Pride' or formerly clone A

This is a resinous plant growing to 1.2 to 1.5 × 1.2 m, but plants 2–4 m high have been reported from South Africa. The stems and the two distinct types of leaves (entire or somewhat toothed at

Figure 21.1 Lavandula x allardii 'African Pride'.

the apex) are grey-green, while the broadly ovate, tomentose bracts and long, linear bracteoles are a greenish-purple. The flower spikes are long and narrow, with a purplish-green calyx and violet-purple flowers which are 90C/D Violet-Purple Group on the RHS Colour Chart, 1989.

Four South African specimens were collected by Yvette van Wijk in $18 = 983 - 84$ from nurseries and gardens in the Grahamstown area, eastern Cape. There it was recognised as having a very strong scent for drying. Recently, while working at the herbarium of the Arnold Arboretum in Boston, USA, I came across a specimen collected outside Nairobi in 1976 at the former nursery of Peter Greensmith. On a recent visit to the site, now the headquarters of IUCN, two plants had recently been transplanted out from pots lurking in a corner for years.

As this clone is so distinct, I have decided to call it 'African Pride' as it appears to be more widespread on southern Africa (Figure 21.1). Herbarium specimens can be seen at the National Botanical Institute, Pretoria and at the Royal Botanic Gardens, Kew.

Clone B

There is another *L. x allardii* clone in Australia which is misleadingly known as the Mitcham lavender (Figure 21.2). This grows to 1.5×1.3 m and has greenish-grey foliage which is more

Figure 21.2 Lavandula x allardii 'Clone B'.

toothed than 'African Pride'. The long, slender flower spikes have a greyish calyx and the bright purple flowers are 87A Violet Group on the RHS Colour Chart 1989.

On a recent visit to Australia, Joan Head (editor of the *Lavender Bag*) visited the National Herbarium in Melbourne and very kindly made xeroxes of all their material of the clone. According to the notes on the herbarium labels, this is a very fragrant lavender and the leaves are full of essential oil. It is robust and quick growing, and in spite of being cut back each year, will still grow to 1.5 m tall in a season (in Victoria). The specimen R.D. Spencer 5295 was described as an 'exceptionally robust "monster" that can be reproduced reliably in cultivation of cuttings'.

Note: The name Mitcham lavender is a mystery, as though lavender was grown in Mitcham near London, UK there was no particular 'mitcham' type. The Clone B could not survive the English climate if grown outside.

Acknowledgements

Grateful thanks to David Chrietie, Ray Harley, Henry Head, Joan Head, Graham Pattison and Brian Schrire. Part of this article first appeared in *The Lavender Bag* in November 1994, vol. 2, 8–14.

References

Balbis, J.B. (1813) *Catalogus stirpium horti academici taurinensis*, p. 46.
Chaytor, D.A. (1937) A taxonomic study of the genus *Lavandula*. *Linn. Soc. J. Bot.*, 51, 170–1.
DeBaggio, T. (1991) Tender lavenders. *The Herb Companion*, pp. 52–7.
Desfontaines, M. (1804) *Tableau de l'ecole de botanique du Museum d'Histoire Naturelle*, 2nd edn, p. 59.
Desfontaines, M. (1815) *Tableau de l'ecole de botanique du Jardin du Roi.*, 2nd edn, p. 71.
Desfontaines, R. (1829) *Catalogue Plantarum horti regii parisiensis*, 3rd edn, p. 98.
Festing, S. (1989) *The Story of Lavender*, 2nd revised edn, Heritage in Sutton Leisure, Sutton, pp. 1–132.
Gingis de la Sarraz, F.C.J. (1826) *Natural History of the Lavenders*, Translated from the French, 1927, Boston. p. 63.
Hy, F. (1895) *Lavandula allardii* Hy, *Bull. Herb. Boiss.* 3, Appendix 1, 16–17.
McLeod, J.A. (1989) *Lavender, Sweet Lavender*, Kangaroo Press, Australia, pp. 1–120.
Mirabel, M. (1810) Memoire sur l'Anatomie et la Physiologie des plantes de la familie des Labiees, *Ann. Mus. His. Nat.*, 50, 13–162, plate 15, fig. 5.
Persoom, C.H. (1806) *Synopsis Plantarum*, 2, 116.
Poiret, J.L.M. (1813) Lavande. *Encyclopedie Methodique Botanique.* Suppl. Lll, pp. 308–9.
Royal Horticultural Society and the Flower Council of Holland (1989), *RHS Colour Chart*.
Viviani, D. (1802) Elenchus plantarum. *Hor. Botanici J. Car. Dinegro.*, p. 23.

22 Comparative study of essential oil quantity and composition from ten cultivars of organically grown lavender and lavandin

Denys J. Charles, Erica N.C. Renaud and James E. Simon

Introduction

Organic agriculture is an ecological production management system that promotes and enhances biodiversity, biological cycles and soil biological activity. It is based on minimal use of off-farm inputs and the promotion of management practices that restore, maintain and enhance ecological harmony. Organic agriculture uses naturally occurring products such as compost, animal manure and crop residue and also natural minerals. Weed and insect pests are controlled using botanical, biological and cultural methods. This method reduces air, soil and water pollution thereby increasing soil fertility, balancing insect populations and enhancing ecological harmony. Fields must be free from synthetic fertilizers and pesticides for a minimum of 3 years. The primary goal of organic agriculture is to optimize the health and productivity of interdependent communities of soil life, plants, animals and people. The holistic vision includes the maintenance of valuable relationships between soil, water, air, plants, animals and people.

Organic fertilizers

Organic farmers use compost or manure to replenish the soil with minerals, as the soil is rich in beneficial soil microorganisms, which in turn slowly and steadily make minerals available to plants. Crop rotation is another method used in organic farming to maintain the soil nutrients without the use of synthetic fertilizers. A field is used to grow crops like wheat or corn for a period of several years followed by a season of legumes, which help to incorporate nitrogen from the air into the soil through the bacteria in their roots. Soil conservation is strictly observed by using cover crops or practicing low till farming or leaving unwanted portions thus preventing soil erosion.

Organic pest control

Organic farming uses a variety of methods to control fungi and harmful insects. Farmers often use the method of intercropping. This is done by planting different crops in alternating rows, thereby interrupting the movement of disease-causing organisms through a field. Sometimes crops are sprayed with bacteria, which then destroy the larvae of harmful insects.

Organic farmers also use pesticides primarily derived from chemically-unaltered plant, animal or mineral substance in which the active ingredient becomes non-toxic after being applied to the crops. For example, pyrethrum extracted from chrysanthemums and oil extracted from neem trees is being widely used.

Organic weed control

In organic farming weed control is mainly done through mulching to smother weeds and by planting cover crops like cereal rye and oats that either inhibit weed seed germination or deprive them of the nutrients they need to grow. Sometimes tractor drawn equipment is also used to uproot weeds.

Ecological balance

Organic farming represents long-term savings and also maintains ecological balance and harmony. The crops are free from synthetic, toxic chemicals and pesticides, being pure and natural and often believed to be more nutritious by organic farmers. Organic farming also preserves top soil so more crops can be grown in future without polluting the environment; also saving on fuels, pesticides and fertilizers make it a more attractive proposition for many people than conventional farming.

Certified organic

'*Certified organic*' refers to agricultural products that have been grown and processed according to strict uniform standards, verified annually by independent state or private organization. Certification includes inspection of farm fields and processing facilities. Farm practices which are inspected include: long-term soil management, buffering between organic farms and any neighboring conventional farms, product labeling, and record keeping. Processing inspections include: review of the facility's cleaning and pest control methods, ingredient transportation and storage, and record keeping and audit control.

The phrase 'certified organic' needs to be understood from the point of origin: organic foods are products using organic agriculture. Engaging in organic agriculture requires a different approach from that used in conventional modern agriculture. In organics, the focus is not so much on pushing the resources to produce tremendous yields and profits. Instead, organics foster the ecological processes that produce resources and add value to the resulting products. There is the added value of the designation, as well as the assurance that the products' consumers are buying not just chemical free products, but that these products are good for them, as well as good for the environment. Organic products help minimize the adverse effects that agriculture can have on soil, water and air.

Organic growing is a challenging, more detailed type of agriculture that requires the farmer to take responsibility for and make a commitment to the land even while pursuing the traditional business goal of making a profit. This commitment is embodied by the organic farming practices of weed, pest and disease control.

Introduction to lavender crops

Lavender (*Lavandula angustifolia* Mill.), a perennial shrub and a member of the Lamiaceae (Labiatae) family are native to southern Europe and the Mediterranean area. Lavender has a long history of cultivation worldwide, but commercial cultivation in the United States is a relatively recent phenomenon. Lavenders flower in summer with many small fragrant flowers clustered in spikes that extend above the green or gray-green narrow lanceolate or toothed leaves. Lavender is one of the most loved and best known of all herbs, and the wide range of lavender products has grown in appeal to many people.

Lavender grows in full sun on dry, well-drained, stony calcareous soils with a soil pH of 6–8 (Hortus Third, 1976). Lavender is cultivated primarily for its aromatic essential oils though its fresh and dried flowers are also marketed. Cultivars of lavender, lavandin and spike lavender are the most commonly grown for commercial production. While lavender plantings can last as long as 30 years (Simon et al., 1971–80), best commercial yields are reported to be derived from 3-and 4-year-old fields. *L. angustifolia* cultivars flower earlier than *L. x intermedia* cultivars, at the beginning of summer, and if pruned lightly immediately following the first flowering, they will often flower a second time before autumn.

Lavender essential oils

Oil production differs for 'true' lavender *(L. angustifolia* Mill.*)* and lavandin *(L. x intermedia* Emeric ex loisel.*)*, a sterile hybrid of lavender *(L. angustifolia)* and spike lavender *(L. latifolia)* (Tucker et al., 1984). Lavender oils contain more than 100 compounds, with the two major constituents being linalool and linalyl acetate (Lawrence, 1994). Other constituents include α-thujene, α-pinene, camphene, sabinene, β-pinene, myrcene, p-cymene, limonene, 1,8-cineole, (Z)- and (E)-β-ocimene, gamma-terpinene, camphor, terpinen-4-ol, lavandulol, lavandulyl acetate, β-caryophyllene etc. *L. angustifolia* is reported to produce the highest quality oil, while lavandin the highest yields, but the latter is considered lower in quality due to its relatively higher camphor and lower linalyl acetate contents (Lawrence, 1994). Because lavandin is such a high yielding oil plant, oil of lavandin is used to adulterate true lavender oil.

Research aims

The goal of this research was to identify sources of lavender and lavandin that produce high quality essential oils from organically field-grown plants in the Midwest, United States. Specifically, we sought to: (1) comparatively evaluate four cultivars of lavandin *(L. x intermedia)* and six cultivars of lavender *(L. angustifolia)* for growth, yield and essential oil yield and quality; (2) assess the uniformity of each source; (3) determine which cultivars could be grown for the dried floral market; and (4) use the pure essential oils from our field grown plants to assess quality and as reference standards.

Experimental conditions

The plants

Seedlings were sourced from three commercial herb companies: Companion Plants (Athens, Ohio); Richter's Seeds (Goodwood, Ontario); Goodwin Creek Gardens (Williams, Oregon). Voucher specimens of each cultivar have been collected and one set was kept at the Frontier Organic Research Farm and a second set deposited at the Purdue University Herbarium. Cultivar names as purchased and marketed in the US horticultural industry are presented. Cultivar identities were in part confirmed by Art Tucker, Delaware State College (personal communication).

The site

The study was conducted at the Frontier Organic Research Farm, Norway, Iowa. The farm is certified organic under Quality Assurance International (QAI). The site is a moderately well drained Dinsdale silty clay loam soil with a 2–5 per cent slope. The soil is gently sloping and moderately well drained. Permeability is moderate with medium surface runoff. Available water capacity is high and organic matter is 3–4 per cent. Reaction is generally slightly acid or medium acid in the surface layer. The subsoil is low in available phosphorus and very low in available potassium. Soil tests of the experimental site were performed and the soil pH was 7.4 at the outset of the study.

Field design

A covercrop of oats *(Avena sativa)* was cut (April 30, 1998) and rototilled (May 15, 1998) into the beds to prepare the site. On 16 June 1998, twenty-four plants (about 15 cm tall) of each lavender and lavandin cultivar were transplanted into raised beds in a randomized block design (RBD). The design was replicated three times with eight plants in each block at 46 cm spacing, 0.61 m between rows. Plants were monitored for plant vigour, plant uniformity, date of flowering, plant height at flower, and the presence of disease or insects. Flowering spikes were harvested at 50 per cent bloom and dates recorded. Freshly harvested flowers were dried in a forced air dryer for 48 h at 40°C. Dry, coded flower samples, which could not be traced to any specific variety or lavender source, were submitted to Frontier's quality assurance labs for visual assessment and organoleptic evaluation. Each sample was scored for color, aroma, flavour, and flower size (0–5 scale) relative to market acceptability. Essential oils, derived from each sample were assessed for yield and chemical composition.

Essential oil isolation

Essential oil was isolated by hydro distillation and steam distillation. For hydro distillation, 75 g of dried flower was placed in a 2000 ml round bottom flask with 1000 ml distilled-deionized water and the essential oil extracted for 1 h using a modified Clevenger trap (ASTA, 1968). The essential oil content was determined on an oil volume to tissue weight basis. Essential oil samples were stored in Varian autosampler vials at 2°C in the dark. For steam distillation, 10 lb partially dried flowers were placed in the glass percolator. Steam was passed through the packed flowers and allowed to condense into a Clevenger trap. The distillation was run for 3 h to completion. The essential oil content was determined on an oil volume to tissue weight basis (Charles and Simon, 1990a). Essential oil samples were stored in Varian autosampler vials at 2°C in the dark.

Essential oil composition and quantification by gas chromatography

Hydro distilled and steam distilled essential oil samples were analyzed by gas chromatography using a Varian 3400 CX gas chromatograph equipped with FID and a Varian Workstation (Varian, Walnut Creek, CA). A fused silica capillary column (30 m × 0.22 mm i.d.) with a DB-5 bonded phase (J & W, polydimethylsiloxane) was used to separate the oil constituents. A 1-µl oil sample was injected with helium as the carrier gas and 100 : 1 split-vent ratio. The oven temperature was held isothermal at 75°C for 2 min and then programmed to increase at 3°C/min to 200°C to give complete elution of all peaks. The injector and detector temperatures were 250 and 300°C, respectively. The relative peak area for individual constituents was determined by a Varian Star Workstation.

Refractive index, optical rotation and specific gravity measurements

Refractive Index was measured using a Leica ABBE Mark II Refractometer. Optical rotation was measured using a Perkin Elmer Polarimeter 343. Specific gravity was measured using a Mettler Toledo DA-110M Specific gravity meter.

Gas chromatography/mass spectroscopy analysis

Pure compounds and essential oil constituents were verified by GC/MS. A Finnigan (San Jose, CA) GC and MS (4000) hooked on-line to a Data General Nova/4 data processing system, used electron impact analysis as previously described (Charles and Simon, 1990a; Charles et al., 1990b; Charles et al., 1995). The GC conditions were as follows: direct injection of 1-μl sample diluted 10:1 with MeOH; fused silica column (30 × 0.25 mm) with DB-1 bonded phase (polydimethylsiloxane) (J & W Scientific, Folsom, CA); helium as a carrier gas with a column pressure of 72.2 kPa (10.5 psi) and split vent of 40 mL/min; oven programme, 80°C at 2 min rising to 180°C at 2°C/min; injector temperature, 225°C. The MS conditions were as follows: ionization voltage, 70 eV; emission current, 40 μA; scan rate, 1 scan/s; mass range, 40–500 Da; ion source temperature, 160°C.

GC-chiral analysis

Column-30 m, 0.32 mm i.d., 0.25 μm Rt-βDEX-se; carrier gas-Helium at 80 cm/sec; FID detector at 250°C; oven temperature-40°C for 1 min and then programmed to rise at 2°C/min to 200°C and hold for 3 min. Enantiomeric pure linalool and linalyl acetate were injected to verify peaks.

Radiocarbon (^{14}C) and stable isotope ratio (^{13}C and D) analysis

This analysis was performed by Randy Culp at Center for Applied Isotope Studies, University of Georgia, Athens, Georgia.

Identification of the compounds

The identification of essential oil constituents was accomplished by retention time and co-injection with authentic standards wherever possible and matching the mass spectra of each compound with standards or different MS compound libraries for best fit (Finnigan; Keller and Milne, 1978; Stenhagen et al., 1987; Adams, 1989).

Results

Flower production

Of the ten cultivars, only seven produced sufficient flowers in the first year to harvest and evaluate. 'Grey Lady' was the only cultivar that produced no flowers in year one. The lavandin cultivars 'Abriallii', 'Grosso' and 'Super' were the earliest to reach full bloom (30 July 1998), while the lavender cultivars 'Hidcote', 'Lady' and 'Munstead' developed flowers four weeks later. Lavandin 'Super' produced the tallest plant at bloom, 41 cm; while lavender cultivars 'Hidcote' and 'Lady'

Table 22.1 Growth and development of ten cultivars of organically grown lavandin and lavender (1999) in Norway, Iowa

Lavandin and lavender cultivars	Plant source	Date of bloom	Plant height (cm)	Flowering plants (%)	Plant vigour[a]	Plant uniformity[b]	Winter survival (%)
Lavandin (L. x intermedia)							
Abriallii	Goodwin Creek	1 July	53.7 ± 1.5	94	4.3 ± 0.6	3.7 ± 0.6	100
Grosso	Richter's	7 July	57.0 ± 5.3	80	3.3 ± 1.5	3.0 ± 1.7	84
Provence	Richter's	7 July	51.0 ± 4.0	50	3.0 ± 0.0	2.0 ± 0.0	80
Super	Goodwin Creek	1 July	67.0 ± 9.9	100	4.7 ± 0.6	4.3 ± 0.6	100
Lavender (L. angustifolia)							
Alba	Companion plants	27 June	47.5 ± 2.3	88	3.7 ± 0.6	3.7 ± 0.6	100
English	Richter's	25 June	42.7 ± 1.5	96	3.0 ± 1.0	2.7 ± 0.6	96
Grey Lady	Companion plants	25 June	41.7 ± 3.1	71	2.0 ± 1.0	2.3 ± 1.5	100
Hidcote	Richter's	1 July	37.0 ± 2.0	74	3.0 ± 0.0	2.3 ± 0.6	92
Lady	Richter's	1 July	25.8 ± 4.4	97	1.3 ± 0.6	2.0 ± 1.0	60
Munstead	Richter's	22 June	41.8 ± 1.2	81	3.3 ± 0.6	2.7 ± 1.2	96

Notes
a Plant vigour, 1 = no vigour, little vigour; 2 = poor, very slow growth; 3 = moderate growth; 4 = vigorous growth; 5 = very vigorous growth.
b Plant uniformity, 1 = lacks uniformity; 2 = low uniformity; 3 = some uniformity; 4 = highly uniform; 5 = totally uniform.

produced the shortest plants, 29 cm. Lavender 'Alba' produced plants that rated above average in both vigour and uniformity. Lavandin 'Super' produced the highest quantity of EO (10 per cent dry floral weight), while the lavenders 'Hidcote' and 'Munstead' produced the lowest (2.3 per cent) (Table 22.1). All cultivars had acceptable colour and aroma, but the calyx size of the lavenders 'Hidcote', 'Lady' and 'Munstead' were unacceptably small (Table 22.2).

In year two, four of the six lavender cultivars were the earliest to flower ('Munstead', 'Grey Lady', 'English' and 'Alba', June 22–27, 1999), 7–14 days earlier than the lavandin cultivars. Lavandin 'Super' produced the tallest plants again in year two (65 cm at harvest) and was the cultivar in the study that exhibited 100 per cent winter survival and bloom from all the plants (Table 22.1). 'Super' also produced the most vigorous and uniform plants. Lavender 'Lady' produced the least vigorous and least uniform plants. Highest oil yields came from the lavandin 'Grosso' (9.9 per cent dry floral weight), a 3.65 per cent increase from year one. Lavender 'Lady' averaged the lowest oil yield across both years (Table 22.2). In year two, most lavandin cultivars produced market-acceptable flowers for the dried market. Among the lavenders, however, only 'Hidcote', 'Munstead' and 'Alba' were acceptable, and 'Hidcote' rated the highest in colour, aroma and size. Winter survival under the cold, high winds of the Midwest varied among cultivars, with the cultivars 'Abriallii', 'Super', 'Alba' and 'Grey Lady' exhibiting 100 per cent. Lavender 'Lady' had the lowest winter survival of 60 per cent.

Essential oil production

The percentages of linalool among the varieties compared well to published results (Lawrence, 1994) (Table 22.3). The camphor content (0.1–0.7 per cent) in the lavender cultivars and 3.5–8.1 per cent in the lavandin cultivars fell within the reported range. However, the linalyl acetate content (max. 16 per cent, year one and 18 per cent, year two) was well below the level of 25 per cent set by International Standards. Whether this could be due to altitude differences,

Table 22.2 Essential oil and quality control evaluation of ten cultivars of organically grown lavandin and lavender over 2 years in Norway, Iowa

Lavandin and lavender cultivars	Essential oil (% dry inflorescence)		Flower quality assessment					
			Color[a]		Aroma[b]		Size[c]	
	Y1	Y2	Y1	Y2	Y1	Y2	Y1	Y2
Lavandin (L. x intermedia)								
Abriallii	8.72	9.0 ± 0.7	3	4	4	5	4	3
Grosso	6.25	9.9 ± 0.2	3	4	4	5	3	3
Provence	—	7.1 ± 0.5	—	1	—	5	—	3
Super	10.00	8.7 ± 0.7	3	3	4	4	3	4
Lavender (L. angustifolia)								
Alba	3.85	3.3 ± 0.2	4	4	5	4	3	3
English	—	3.2 ± 0.5	—	3	—	4	—	3
Grey Lady	—	5.0 ± 0.7	—	1	—	3	—	4
Hidcote	2.30	3.5 ± 0.2	5	5	5	5	2	5
Lady	2.70	2.8 ± 0.0	5	4	5	4	1	1
Munstead	2.30	3.6 ± 0.6	5	4	5	5	2	4

Notes
a Flower colour, scale 1–5: 1 = poor colour, unmarketable; 3 = acceptable colour; 5 = highest quality.
b Flower aroma, scale 1–5: 1 = lack of or off-aroma; 3 = acceptable aroma; 5 = excellent and strong aroma.
c Flower size, scale 1–5: 1 = very small, unacceptable; 3 = acceptable size; 5 = very large.
— Data not available due to insufficient or no flowering in first year.

Table 22.3 Comparative evaluation of the oil composition of ten cultivars of organically grown lavender, 1999, Norway, Iowa

Lavandin and lavender cultivars	1,8-cineole (%)[a]	Linalool (%)[a]	Camphor (%)[a]	Linalyl acetate (%)[a]	Lavandulyl acetate (%)[a]	β-caryophyllene (%)[a]
Lavandin (L. x intermedia)						
Abriallii	8.8 ± 1.4	31.1 ± 3.7	7.5 ± 1.1	17.2 ± 2.2	2.9 ± 0.1	1.9 ± 0.0
Grosso	10.7 ± 0.9	27.9 ± 2.3	8.1 ± 0.1	17.8 ± 2.6	3.1 ± 0.1	1.9 ± 0.1
Provence	21.1 ± 0.8	29.1 ± 1.8	3.8 ± 0.3	1.5 ± 0.4	0.49 ± 0.1	0.8 ± 0.2
Super	6.8 ± 0.5	38.5 ± 1.9	3.5 ± 1.1	17.7 ± 1.0	2.2 ± 0.2	1.1 ± 0.3
Lavender (L. angustifolia)						
Alba	1.0 ± 0.1	2.8 ± 4.0	0.1 ± 0.0	16.8 ± 1.8	3.8 ± 0.2	4.3 ± 0.7
English	1.0 ± 0.2	37.8 ± 1.5	0.1 ± 0.1	14.7 ± 4.3	6.3 ± 0.6	3.8 ± 0.9
Grey Lady	0.9 ± 0.3	34.5 ± 3.0	0.5 ± 0.7	14.3 ± 3.7	3.2 ± 0.4	3.7 ± 1.5
Hidcote	1.2 ± 0.6	29.1 ± 2.5	0.7 ± 1.0	17.3 ± 6.4	5.3 ± 0.9	3.0 ± 1.2
Lady[b]	0.2	53.4	0.3	5.1	1.2	1.5
Munstead	1.5 ± 0.7	38.3 ± 2.9	0.1 ± 0.1	15.0 ± 1.5	5.4 ± 0.7	3.5 ± 1.6

Notes
a Percentage of constituent relative to percentage of total essential oil.
b Lady lavender did not produce sufficient flower samples to run three replicates of the oil analysis.

extraction processes, or because oil composition could change over the years is unclear (Table 22.3). Lawrence (1994) also reported a wide range in linalool and linalyl acetate percentages depending on plant genotype and cultivation area. Tucker (1984) reported a very wide range of linalyl acetate from 7.46 per cent to 56.6 per cent among the cultivars they evaluated.

Enantiomeric analysis

Results for the enantiomeric distribution of linalool and linalyl acetate from four cultivars of lavandin, six cultivars of lavender, and two commercial samples of L. angustifolia is presented in Table 22.4. The (R)-(−) enantiomer of linalyl acetate is predominant in this study (>99 per cent) in pure authentic samples. The (R)-(−) linalool appears to be present in at least 85 per cent enantiomeric excess.

Two authentic samples of essential oil obtained by steam distillation of organically grown lavender at Iowa, and two commercial samples of lavender essential oil were also analyzed (Table 22.5). The percentages of linalool, linalyl acetate and camphor, and the values for refractive index, optical rotation and specific gravity fell within the ranges specified by ISO 2000 for pure lavender essential oil. However, the percentage of camphor and optical rotation value for the two commercial lavender oils was way off the ISO 2000. The (R)-(−) linalool enantiomer (>93 per cent) and the (R)-(−) linalyl acetate enantiomer (>99 per cent) were predominant in pure authentic lavender essential oils obtained by steam distillation from organically grown lavender. Enantiomeric distribution of linalool and linalyl acetate in the two commercial lavender oils ranged from 68–72 per cent in linalool R-(−) and 67–80 per cent in R-(−) linalyl acetate. Sample 1 of commercial lavender oil that appeared to be adulterated based on optical

Table 22.4 Enantiomeric distribution of linalool and linalyl acetate in hydrodistilled oils of lavandin and lavender grown in Norway, Iowa

Lavandin and lavender varieties	Linalool		Linalyl acetate	
	R-(−) (%)	S-(+) (%)	R-(−) (%)	S-(+) (%)
Lavandin (L. x intermedia)				
Abriallii	89.8	10.2	100	0
Grosso	89.8	10.2	100	0
Provence	97.3	2.7	99	1
Super	87.2	12.8	100	0
Lavender (L. angustifolia)				
Alba	84.8	15.2	100	0
English	86.6	13.3	100	0
Grey Lady	86.3	13.7	100	0
Hidcote	80.9	19.1	100	0
Lady	92.7	7.3	100	0
Munstead	86.6	13.4	100	0
Commercial Lavender Samples[a]				
Sample 1	68.1	31.9	66.4	33.6
Sample 2	70.2	29.8	66.5	33.5

Note
a Both samples exhibiting adulteration.

Table 22.5 Analysis of authentic and commercial lavender oils

Origin	RI	OR	SG	Camphor	Linalool	Linalyl acetate	R-(−)-linalool	S-(+)-linalool	R-(−)-linalyl acetate	S-(+)-linalyl acetate
Authentic samples *(Steam distilled)*										
1 USA	1.458	−6.78	0.888	0.1	25	35	93	7	100	0
2 USA	1.461	−9.12	0.887	0.5	30	40	95	5	99	1
Commercial samples										
1 Unknown	1.455	−4.46	0.886	2.2	44	41	68	32	66	34
2 Unknown	1.456	−1.09	0.884	2.1	47	29	70	30	67	33

rotation and chiral analysis was analyzed for radiocarbon and stable isotope ratio. The results (Table 22.6) clearly indicate adulteration with synthetic linalool and linalyl acetate.

These results clearly demonstrate that organically grown lavandin and lavender can be successfully grown in Midwest, United States, though the commercial feasibility and marketing of such oils produced organically was not evaluated.

Discussion

Our results demonstrate that acceptable lavender and lavandin oil can be organically produced, and that the use of fresh and pure lavender oils are useful in assessing possible adulteration. First year lavandin cultivars produced high essential oil contents of 6.25–10 per cent (dry floral weight). All lavandins produced oils with acceptable camphor contents between 2–15 per cent. Lavandin 'Super' was the overall best performer in oil yield and flower quality (colour, aroma and floral size). All lavenders evaluated produced 2.8–4.96 per cent essential oil, and had camphor contents <1 per cent.

Second year lavandin cultivars produced essential oil contents of 7.10–9.9 per cent, an overall 0.35 per cent increase from year one. Camphor contents in the lavandins remained about the same and averaged 3.5–8.1 per cent. Lavandins 'Grosso' and 'Super' produced the highest oils (9.9 and 8.7 per cent, respectively), but only 'Super' exhibited 100 per cent winter survival under our harsh winter conditions and exhibited a 100 per cent bloom among all its plants. 'Super' also rated very high on plant vigour and uniformity. Lavender cultivars produced >1–2 per cent essential oil, with an average yield of 3.55 per cent, only a minor increase of 0.76 per cent from year one. Although lavenders 'Grey Lady' and 'Hidcote' produced high quantities of oil (5 and 3.50 per cent, respectively), 'Alba' and 'Munstead' performed best relative to plant vigour, uniformity. Relative to dried floral quality, 'Hidcote', 'Alba', and 'Munstead' were quite promising.

Chiral evaluation was an effective tool for authenticity control of essential oil purity. A quantification profile by gas chromatography alone is not sufficient for authenticity control. Our results on chiral analysis of linalool and linalyl acetate clearly show a profile for lavender oils (Tables 22.4 and 22.5). The enantiomeric distribution of (R)-(−) and (S)-(+) forms of linalool and linalyl acetate can be used for testing the authenticity of lavender essential oils. This study was further confirmed by Radiocarbon and Stable Isotope Ratio Analysis (Table 22.6). One commercial sample of lavender oil was shown by Radiocarbon and Stable Isotope Ratio Analyses to be diluted with synthetic chemicals.

Table 22.6 Radiocarbon and stable isotope ratio analyses

	^{14}C	^{13}C	D
Sample 1	10.98 ± 0.45	−28.16 ± 0.01	−220 ± 5

^{13}C ($^{13}C/^{12}C$) is in ppm (0/00) relative to the International Standard PDB (±1).
D (D/H) is in ppm (0/00) relative to the International Standard V-SMOW (±1).
^{14}C activity is in disintegrations per min per gram carbon (dpm/gC) (±1).
The ^{14}C activities of Sample 1 are equivalent to 73 per cent of the 1997 or present day ^{14}C reference activity of 15.0 dpm/g carbon. This indicates the addition or dilution with fossil fuel derived material of approximately 27 per cent.

The enantiomeric distribution of (R)-(−) and (S)-(+) forms of linalool and linalyl acetate is important indicators of the purity of lavender essential oils. Oil samples that fall outside the expected range of the expected enantiomeric values should be suspect relative to their purity and quality. While these values represent only lavender samples from 2-year old fields in the Midwestern United States, this technique has long been recognized by industry as a valuable tool to ascertain purity. The results we present here clearly show that the enantiomeric distribution in two commercial samples obtained at random as a comparative screen clearly fall outside expected range.

As higher essential oil yields are expected after the third and fourth year of growth, this study will next examine the annual growth, oil yield, and oil quality of these varieties over a 4-year time period.

Acknowledgements

We thank Jeanie Abi-Nader for her support and review of this chapter, Phil Forbes for his assistance in the field trials, Kendall Ortberg for her assistance in the quality ratings of the lavender flowers, and Kathy Krezak-Larson and Tom Havran for their ideas and contributions.

References

Adams, R.P. (1989). Identification of Essential Oils by Ion Trap Spectroscopy. Academic Press, San Diego, CA.
AFNOR (1996). Huiles Essentialles. AFNOR, Paris, France, 5th edn.
ASTA (1968). Official Analytical Methods of the American Spice Trade Association. ASTA, Engelwood Cliffs, NJ, 8–11.
Charles, D.J. and Simon, J.E. (1990a). Comparison of extraction methods for rapid determination of essential oil composition of Basil (*Ocimum spp.*). *J. Amer. Soc. Hort. Sci.*, 115, 458–62.
Charles, D.J., Simon, J.E. and Wood, K.V. (1990b). Essential oil constituents of *Ocimum micranthum* Wild. *J. Agric. Food Chem.*, 38, 120–2.
Charles, D.J., Simon, J.E. and Widrlechner, M.P. (1995). Characterization of essential oil of dill (*Anethum graveolens* L.). *J. Essential Oil Res.*, 7, 11–20.
Finnigan.400 GC/MS; NBS Library 31, 331 entries.

ISO/DIS 3515 (2000). International Organization for Standardization.

Keller, R.S. and Milne, G.A.W. (1978). EPA/NIH Mass Spectral Data Base. Department of Commerce, Washington, DC.

Lawrence, B.M. Essential Oils (1991–94). Allured Publ. Corp., Wheaton, IL.

Liberty Hyde Bailey Hortorium (1976). Hortus Third. Macmillan Publishing, New York, NY.

Simon, J.E., Chadwick, A.F. and Craker, L.E. An indexed bibliography. (1971–80). The Scientific Literature on Selected Herbs and Aromatic and Medicinal Plants of the Temperate Zone. Archon Books, Hamden, CT.

Stenhagen, E., Abrahamson, S. and McLafferty, F.W. (1987). Registry of Mass Spectral Data. J. Wiley & Sons, New York, NY.

Tucker, A.O., Maciarello, M.J. and Howell, J.T. (1984). A preliminary analysis of some lavender and lavandin cultivars. *Perf. & Flav.*, **9**, 49–52.

23 Chemical profiles of lavender oils and pharmacology

Maria Lis-Balchin and Stephen Hart

Introduction

Lavender oil, like other plant essential oils have been used for centuries in the manufacture of perfumes and these have been designed to either stimulate or relax the wearer or possibly even to attract the opposite sex. Perfumes are made up of a large number of essential oils and nowadays also numerous synthetic chemical components, but the basis of a good perfume is to have a specific blend of 'top notes', 'middle notes' and 'base notes'. Perfumes were originally designed for either men or for women individually, but nowadays there is a tendency for unisex perfumes, which suggests that the emphasis lies nowadays more on the stimulant or relaxant nature of the perfume rather than sexual attraction.

Essential oils have also been individually categorised by aromatherapists as been either relaxant (sedative) or stimulant. It is not clear whether this refers to the action on the brain or to some or all of the muscles, as many essential oils are also classified as anti-spasmodic. Most essential oils are used by aromatherapists in mixes of three or more, diluted by 95% with almond or other carrier oil and then massaged into the skin. The mixture almost inevitably has the three 'notes' as its theme, in order to give the correct 'balance'.

The overall effect of this 'perfume' together with the massage is supposed to relieve stress and in so doing may also 'cure' or at least alleviate many of the stress-related conditions like eczema, stomachache, backache, headache. Whether or not the aroma of the perfume mix has any effect is unclear as clinical studies have not provided any statistically significant results (see Chapter 16).

There are many studies on the effects of aroma on psychology as well as physiology of the 'recipient', however, there is so far no published evidence for the role of the chemistry of the essential oil, component and mixture of these in a perfume or 'aromatherapy mix'. These parameters were therefore studied using smooth muscle.

Stress: biochemical and physiological implications

Changes in the body which occur outside of the brain, as a result of stress, are not under conscious control but are mediated by the sympathetic branch of the autonomic nervous system. The activity of most organs of the body is controlled by the autonomic nervous system and as a general rule the sympathetic system may be considered to be activated in times of flight or fight which will include stress. Stress-related changes in the body will also be mediated by hormones, such as those released from the adrenal gland. Stimulation of the sympathetic system, and adrenaline released from the adrenal gland, will increase heart rate and stroke volume and by dilating and contracting different blood vessels will cause blood to be distributed to those organs such as skeletal muscle, heart and lungs which are involved in exercise.

Smooth muscle will also be either contracted or relaxed such that the body is prepared for exercise, thus bronchial muscle relaxes and sphincters of the gastro-intestinal system contract. If one considers the fight response in animals, smooth muscle contracts to give dilated pupils and make hair stand on end. In both man and other animals, stimulation of the sympathetic system will cause metabolic changes which favour activity, such as an increase in blood glucose. The nerves of the sympathetic system which innervate smooth, cardiac and vascular smooth muscle all release noradrenaline as their neurotransmitter and the differential response, either contraction or relaxation, is brought about by the presence of different adrenoceptors on the innervated tissue.

In general, α-adrenoceptors mediate contraction and β-adrenoceptors relaxation, but of course there are exceptions to this rule. Further differentiation and control of the system is obtained by the presence of sub-types of α- and β-adrenoceptors. Occupation of a receptor by an appropriate agonist results in a change in cell activity (such as contraction or relaxation) which is mediated via a secondary messenger within the cell. Alpha-2 adrenoceptors mediate their actions via a fall in cyclic AMP (cAMP), while beta-adrenoceptor activation is associated with a rise in cAMP. Alpha-adrenoceptors are linked to the phosphoinositide pathway. In general, contraction is associated with an increase in the concentration of calcium ions within the muscle fibre while relaxation involves either a removal of calcium, the blocking of calcium channels or the opening of potassium channels.

Many tissue of the body receive a dual innervation from the two branches of the autonomic nervous system (sympathetic associated with activity and parasympathetic with feeding and the restoration of energy). In the gastro-intestinal tract we have this dual innervation plus an additional plexus of nerves in the wall of the intestine, often called the enteric nervous system, which involves several other neutotransmitters.

It is on account of this rich innervation of the intestine that we have studied the action of essential oils on the smooth muscle of the guinea-pig ileum *in vitro*.

Smooth muscle preparation

The preparation of the smooth muscle of the guinea-pig ileum will remain viable for several hours after removal from the animal and will respond to electrical field stimulation with reproducible contractions which are due to the stimulation of the parasympathetic nerve with the release of acetylcholine.

Essential oils which stimulate smooth muscle contraction can be recognised immediately while the site of action of those which reduce the size of the electrically-induced contraction can be determined. Possible sites of action include inhibition of the release of acetylcholine, or relaxation of the tissue via stimulation of adrenoceptors, action on secondary messengers or on calcium or potassium channels. This preparation thus allows us to recognise spasmogenic and spasmolytic activity, to determine whether or not the activity is dose-related, to measure duration of action and also attempt to determine the mechanism of action.

The question arises whether the knowledge of the activity of essential oils on smooth muscle gives us any clues about the likely actions of these compounds if and when they enter the central nervous system (CNS).

A famous English pharmacologist suggested that the intestine could be considered a paradigm of the CNS, but it still remained almost impossible to infer action in the CNS from activity on isolated smooth muscle.

The reason for this is simply the complexity of the CNS, with the interaction between excitatory and inhibitory fibres being such that reduced activity in one neurone can lead either to sedation or excitation. Thus, alcohol appears to stimulate some behaviour although it is a CNS depressant, the explanation being that the inhibition of inhibitory pathways removes a normal break and behaviour therefore changes.

Another aspect of the complexity of the CNS is that any one particular behaviour is controlled by several neurotransmitters, each of which is likely to be able to bind to different sub-groups of receptors.

If one, for example, considers pain, this involves neurotransmitters in the afferent pathway such as Substance P, glutamate and nitric oxide and this afferent pathway can be modulated by neuronal pathways releasing a range of neurotransmitters including opioid peptides, acetylcholine, histamine, 5-hydroxytryptamine and cholecystokinin.

With so many neurotransmitters involved in the pain pathway it is not surprising that the experience of pain can be influenced by many different compounds. For example, monoterpenes like menthone and α-terpineol (administered by the subcutaneous route) showed activity similar to that of accepted analgesics, for example, indomethacin and naproxen in reducing the behavioural activity of the mouse to a noxious stimulus (Hart et al., 1994).

In experiments studying the motor activity of mice after exposure to the aroma of various essential oils (Buchbauer, 1991; 1992; Jager et al., 1992) rosemary, jasmine and ylang ylang increased activity while lavender, neroli, lime-blossom, passiflora, citronellol and linalool decrease motor activity. The presence of components in the blood when applied by inhalation has also been demonstrated (Jager et al., 1992). The effect on the motor activity has been shown to be similar to that when the essential oil was injected. It has been assumed that changes in motor activity are a central effect but the possible action on neuromuscular transmission has not been investigated.

However, recent experiments on the motor-nerve skeletal muscle preparation (rat phrenic-nerve diaphragm) by the authors has shown that lavender and tea tree oils cause a reduction in the size of the twitch of the skeletal muscle in response to electrical stimulation of the motor nerve.

Linalool, which was shown to reduce motor activity in the mouse has also been shown to have an action within the brain itself: using membranes from rat cerebral cortex, linalool exhibited a dose-related inhibition of the binding of glutamate, a main excitatory neurotransmitter of the CNS (Elisabetsky et al., 1995).

The effect of essential oils in man has been studied in several different ways including measuring the alertness and reaction times (Manley, 1993) and human brain activity (Torii et al., 1988; Kubota et al., 1992) using Contingent Negative Variation (CNV). The latter is the brain potential which occurs between a warning stimulus and an imperative stimulus, that is, when the subject is expecting something to happen. The CNV amplitude is increased by caffeine, jasmine and peppermint and decreased by chlorpromazine, lavender and marjoram. There is some discrepancy between results from different groups regarding many oils.

The present experiments were designed to see whether the effect of stress on man could be mimicked using isolated segments of small intestine and by monitoring their spasmogenic or spasmolytic effects, it could be possible to assess their relaxant or stimulant nature.

Materials and methods

Essential oils were obtained from various commercial sources and each oil was analysed by GC using a Shimadzu GC 8A with a 50 m $\times$ 0.32 mm OV101 column; the temperature program was set at $4°C\,min^{-1}$ from 100 to 230°C.

The percentage of all the components was calculated in each selected Retention Time interval of under 10 min, 11–15 min, 16–20 min, 21–30 min and 30+ min. The main components present in each RT interval was also determined. The essential oils were diluted in methanol (usually ×1,000) and 0.1–0.2 ml was applied to the tissue preparations in the organ bath giving a final dilution of ×200,000 to ×400,000 (a concentration of 2.5×10^{-6} to 5×10^{-6}).

Pharmacological studies, carried out on guinea-pig ileum were contrasted against many practising aromatherapists' predictions of the effect of essential oils on the patient (alone or as mixtures).

Results of the studies

Monoterpenes versus contractions of smooth muscle

Previous comparisons of the pharmacological activity of many components and essential oils suggested that monoterpenes were responsible for contractions in the guinea-pig ileum *in vitro* (Lis-Balchin *et al.*, 1996a,b).

This was best illustrated by work on two New Zealand essential oils Manuka and Kanuka. The former was largely composed of sesquiterpenes and produced a relaxation in the gut, while the latter was composed largely of monoterpenes and produced a contraction (Lis-Balchin *et al.*, 1996a). Further, work on over seventy essential oils suggested that there was a considerable correlation of contraction of the small intestine with a high percentage of monoterpenes (but not sesquiterepenes) (Lis-Balchin *et al.*, 1996b; 1998).

Table 23.1 The predicted effect on guinea-pig ileum based on the percentage of components at different retention time intervals

RT	>10	11–15	16–20	21–30	30+	Effect
a	19	66	—	5	—	R
b	13	64	12	7	—	R
c	—	3	—	47	45	R
d	82	1	—	5	—	S
e	97	3	—	—	—	S
f	57	—	—	32	—	S/R
g	69	25	2	2	—	S/r
h	2	—	85	2	—	R
i	82	—	—	4	—	S
j	6	41	45	5	—	R
k	46	22	30	—	—	S/R
l	8	30	5	39	10	R
m	—	—	—	59	48	R
n	—	—	—	78	34	R
o	7	29	58	—	—	R
p	—	92	2	—	—	R
q	—	52	37	6	—	R
r	99	—	—	—	—	S
s	1	30	51	5	—	R
t	14	—	1	69	—	S/R
u	16	—	76	—	—	S/R
v	67	7	2	9	—	S
w	1	—	34	56	—	R
x	—	—	—	89	4	R
y	58	32	—	—	—	S/R
z	4	—	1	76	1	R

Chemical profiles of lavender oils and pharmacology 247

These results therefore suggested that it was simply the actual percentage composition of the monoterpenes which determined whether the effect on the smooth intestinal muscle would be contractile or relaxant. This was therefore investigated further.

Retention times versus percentage of components

This hypothesis was put to the test, using essential oils alone or in mixes, by calculating the total per cent of components in different RT intervals (Table 23.1) and predicting what the effect on the smooth muscle would be.

It was noted that monoterpenes were in the RT >10 interval, with the exception of 1,8-cineole which was also found here, while alcohols, ketones and aldehydes occurred in the 11–15 min interval, esters and phenols in the 15–20 min interval and sesquiterpenes thereafter.

Chemical predictions versus actual effect on ileum

Whenever there was a considerable percentage of components in the >10 min interval, this would be associated with a small to large contraction of the ileal muscle (depending on the actual percentage). Predictions of pharmacological activity could therefore be easily made based on the chemical composition, with the exception of essential oils containing 1,8-cineole, for example, *Eucalyptus globulus* (Table 23.2). Lavender showed a very high proportion of the

Table 23.2 Comparison of the actual effect of essential oils on guinea-pig ileum and the predicted effects using chemical composition and aromatherapists' predictions

	Actual effect on tissue	Chemical prediction	Aromatherapists' prediction
a. Tea tree	R & s/R	s/R	s/r
b. Neroli	R	s/R	S/r
c. Camomile German	R	R	R
d. Frankincense	S	S	R
e. Camphor	S	S	S
f. Black Pepper	S/r	S/r	S/r
g. Rosemary	S/r	S/r	S
h. Lemongrass	R	R	S/r
i. Juniper	R	S	S
j. **Lavender**	R & s/R	R	R
k. Bergamot	s/R	S/R	S/R
l. Ylang Ylang	R	R	r
m. Sandalwood	R	R	r
n. Vetivert	R	R	R
o. Petitgrain	S/r	R	S/R
p. Rosewood	R	R	R
q. Geranium Bourbon	R	R	R
r. Eucalyptus globulus	R	S	S
s. Clary Sage	S & S/r	R	r
t. Ginger	R	s/R	s/r
u. Dillweed	S/r	s/R	r
v. Nutmeg	S	S	r
w. Manuka	R	R	R
x. Spikenard	R	R	r
y. Camomile Roman	S/R & s/R	S/R	R
z. Valerian	R	R	R

non-stimulating components and therefore was stated to be relaxant according to the chemical prediction.

Chemical predictions versus aromatherapists' predictions of effect on patient

The chemical predictions were largely similar to both the actual observed effect on the smooth muscle and also similar to the aromatherapists' prediction on the 'patient'. The latter effect was either a relaxant effect or a stimulant effect on the 'patient'; the stimulant effect could be directly related to a contraction on the isolated muscle. The chemical prediction for lavender was confirmed by the actual effect and that of the aromatherapists' predictions.

Effect of mixtures of essential oils

Mixtures of two or more essential oils also showed the same trend, some of which are shown in Table 23.3. This proves that contractions of smooth muscle are largely as a result of a high monoterpene concentration, regardless of the actual monoterpene component. Lavender added to other components caused a swing towards the chemical readjustment towards a greater concentration of relaxant components and therefore it was predicted to be a relaxant mixture.

As before, the correlation broke down if 1,8-cineole was involved, for example, in mixtures with rosemary or *E. globulus*. There is no easy explanation for this discrepancy. It is also of interest that if *E. globulus*, containing 95% of 1,8-cineole is presented to the smooth muscle preparation it will cause a relaxation, whereas if 1,8-cineole alone is presented it causes a contraction.

Studies on well-known commercial perfumes were only effected on their chemical composition. There seemed to be a very positive correlation between the chemical distribution and the product's intention, as determined by the publicity information. Thus, the lavender-containing eau de colognes, were refreshing and stimulating only because there was a predominance of

Table 23.3 The predicted effect on guinea-pig ileum based on the percentage of components at different RT intervals

RT	>10	11–15	16–20	21–30	30+	Effect
1	—	85	2	—	7	R
2	41	5	49	—	—	S/R
3	85	1	—	6	—	S
4	78	22	1	14	—	S/r
5	86	2	2	1	—	S
6	44	45	—	5	—	S/R
7	37	47	8	4	—	S/R
8	4	90	—	2	—	R
9	9	56	12	15	—	s/R
10	—	—	—	58	37	R
11	98	—	—	—	—	S
12	99	—	—	—	—	S
13	—	—	—	71	19	R
14	22	69	—	—	—	S/R
15	10	26	4	41	6	s/R
16	10	39	36	9	—	s/R
17	78	9	3	5	—	S

Table 23.4 Comparison of the predicted effect of essential oil blends on clients/patients by aromatherapists and by the chemical composition with their actual effect on guinea-pig ileum

Blend	Actual effect	Aroma-therapist	Chemical prediction
1. Orange 2: nutmeg 1: dill 1	S	S	s/R
2. Lemongrass 1: juniper 1: rosemary 2	S	S	R
3. Frankincense 1: rose abs. 1: clary sage 2	R/S	R/S	s/R
4. Eucalyptus glob.1: black pepper 1: ginger 1	S	S	s/R
5. Ginger 1: tea tree 1: rosemary 2	S	S/R	S/R
6. Frankincense 2: ylang ylang 1: geranium 1	R/S	S/R	R
7. Frankincense 1: geranium 1: bergamot 2	S/R	S/R	S/R
9. Camomile roman 1: **lavender** 1: geranium 1	R/S	s/R	S/R
10. Frankincense 1: mandarin 2: scotch pine 1	S/R	S	S/R
11. Camomile roman 1: valerian 1: rose abs. 1	R/S	s/R	R
12. Ylang ylang 1: marjoram 1: thyme red 1	R	s/R	R
13. Petitgrain 1: melissa 1: sage dalmatian 1	s/r	R	R
14. Kanuka 1: **lavender** 1: frankincense 1	R	S/r	R
15. Manuka 1: **lavender** 1: frankincense 1	S	s/R	R
16. Fennel 1: orange 1: bergamot 1	S/r	S/R	S/R
17. Basil 1: bergamot 1: clary sage 1: jasmine 1	S	s/R	S/R

monoterpenes (largely limonene) from the citrus essential oils used in greater concentration than lavender in the formulation. Paris by Kenzo, showed a similar over-preponderance of monoterpenes due again to the limonene of its citrus components. These are obviously stimulating oils and their predicted effect would be contractile on smooth muscle. Lavender perfume, with its high concentration of relaxant components, would, however, be predicted to be relaxant on smooth muscle. Its holistic effect would also be of a relaxant nature, unless the 'patient' had a great dislike for that particular odour ... but that is an unpredictable psychological effect, which could completely contradict the chemical findings.

Conclusion

The present results indicate that there is a very close correlation between the pharmacological activity of essential oils on the isolated smooth ileal muscle of the guinea-pig and the predicted effect on the human psyche by aromatherapists, as with the use of lavender (Table 23.4).

This 'holistic' effect would probably have originated from the direct effect of essential oils on the CNS with the concomitant effect of the massage (and of course counselling and possible placebo effect. The actual effect on the isolated smooth muscle is less complex and probably involves various adrenoceptors, but there could also be a simple direct action of components on the membrane with all monoterpenes initiating a rise in calcium levels which cause contraction of the muscle.

References

Buchbauer, G., Jirovetz, L. and Jager, W. (1992). Passiflora and limeblossoms: motility effects after inhalation of the essential oils and some of the main constituents in animal experiments. *Arch. Pharm. (Weinheim)*, 325, 247–8.

Buchbauer, G., Jirovetz, L., Jager, W., Dietrich, H., Plank, C. and Karamat, E. (1991). Aromatherapy: evidence for sedative effects of the essential oils of lavender after inhalation. *Z. Naturforsch.*, 46, 1067–72.

Elisabetsky, E., Marschner, J. and Souza, D.O. (1995). Effects of linalool on glutamatergic system in the rat cerebral cortex. Neurochem. Res., 20, 461–5.

Hart, S.L., Gaffen, Z., Hider, R.C. and Smith, T.W. (1994). Antinociceptive activity of monoterpenes in the mouse. *Can. J. Physiol. Pharmacol.* 72, Sup.1, 344.

Jager, W., Buchbauer, G., Jirovetz, L. and Fritzer, M. (1992). Percutaneous absorption of lavender oil from a massage oil. *J. Soc. Cosmet. Chem.*, 43, 49–54.

Kubota, M., Ikemoto, T., Komaki, R. and Inui, M. (1992). Odor and emotion-effects of essential oils on contingent negative variation. *Proc.12th Int. Congress on Flavours, Fragrances and Essential Oils*, Vienna, Austria, Oct. 4–8, p. 456–61.

Lis-Balchin, M., Deans, S.G. and Hart, S.L. (1996a). Bioactivity of New Zealand medicinal plant essential oils. *Acta Hort.* 426, 13–30.

Lis-Balchin, M., Hart, S.L., Deans, S.G. and Eaglesham, E. (1996b). Comparison of the pharmacological and antimicrobial action of commercial plant essential oils. *J. Herbs, Spices Med. Plants.*, 4, 69–86.

Lis-Balchin, M., Deans, S.G. and Eaglesham, E. (1998) Relationship between the bioactivity and chemical composition of commercial plant essential oils. *Flav. Fragr. J.*, 13, 98–104.

Manley, C.H. (1993). Psychophysiological effect of odor. *Crit. Rev. Food. Sci. Nutr.*, 33, 57–62.

Torii, S., Fukuda, H., Kanemoto, H., Miyanchio, R., Hamazu, Y. and Kawasaki, M. (1988). Contingent negative variation and the psychological effects of odor. In: *Perfumery: The Psychology and Biology of Fragrance* (S. Toller and G.H. Dodd, eds) New York: Chapman and Hall.

24 Chemical composition of essential oils from different species, hybrids and cultivars of *Lavandula*

Maria Lis-Balchin

Introduction

The chemical composition of the essential oils from different *Lavandula* species, hybrids and cultivars show not only interspecific differences, but also intraspecific differences, which may sometimes be due to climatic, geographical or seasonal differences or due simply to the amount of watering or fertilization used. It may also depend on variation due to genotypes, which can occur either in plants growing in close proximity or in plants some considerable distance away, for example, in different countries. It can also be due to the method of essential oil extraction and, in the case of commercial oils, to the degree of blending and adulteration. The many species, hybrids and cultivars of *Lavandula* have only been studied to a limited extent and many of these show wide differences in their compositions, while some are apparently very consistent.

The commercial hybrids of lavandin have variable concentrations of 1,8-cineole and camphor, which are absent from the most favoured species, *Lavandula angustifolia* P. Miller, but which give the harsher notes to the lavandins (Table 24.1). The 'rhodinol content', consisting of citronellol, geraniol, nerol, neryl acetate and geranyl acetate, which together amount to a very small percentage of the total composition, give a sweet, rose-like odour to the lavandin oils. The lavandin hybrids (Table 24.1), originated from France, but were grown in the Po Valley, Italy, and were found to have a similar composition to those grown in France.

In 1981, Tucker gave the following average percentage composition of the main components of cultivars of lavandin (Table 24.2). This showed small differences between the cultivars, the

Table 24.1 Composition of French lavender (ISO 3515) (*L. angustifolia* P. Miller) compared to lavandin hybrids

Constituent	*L. angustifolia* proportion		'Abrialis' Average	'Grosso' Average	'Super' Average
	Minimum	Maximum			
1,8-cineole	—	1.5	8.6	5.2	2.0
Camphor	—	0.5	8.2	5.9	3.0
Lavandulyl acetate	2	—	—	—	—
Linalool	25	38	30.4	28.4	29.3
Linalyl acetate	25	45	20.8	27.6	30.4
'rhodinol content'	not assesed		1.5	1.7	2.2

Source: Piccaglia (1998).

Table 24.2 Chemical composition of lavandin and cultivars

Component	Composition (%)		
	'Abrialii'	'Super'	'Grosso'
α-pinene	0.4	0.2	0.6
Camphene	0.3	0.1	1.2
Linalool	33.1	31.7	32.3
Linalyl acetate	29.4	46.0	—
Camphor	9.4	5.0	7.0
Terpinen-4-ol	0.5	—	2.8
Cis-ocimene	2.6	—	1.2
Trans-ocimene +3-octanone	4.0	—	0.6

Source: Tucker (1982).

main one was in the high concentration of linalyl acetate in 'Super' and the apparent absence of terpinen-4-ol, cis-ocimene and trans-ocimene in this cultivar, however, the analyses had been done by different groups, which may account for the slight differences.

Spike lavender

L. latifolia Medicus is the spike oil of commerce. Different authors have presented variable compositions for spike oil (Lawrence, 1976–8; 1979–80; 1981–7). This is the most important of the Spanish oils produced commercially and the yield is high, 0.8–1.2 per cent. More than 300 components have been identified (Table 24.3) and the main ones are linalool (19–48 per cent), 1,8-cineole (21–42 per cent) and camphor (5–17 per cent).

Table 24.4 shows the similarities in the composition of *L. latifolia* (as averages) published by different authors in different years, with the exception of linalool.

L. angustifolia *and L.* angustifolia *ssp.* pyrenaica

Data is also given in Table 24.3 for *L. angustifolia, L. angustifolia* ssp. *pyrenaica* and *L. lanata*, for comparison. *L. angustifolia* of commerce (Naef and Morris, 1992), whose main components are linalool (25–38 per cent) and linalyl acetate (25–45 per cent) shows some considerable differences between the subspecies *L. angustifolia* ssp. *pyrenaica* (DC.) (Garcia-Vallejo et al., 1989) whose three main components were: linalool (20–66 per cent), borneol (6–32 per cent) and camphor (2–14 per cent). The subspecies grows wild in the mountains in NE Spain on calcareous soil at 700–1400 m altitude. This subspecies therefore would not be acceptable as normal lavender oil.

L. lanata

L. lanata Boisse (Garcia-Vallejo et al., 1989) is also shown in Table 24.3. This lavender is morphologically similar to *L. latifolia* but is more tomentose. *L. lanata* had a very high concentration of camphor (43–59 per cent) and variable amounts of lavandulol (3–27 per cent) as its main components.

Table 24.3 Chemical composition of *Lavandula* species and cultivars

Component	Composition (%)			
	L. angust.	L. ssp. pyr.	L. lanata	L. latifolia
Linalool	17.8	38.0	41.7	19.5–47.8
Linalyl acetate	21.5	1.1	1.1	0–1.8
Camphor	0.5	8.1	12.8 (59.2)	5.3–16.6
1,8-cineole	0.9	3.0	26.3	20.5–42.4
Terpinen-4-ol	6.4	3.0	0.6	0.2–0.7
(Z)-b-ocimene	8.2	0.8	0.1	0–0.5
(E)-b-ocimene	6.2	0	0.2	0–0.5
β-caryophyllene	8.0	0.9	1.4	0.3–1.9
Lavandulyl acetate	7.3	0.3	0	0–0.3
Lavandulol	1.2	0.7	0.6 (26.7)	0.1–1.5
Borneol	1.0	19.7 (32)	0.8	0.4–4.9

Notes
L. angust. = L. angustifolia (Naef and Morris, 1992).
L. ssp. pyr. = L. angustifolia ssp. pyrenaica (DC.) (Garcia-Vallejo et al., 1989).
L. lanata Boisse (Garcia-Vallejo et al., 1989).
L. latifolia (Garcia-Vallejo et al., 1989).
L. angustifolia (Main components are linalool (25–38 per cent) and linalyl acetate (25–45 per cent).
The data in brackets show the maximum concentration found.

Table 24.4 Chemical composition of L. *latifolia*

Component	Composition (%)	
	L. latifolia '86	L. latifolia '92
Linalool	41.7	27.6
Linalyl acetate	1.1	1.1
Camphor	12.8	16.3
1,8-cineole	26.3	22.9
Terpinen-4-ol	0.6	0.4
(Z)-b-ocimene	0.1	0.5
(E)-b-ocimene	0.2	0.1
β-caryophyllene	1.4	2.2
Lavandulol	0.6	0.5
Borneol	0.8	1.7
Myrcene	0.2	0.8
b-farnesene	0	0.3
Limonene	1.1	3.1

Sources: L. *latifolia* (Boelens, 1986); L. *latifolia* (Naef and Morris, 1992).

L. dentata

L. *dentata* L. grows wild along the Mediterranean coast of Spain at 50–100 m (Garcia-Vallejo et al., 1989). Two chemotypes have been found with the following main components: 1,8-cineole/β-pinene and β-pinene/α-pinene.

L. multifida

L. multifida was also studied by (Garcia-Vallejo *et al.*, 1989) and was found to have carvacrol and β-bisabolene as the main components. These are very unusual components in lavenders.

L. stoechas

L. stoechas L. ssp. *pedunculata* (Miller) Samp. ex Roziera (*L. pedunculata* Cavanilles) shows two chemotypes (Table 24.5). Chemotype 1 was the most common, comprising 34/37 of the oils studied in fourteen provinces in Spain.

Both this subspecies and ssp. *sampaioana* (*L. stoechas* L. ssp. *sampaioana* Roziera) had two chemotypes: one had camphor/fenchone as its main components and the other had β-pinene/camphor/fenchone. *L. stoechas* L. ssp. *stoechas* included *L. stoechas* ssp. *caesia* Borja et Boday. The main components were camphor and fenchone (with high 1,8-cineole).

Four wild populations of *L. stoechas* L. ssp. *stoechas* growing wild in Crete were analysed by Skoula *et al.* (1996). The essential oils from leaves and flowers consisted mainly of α-pinene, 1,8-cineole, fenchone, camphor and myrtenyl acetate. Considerable variation was, however, found in the percentage of these components. In general, the flowers contained more fenchone, myrtenyl acetate and α-pinene than the leaves, while camphor and 1,8-cineole was higher in the leaves. Flowers also produced a larger quantity of the EO. Two chemotypes were noticed in the four populations: a fenchone/camphor type (three populations) and a 1,8-cineole/fenchone type for the other population.

L. lusieri

L. lusieri (Rozeira) Rivas-Martinez (*L. stoechas* ssp. *luisieri* (Rozeira) Rozeira) had two chemotypes (Garcia-Vallejo *et al.*, 1989) (Table 24.6). Chemotype 1 and chemotype 2 were roughly equally distributed in the seven provinces in Spain (total thirty samples).

Both had an unidentified ester as their main component with chemotype 1 having a high 1,8-cineole concentration as well.

Table 24.5 Chemical composition of *L. stoechas* species and their chemotypes

Constituent	Main components (%)				
	Pedunculata		Stoechas	Sampaioana	
	Chem 1	Chem 2		Chem 1	Chem 2
α-pinene	9.7 (21.1*)	14.7	1.8	7.4 (18.6*)	13.5 (23.0*)
β-pinene	1.0	22.6 (26.0*)	0.1	0.8	21.0 (31.0*)
1,8-cineole	16.7 (67.7*)	15.1	9.4 (52.7*)	7.2 (25.8*)	9.4
Limonene	1.5	1.5	1.3	1.5	2.1
Fenchone	20.1 (44.5*)	7.3	42.1 (68.2*)	20.2 (56.0*)	15.8
Camphor	23.5 (55.7*)	7.2	23.0 (51.6*)	38.0 (84.4*)	5.2
Linalool	3.6	4.1	0.9	4.9 (12.2*)	3.5
Pinocarvone	0.1	1.6	0	0.1	1.8

Source: *(data after Garcia-Vallejo *et al.*, 1989).

Notes
pedunculata = *L. stoechas* L. ssp. *pedunculata* (Miller) Samp. ex Roziera (*L. pedunculata* Cavanilles).
sampaioana = *L. stoechas* L. ssp. *sampaioana* Roziera.
stoechas = *L. stoechas* L. ssp. *stoechas* (included *L. stoechas* ssp. *caesia* Borja et Boday).

Table 24.6 Chemical composition of L. luisieri and its chemotypes

Constituent	Main components (%) L. luisieri	
	Chem 1	Chem 2
α-pinene	2.0	2.0
β-pinene	1.8	1.3
1,8-cineole	22.2 (43.2*)	2.4
Fenchone	1.6	0.7
Camphor	2.1	2.7
Linalool	2.1	3.1
Unknown ester A	15.4 (26.8*)	22.5 (28.4*)
Verbenone	1.7	2.1

Source: Garcia-Vallejo et al., 1989.

Notes
L. lusieri (Rozeira) Rivas-Martinez (L. stoechas ssp. luisieri (Rozeira) Rozeira).
Chem = Chemotype 1 and chemotype 2.

Table 24.7 Chemical composition of L. viridis oil

Constituent	Main components (%) L. viridis	
	Minimum	Maximum
α-pinene	3.2	25.1
β-pinene	0.1	0.4
1,8-cineole	17.8	57.8
Limonene	0.2	0.7
Fenchone	0	0
Camphor	10.3	17.2
Linalool	1.2	5.8
Cis-verbenol	0.5	1.2
Trans-verbenol	0.8	0.4
Borneol	1.6	4.2

Source: Garcia-Vallejo et al., 1989.

Apart from ester A, four other esters were present, together with an unknown ketone and an 'alcohol A'.

L. viridis

L. viridis (Garcia-Vallejo et al., 1989) (Table 24.7) has a high concentration of 1,8-cineole, camphor and pinene.

Table 24.8 L. pinnata L. il. var. pinnata grown on Madeira

Monoterpenes	37–80
Oxygenated monoterpenes	2–4
Sesquiterpenes	13–22%
Main monoterpenes	
β-phellandrene	12–32
α-phellandrene	6–16
Main sesquiterpenes	
β-caryophyllene	11%
Phenylacetaldehyde	6–9%

Source: Figueiredo et al., 1995.

L. pinnata

The main components for *L. pinnata* L. il. *var. pinnata* grown on Madeira (Figueiredo et al., 1995) are shown in Table 24.8.

L. pinnata L. il. *var. pinnata* grown on Madeira (Figueiredo et al., 1995) has a percentage of monoterpenes (37–80) and a relatively small proportion of sesquiterpenes (13–22 per cent). Of the monoterpenes, the highest concentration is of β-phellandrene at 12–32 per cent with α-phellandrene at 6–16 per cent.

Aroma profiles

The aroma profile is determined by analyzing the headspace of a growing plant, which is a non-destructive technique compared to the essential oil produced by steam distillation. The odour profile of the living plant would therefore be expected to be different than that of the processed oil.

The aroma profile may be thought of the perceived aroma of the living plant. The oil, however, reflects the composition of volatiles and semi-volatiles present in the plant, with molecular transformations occurring during the distillation process and during storage. Several different species of *Lavandula* were investigated by Wiesenfeld (1999) derived by dynamic headspace analysis and compared to that of commercial Bulgarian lavender oil obtained by steam distillation.

Experimental conditions

Botanicals used in the survey

One-year-old *Lavandula* plants were obtained from a supplier and were repotted and acclimatised in the laboratory for several weeks. The specimens were sampled after each put showed active growth and all sampling took place within a three-month period, *prior to flowering*: this means that in this survey, no flowers were included, and that this is not a true aroma profile of the plant but only the leaves (present author's note).

Each specimen was sampled at least twice, and the results averaged. The lavender species involved in this study were: *L. officinalis* or *angustifolia* (English lavender), *L. dentata* (French or fringed lavender), *L. stoechas* (Spanish lavender), *L. spica*, *L. viridis* (green lavender), *L. lanata* (wooly lavender) (11), *L. pinnata*, *L. multifida*, and *L. x heterophylla* 'Goodwin Creek'.

Sampling and analysis technique

In preparation for sampling, several stems of each plant are enclosed in a two-piece spherical, custom-designed glass vessel, taking care to not damage or stress the plant. The two sections of the vessel were clamped around the stems and, over the course of several hours, the headspace is sampled by means of a low-flow vacuum pump. The volatile components were trapped on an adsorbent resin (tenax or equivalent), which was packed within a desorption tube attached to the pump. Subsequently, these trapped components were desorbed by means of a short path thermal desorption unit into a gas chromatograph equipped with a cryo trap. A blank was purged and analyzed to determine artifacts from the system.

Instruments and conditions used were: a GC/MS (Hewlett Packard 5890/5971) was configured with two matched (50 m, methyl silicone) capillary columns, one inlet, and two detectors (MSD and FID). The inlet was modified to accept the thermal desorption unit. The cryo trap was installed at the head of the columns and cooled to $-50°C$ by carbon dioxide. Desorption took place at 220°C for 4 min. After desorption was completed, the cryo trap was heated to match the inlet temperature, while the GC temperature was programmed from 35 to 240°C at 3°C/min. Data analysis was by means of a proprietary mass spectrum library and quantitation was determined by FID area normalization. Because of the complexity of the chromatograms, no corrections were made for response differences of individual components.

Weisenfeld (1999) had a problem with coelution (which is normally circumvented by the use of suitable conditions and also by diluting the samples appropriately). This occurred in this case with a multiple elution consisting of β-phellandrene, limonene, cis-ocimene and 1,8-cineole. An exact quantitation of their distribution within the peak could apparently not be determined, but an approximation can be made by mass spectrometry (Weisenfeld, 1999) and appears in the Tables 24.8–24.10.

An analysis of this peak showed that in these profiles: limonene dominates in *L. multifida*, *L. pinnata* and *L. heterophylla*; cineole dominates in *L. viridis*, *L. dentata*, *L. lanata* and *L. spica*. *L. angustifolia* contains a greater percentage of β-phellandrene and *L. stoechas* seems to contain almost equal amounts of limonene and cineole. Cymene most commonly occurs as the para isomer. However, *L. angustifolia* shows two other isomers of cymene in addition to the para form.

The monoterpenes concentration identified in each species could be correlated with the actual odour of the plant (Weisenfeld, 1999). Beta-phellandrene is 'peppery-minty and slightly citrusy', contributing to the herbaceous scent of *L. angustifolia*. Cineole is 'camphoraceous, cool', adding a substantial sharp, penetrating quality to *L. viridis*, *L. dentata*, *L. lanata* and *L. spica* species. Limonene is either dextro or laevo: D-limonene being 'sweet citrusy', whereas L-limonene is 'very clean, not reminiscent of citrus fruits'. Limonene dominates the terpene fraction of the *L. heterophylla* and contributes to the scent of *L. stoechas*. Trans-ocimene, the dominant terpene in *L. pinnata* and *L. multifida*, has a 'warm-herbaceous odor', enhancing the balsamic qualities of the carvacrol component.

The sesquiterpenes were shown in the tables as a sum of the various isomers (beta, gamma, delta etc.) for that particular sesquiterpene. Those sesquiterpenes listed under 'other sesquiterpenes' are mostly unknown, but are usually not the same unknown. The total sesquiterpene content of the different species varies notably, from 1.7 per cent in *L. stoechas* to 35.9 per cent in *L. angustifolia*. The high amount of caryophyllene, along with a considerable amount of cadinenes, in *L. angustifolia* contributes to the distinctive woody note found in this species. It is not surprising to see a large amount of bisabolenes (sweet-spicy-balsamic) in *L. pinnata* and *L. multifida*, enhancing their warm, sweet, spicy scent.

Table 24.9 Aroma profiles of selected *Lavandula* species (% composition)

Species	Angustifolia	Dentata	Stoechas	Lanata	Spica	Viridis
Component						
α-pinene	0.5	1.4	2.1	1.4	0.8	2
Sabinene	0.7	3.6	0.7	0.8	5.4	2.8
β-pinene	0.2	5.1	0.4	1.1	1.2	1.9
Myrcene	2.8	1.2	0.8	1.1	2.9	2.6
Cymenes*	3.8	0.4	0.7	0.2	0.7	7.5
Multi-component peak**	20.5	43.4	20.8	10.8	27.1	24.9
Ocimene	0.5	—	—	1.4	0.1	0.3
Fenchone	—	3.4	33	<0.1	—	—
α-terpinolene	1.1	—	0.8	2.1	0.6	0.3
Linalool	0.5	3	0.6	4.2	2	14.7
Camphor	1.7	2	26.2	39.2	10.9	12.4
Polycyclic ketones*	—	1.5	0.6	—	1.9	1.3
Borneol	4.6	—	—	1.2	0.5	1.3
Lavandulol	—	—	—	0.4	—	—
Terpineols*	—	1.4	0.2	0.6	6.7	4.7
Methyl thymyl ether	—	—	—	—	—	
Linalyl acetate	0.2	—	—	—	—	1.5
Bornyl acetate	0.4		0.3	0.3	0.6	0.1
Lavandulyl acetate	—	—	—	—	—	—
Carvacrol	—	—	—	<0.1	—	—
Neryl/geranyl acetate	0.2	—	—	—	0.1	1.3
Caryophyllene	15.9	0.4	0.5	7.2	1.1	—
Bergamotenes*	1.7	1.2	—	—	1.5	—
Germacrenes*	—	—	—	1.9	4.3	0.6
Selinenes*	—	1.6	0.2	—	0.3	1.8
Farnesenes*	—	1	—	—	1.3	—
Bisabolenes*	—	0.5	0.3	0.2	0.5	—
Cadinenes*	7.1	0.4	0.2	—	1.2	—
Selinadiene	—	—	—	—	—	4.9
Other sesquiterpenes	3.5	3.4	0.4	3	4.5	8.3
Caryophyllene oxide	2	0.3	—	0.7	0.7	0.1
Unknown diterpenes	—	—	<0.1	0.5	—	1.2
Total sesquiterpenes	35.9	8.5	1.7	12.3	14.7	15.6

Source: Weisenfeld (1999).

Notes
* Sum of isomers.
** May include β-phellandrene, cis-ocimene, limonene, cineole.

L. x heterophylla and *L. lanata* contain high concentrations of lavandulol. Lavandulol, an isomer of geraniol, has a 'warm-rosy odor', 'with a slightly spicy note'. This, along with the floral note of the germacrenes, likely contributes to the uncharacteristically floral note in *L. heterophylla*.

Borneol, more widely distributed throughout the genus than lavandulol, has a 'dry-camphoraceous, woody-peppery odor', more characteristic of the lavenders. This terpene alcohol, along with its acetate and ketone form (camphor), adds to the distinctive warm, minty, herbaceous aroma of the typical lavender.

Comparison of aroma profile to distilled oil

Since these aroma profiles are produced by headspace analysis of the leaves and stems of living plants, while the essential oils are prepared by distillation of the leaves, stems and flowers of

Table 24.10 Aroma profiles of selected *Lavandula* species (% composition)

Species Component	Heterophylla	Pinnata	Multifida
α-pinene	0.9	0.1	0.3
Sabinene	1.5	<0.1	—
β-pinene	1.3	<0.1	—
Myrcene	0.7	5.8	2
Cymenes*	0.3	0.5	0.2
Multi-component peak**	7.7	2.2	1.2
Ocimene	—	11	28.2
Fenchone	0.4	<0.1	—
α-terpinolene	—	12.8	8.3
Linalool	9.3	0.1	—
Camphor	14	0.1	—
Polycyclic ketones	0.5	—	—
Borneol	<0.1	—	—
Lavandulol	16.8	—	—
Terpineols*	—	0.3	—
Methyl thymyl ether	—	0.5	3.3
Linalyl acetate	—	—	—
Bornyl acetate	0.3	—	—
Lavandulyl acetate	—	—	—
Carvacrol	<0.1	27	12.8
Neryl/geranyl acetate	0.3	—	—
Caryophyllene	4.3	3.7	2.8
Bergamotenes*	—	0.1	—
Germacrenes*	2.1	—	1.5
Selinenes*	1	—	—
Farnesenes*	0.4	7	3
Bisabolenes*	0.1	13.6	9.9
Other sesquiterpenes	9.4	1.2	—
Caryophyllene oxide	4.4	—	—
Diterpenes	—	—	—
Total sesquiterpenes	17.3	25.6	17.2

Source: Weisenfeld (1999).

Notes
* Sum of isomers.
** May include β-phellandrene, cis-ocimene, limonene, cineole.

harvested plants, comparisons between the oil and headspace compositions are of limited value. We can assume that inclusion of the flowers in the oil greatly influences its composition (Weisenfeld, 1999).

Table 24.11 shows a comparison of the composition of the aroma profile of *L. angustifolia* with that of the steam distilled oil of Bulgarian lavender, presumably derived mostly from this same species.

The most obvious differences between the headspace and the distilled oil are in the amounts of linalool and linalyl acetate. The level of linalyl acetate is of great interest in a lavender oil, because the quality of the oil is evaluated by its ester content; the higher the ester content, the finer the oil. Therefore, the low percentage of these two components in the headspace of the leaves is of particular interest. Again, exclusion of the flowers from the aroma profile, is an

Table 24.11 Lavandula angustifolia aroma profile compared to Bulgarian Lavender Oil (results expressed as per cent)

Component	Aroma profile	Lavender oil
α-thujene	0.3	0.1
Cymenes (3 isomers)	3	0.2
Sabinene	0.7	0.1
Myrcene	2.8	0.6
α-phellandrene	1.5	—
Delta-3-carene	9.2	0.1
Trans-ocimene	0.5	2.3
Compound peak*	21	5.5
Sabinene hydrate	0.3	—
Linalool oxide	trace	0.2
α-terpinolene	1.1	—
Linalool	0.5	27.7
Octenyl acetate	—	1
Camphor	1.7	0.3
Lavandulol	—	0.7
Borneol	4.6	0.6
Para-cymenol (2-isomers)	1.8	—
Terpinene-4-ol	—	3.6
α-terpineol	—	1.3
Linalyl acetate	0.2	39.1
Lavendulyl acetate	0.4	3.7
Neryl acetate	trace	0.4
Geranyl acetate	0.2	0.8
Coumarin	—	0.1
Caryophyllene	13**	3.9
Other sesquiterpenes	22.9	2.3
Caryophyllene oxide	2	0.4

Source: Wiesenfeld (1999).

Notes
* May include β-phellandrene, cis-ocimene, limonene, cineole.
** Contains approximately 30 per cent of a tricyclo sesquiterpene.
Aroma profile is from dynamic headspace purge.
Bulgarian oil is steam distilled.

important point and may account for most of the disparity. However, the inherent differences between a headspace sampling and a steam distillation method cannot be discounted as a factor (Weisenfeld, 1999).

Another interesting difference between the composition of the headspace and the oil is the distribution of the sesquiterpenes. Both contain caryophyllene, a sesquiterpene found widely in nature. However, the headspace contains approximately 4 per cent of a tricyclo sesquiterpene eluting on the back of the caryophyllene peak. This peak does not appear in the oil. In addition, the oil contains farnesene and germacrene isomers, while the headspace contains considerable amounts of cadinene and bergamotene isomers (listed as 'other sesquiterpenes' in the legend).

In general, the headspace is dominated by terpenes and sesquiterpenes, while the oil contains mostly alcohols and esters.

Reasons for differences: overall view

The main reasons for the differences seen in the two modes of odour 'extraction' is probably due to the fact that only leaves were used in the headspace analysis. Another reason could be that of the time of day the 'sample' headspace was taken, which can be greatly influenced by the temperature and amount of sunlight at different times of the day.

The odour of the flowers bears a distinct effect on the whole odour of the plant. Recent analyses of the essential oils of flowers compared to the leaves. Also show that there is a distinctive difference in the action of flowers and leaves on smooth muscle which indicates a difference in chemical composition (Lis-Balchin and Hart, 2000).

Lavender extracts: action on smooth muscle

The guinea-pig ileum was again used to study the effect of Lavender extracts: action on smooth muscle. Extracts were made from:

- Dried lavender flowers (commercial lavender, possibly *L. angustifolia*)
- Freshly-picked *L. angustifolia* flowers
- Freshly-picked *L. angustifolia* leaves
- Freshly-picked *L. stoechas* leaves.

Three different hydrophilic extracts were made from each plant part: (a) cold methanolic extract; (b) a tea, made with boiling water; (c) hydrosol, that is, the water remaining after steam/water distillation. And this was compared against the steam/water-distilled essential oils.

The extracts were tested on guinea-pig ileum, with and without electrical stimulation *in vitro*, using the same conditions as described previously. It was found that there was a distinctive difference between the effect of the flowers and leaves of the fresh *L. angustifolia* water extracts on the smooth muscle. The tea and hydrosol of the flowers had a relaxant effect, while the leaves had a contractile effect at similar concentrations. Surprisingly though, the same leaf extracts of *L. stoechas* showed only a relaxant effect. However, the methanolic extracts and essential oils of both the *L. angustifolia* leaves and flowers had a relaxant effect, as did the *L. stoechas* essential oil.

Table 24.12 Composition of different floral parts of a lavender (*L. angustifolia*) clone

Component	Flower spike	Corolla	Calyx
(Z)-b-ocimene	1.8	3.8	1.4
(E)-b-ocimene	0.7	trace	0.6
3-octanone	0.8	0.1	0.2
Camphor	0.3	1.7	0.2
1,8-cineole	0.7	0.9	0.2
Linalool	31.1	20.7	34.2
Linalyl acetate	45.2	30.0	47.2
Terpinen-4-ol	0.2	0.1	0.2
Lavandulyl acetate	2.4	3.5	2.2
Lavandulol	0.4	6.9	0.4
α-terpineol	0.1	0.6	trace

Source: Lawrence, 1995.

This does not therefore correlate with the essential oil composition, which is very different in the two species. The composition of the water-soluble extracts appears to be responsible for the differences between the leaves and flowers of *L. angustifolia*. However, the leaves of a totally different genus, that of most scented *Pelargonium* species and cultivars studied produced a similar contractile effect in contrast to the relaxant effect of the corresponding essential oils. *Geranium robertianum* leaf extracts, including the methanolic one all produced a relaxant effect, although from the same family, with similar water-soluble components. The reason(s) for this differing effect is therefore at present obscure.

There is also an apparent difference within the flower itself (Table 24.12). Data presented by Lawrence (1995) shows that there are substantial differences in the percentage of all the main components within the corolla and calyx compared to the flower spike in *L. angustifolia* clone.

References

Boelens, M.H. (1995) Chemical and sensory evaluation of lavandula oils, *Perf. Flav.*, 20, 23–51

Boelens, M.H. (1986) The essential oil of spike lavender (*Lavandula latifolia* Vill. (syn. *L. spica* DC) *Perf. Flav.*, 11, 43–63

Figuereido, A.C., Barroso, J.G., Pedro, L.G., Sevinate-Pinto, I., Antunes, T., Fontinha, S.S., Looman, A. and Scheffer, J.J.C. (1995) Composition of the essential oil of *Lavandula pinnata* L. fi, var. *pinnata* grown on Madeira. *Flav. Frag. J.*, 10, 93–96

Garcia-Vallejo, M.C., Garcia-Vallejo, I. and Velasco-Negueruela, A. (1989) Essential oils of the genus *Lavandula* L. in Spain. *Proc. ICEOFF*, New Delhi, 1989, vol. 4, pp. 15–26

Garcia-Vallejo, M.C., Garcia-Vallejo, I. and Velasco-Negueruela, A. (1989) Essential oils of the genus *Lavandula* L. in Spain. *Proc. ICEOFF*, New Delhi, 1989, vol. 4, pp. 15–26

Lawrence, B. (1995) Progress in essential oils, *Perf. Flav.*, 20, 30

Lawrence, B. (1976–8; 1979–80; 1981–7) Progress in essential oils, Reprints from *Perf. Flav.*, Allured Pub. Corp.

Lis-Balchin and Hart (2000) To be published.

Naef, R. and Morris, A.F. (1992) Lavender–Lavandin: A comparison. *Rivista Ital. EPPOS*, Numero spaciale, 364–77

Piccaglia, R. (1998) Aromatic plants: a world of flavouring compounds. *Agro Food Industry Hi-Tech*, 93, 12–15

Skoula, M., Abidi, C. and Kokkalou, E. (1996) Essential oil variation of *Lavandula stoechas* L. ssp. *stoechas* growing wild in Crete (Greece). *Biochem. System. Ecol.*, 24, 255–60

Tucker, A.O. (1981) The correct name for lavandin and its cultivars Labiatae. *Baileya*, 21, 131–3

Tucker, A.O. and Lawrence, B.M. (1987) Botanical nomenclature of commercial sources of essential oils, concretes and absolutes. In: *Herbs, Spices and Medicinal Plants: Recent Advances in Botany, Horticulture and Pharmacology*. Oryx Press, Phoenix, A.Z., L.E. Craker and J.E.Simon (eds), pp. 183–220

Wiesenfeld, E. (1999) Aroma profiles of various *lavandula* species, Noville, South Hackensack, NJ, USA, http://www.sisweb.com.referenc/applnote/noville.htm

Index

abortifacient, 39, 49
abortion of dead child, 39, 49, 140
abrialis, 81, 82, 85, 117, 118, 140, 236–40, 251–2
absolute, 119–21
acaricidal, 1, 210
acetylcholine, 134, 145, 163, 244
acne, 132
action on uterus, 149
adrenaline, 243
adrenoceptor, 145, 244, 249
adulteration, 1, 117, 119, 121–3, 133, 176, 189, 207, 234, 240, 251
advertising, 61, 62
afterbirth expulsion, 39, 40, 49, 140
alertness, 124, 185
alkaloid, 86
Allelopathy, 211
allergic airborne contact dermatitis, 1, 130, 190
allergic reaction, 124, 130–1, 190
Alopecia areata, 1, 132, 189
α-wave, 125, 129, 133, 164, 185, 209–10
alternative medicine, 188, 190
Alzheimer's disease, 168
anaesthetic, 131, 134, 162, 208
analgesic, 245
anosmic, 185
anthocyanins, 86, 96
antibacterial activity, 122, 124, 173–6
anticancer, 124, 131–2
anticonvulsive, 124, 127–8, 130, 133, 134, 142, 162–3, 168
antifungal activity, 1, 122, 124, 173, 176–7
anti-inflammatory, 132, 208
antimicrobial action, 171–2, 174
antimicrobial properties, 1, 43, 171–7
antioxidant action, 122, 171, 186, 215, 224
antiseptic, 43, 76, 86, 130, 132, 202
antispasmodic: *see also* spasmolytic effect, 130

anxiety-relieving, 125, 129–30, 167, 185, 187
arithmetic: *see* mathematical
aroma: *see also* odour, 140, 141, 168, 203, 235, 237–8, 243
aroma profiles, 256–8
aromachology, 134
aromatherapists, 1, 40, 46, 154, 173, 184, 189–91, 243, 246–9
aromatherapy uses, 1, 124–5, 130, 134, 152, 154, 168, 180–91, 198, 204
asthma, 208, 211
atropine-like, 129, 145, 163
authentification, 207, 240
autonomic nervous system, 125, 126, 129, 209, 243–4

bactericidal: *see also* antimicrobial action, 130
bath, 36, 127, 132, 168, 180–1, 187, 195, 201, 204
bed sheets, 155, 157–8, 163
behaviour, 164, 168, 187, 245
β-wave, 124, 127, 133, 164, 185, 209
binding sites, 141
bioactivity, 122
blood–brain-barrier, 126, 163, 186
blood flow, 142, 185
Blood Pressure (BP), 133, 168, 185, 201, 208
brain function (human), 129, 162–3, 168, 185, 243, 245
brain waves: *see also* α, β, delta, theta waves, 142
brain stimulant, 128
bronchial muscle: *see also* tracheal muscle, 244
burns, 1, 43
business plan, 60–1, 78

calcium channels, 147–8, 244
callus cultures, 214
calming, 155–7, 162–4, 169, 201
cAMP, 1, 129, 146, 163, 244

camphor, 1, 87–9, 120–1, 209, 234, 237–9, 251–6, 258–9, 261
camphoraceous, 77, 152
carbon dioxide (CO_2) extraction, 120–1, 175
carrier oil, 180–2, 204, 243
cell action potentials, 162
cell cultures, 141, 214–26
central nervous system: *see* CNS
certified organic, 233
cGMP, 129, 146, 163
Chaytor, D.A., 51, 52, 53, 56
chemical composition, 117–24, 133, 171–3, 189, 243–50, 251–62, 235, 238, 247–9, 251–61
chemical profiles, 243
chemo-preventative, 131
chemotherapy, 173, 187–8
chemotypes, 189, 255
chick biventer muscle, 150–1
childbirth, 39, 40, 49, 168, 187, 190
childhood atopic eczema, 189
chiral, 126, 133, 207, 236, 239–41
chiral columns, 1, 122–3, 207
cleanliness, 44
clinical study, 167–9, 181, 186–90, 210, 243
clinical trials: *see* clinical study
CNS, 124, 128–9, 134, 142, 181, 208, 244–5, 249
CNV, 142, 163–4, 185, 245
cognitive skills, 125
cold distempers, 50
colic, 39, 40, 42, 43, 49, 140, 201
commercial essential oils, 122, 129, 140, 143, 184, 189, 207, 251, 256
complementary therapy, 189
concrete, 119–20
condensers, 107–12
conidial germination, 177
contact allergic response, 130
contact dermatitis, 212
Contingent Negative Variation: *see* CNV
contraction, 1, 140–1, 143, 144–5, 148–50, 184, 246–9, 261
convulsions, 39, 40, 41, 43, 155, 162–3
corolla, 7, 261
corolla colour, 81–4
cosmetics, 171, 200, 214
coumarins, 92, 97, 208
crop rotation, 232
Culpeper, Nicholas, 39, 40, 43, 46, 48, 49, 140, 155, 201
cultivars round the world, 80–5, 232, 234
cuttings, 58, 65, 78, 118

cycles, 8
cylindrical still system, 102

dead foetuses, 40
delta waves, 129
dementia, 187
depression, 189, 211
dermal application, 126, 140, 204
dermatitis, 130–1, 190
dermatomycoses, 204
diabetes, 208
dietary supplements, 211
distillation, 100–16, 37, 42, 45, 46, 48, 59, 256
distilled waters, 37, 40, 50, 155
dog, 141, 143
double-blind studies, 188
dry lavender, 1, 78, 144–5, 148, 161, 200–4, 212, 234–5, 261

eau de Cologne, 46, 119, 190, 197, 203, 248
eczema, 243
EEG, 124, 125, 129–30, 163–4, 167, 169, 209–10
electrical activity, 168
electroencephalography: *see* EEG
electromyography, 185
enantiomeric columns: *see* chiral
enantiomers, 122–3, 126–7, 133, 148, 175–6, 258–60
epilepsy, 37, 38, 40, 41, 48, 49, 50, 162, 188
esoteric, 182
essential oils, 1, 86–9, 110, 101, 115, 124–33, 171–4, 180–90, 212, 232, 234, 237–9, 243–9, 261–2
extraction process, 119–21
extracts, 130, 144–5, 148, 171, 173–4, 174, 207, 261–2

fainting, 38, 39, 40, 49, 50, 140, 155, 157, 200–1
fertilizers, 206, 232, 251
fish, 128
fixed oils, 121–2
flavones, 86, 91, 93, 95, 96
flavonoids, 86, 91, 92, 98, 99
flowers, 92, 144–5, 148, 256, 258–9, 261–2
food flavouring, 200–2
food preservation, 1, 171
fougère, 198
fragrance: *see also* odour and perfume, 126, 155, 185
French Lavender, 117

frost-hardy lavenders, 69–71
functional imaging in brain, 126
fungus, 55, 118

GABA, 127, 133, 163, 210
GC and GC/MS analysis, 122, 207, 235–6, 245, 257
Geller-type-conflict-test, 133
Gerard, 37, 38, 39, 42, 43, 44, 46, 47, 48, 140, 201
germination of seeds, 207
glutamate, 133–4, 152, 245
glutamate-binding, 128, 130, 142, 162–3
gram-positive and negative bacteria, 172
Grosso, 1, 14, *15*, 55, 69, 81, 82, 85, 117, 118, 120, 121, 140, 144, 236–40, 251–2
growing conditions, 57–9, 76–7
guinea-pig ileum, 123, 129, 141, 143–8, 184, 244, 246, 248–9, 261
gut microflora, 173

hair growth, 132
half-hardy lavenders, 71–2
hardy lavenders, 69
headache: *see* pain in head
head lice, 36
headspace, 131, 256–61
healing, 1, 132, 134, 181, 190, 204
Heart Rate (HR), 129, 164, 168, 209, 243
hedonic uses, 160–2, 165, 167
heel cuttings, 58
herbal pillows: *see* pillows
Hidcote, 12, *13*, 14, 54, 55, 68, 82, 85, 202, 236–40
Hildegard, Abbess, 35, 36, 39, 43, 44, 46, 47, 140
History of nomenclature, 51–6
History of usage, 35–50
Homes for the Aged, 128
honey bee, 133
hospices, 128, 190
hospitals, 128, 180, 190
household cleaning, 203
human, 1, 127, 141, 245
husbandry: *see also* growing conditions, 55, 74, 118–19, 206
hydrodiffusion, 121
hydrodistillation, 235
hydrosol, 261
hydroxycinnamic acid, 92, 96, 97
hypoglycaemic effect, 132, 208
hysteria, 43

inhalation, 1, 125–8, 130–3, 162, 186, 188, 190, 208
insect repellent, 203–4
insects, 1, 55, 132, 211, 235
insomnia, 128, 201
intercropping, 232
in vitro (pharmacology), 1, 141, 143–52, 184, 185–6, 261
in vivo (pharmacology), 1, 140, 142
irritable bowel syndrome (IBS), 181
irritation, 131, 204
ISO, 117, 118, 207
isomers: *see* enantiomers
isotopes, 207

L. angustifolia, 1, 2, 5, 8, 10, 11, 12, *13*, 33, 53, 54, 55, 58, 67, 68, 77, 80–5, 86–93, 97–8, 117–20, 134, 140, 142, 144, 148, 152, 182, 194–5, 200–2, 206–8, 210–11, 214, 228, 233–4, 237–9, 251–3, 257–8, 260–2
L. angustifolia x L. latifolia, 117, 118, 120
L. antineae, 5, 10, 20, 23, 24, 95
lavandin 119–21, 206, 234, 236–40, 251–2
lavandin absolute, 119–21
lavandin concrete, 119–21
lavandin oil, 119–21, 201–2
Lavandula species/hybrids, 2–34
lavender absolute, 119, 195, 201–2
lavender bags, 46, 158, 200, 203
lavender concrete, 119, 201
lavender cultivars, 59, 80–5
lavender distillate, 37
lavender drops, 156–7, 200
lavender essence, 142
lavender essential oil, 1, 37, 42, 43, 44, 45, 46, 57–9, 117–23, 119, 121, 133, 141–52, 162–4, 168, 174–6, 180–90, 194–8, 200–4, 206–9, 214, 243
lavender oil: *see* lavender essential oil
lavender oil glands, 100
lavender products, 59
lavender spirit, 201
lavender straw, 127, 167
lavender tincture, 200–1
lavender vinegar, 200
lavender water, 37, 143, 156–7, 161, 195–6, 200
L. bippinata, 5, 7, 8, 9, 10, 30, *31*
L. canariensis, 5, 7, 9, 10, 20, 21, *22*, 25, 26, 73, 96
L. dentata, 1, 3, 5, 7, 8, 9, 10, *15*, 17, 33, 51, 52, 71, 85, 89–93, 97, 144, 150, 152, 208, 227–8, 253, 256–8
L. dentata x L. angustifolia, 227

L. dentata x *L. latifolia*, 227–8
leaf shapes, 5
leaves, 91–2, 144–5, 148, 256, 258–9, 261–2
L. hybrida, 228
lice, 36, 43, 45, 47, 140
limbic system, 1, 181, 186
linalool, 1, 87–9, 151, 117–21, 124–8, 130–4, 140, 142–4, 146, 149–53, 162–3, 175–6, 185, 208, 234, 237–41, 245, 251–5, 258–61
linalyl acetate, 1, 87–9, 151, 117, 118, 119, 120, 121, 140, 142, 146, 149–52, 162–3, 175–6, 185, 208, 234, 237–41, 251–4, 258–61
lipid peroxidation, 207
Listeria monocytogenes, 122, 175–6
liver metabolising enzymes, 142
L. lanata, 1, 13, 14, 51, 52, 71, 91, 93, 252–3, 256–8
L. lanata x *L. angustifolia*, 59, 69
L. latifolia, 1, 5, 7, 9, 10, 12, *14*, 33, 51, 52, 80, 86, 89–93, 97, 117, 120, 133, 140, 194, 208, 214, 228, 234, 252–3
L. lusieri, 18, 255
L. minutolii, 5, 10, 20, 23, 73, 94, 95
L. multifida, 5, 7, 8, 9, 10, 20, *21*, 26, 51, 52, 91, 94, 96, 208, 254, 256–7, 259
L. officinalis, 53, 54, 119, 256
L. pinnata, 1, 5, 10, 20, 22, 25, 26, 51, 52, 73, 95, 96, 256, 256–7, 259
L. spica, 1, 11, 12, 35, 36, 42, 51, 52, 53, 144, 152, 227, 256–8, 3, 5, 8, 10, 17, 18, 19, 35, 37, 38, 40, 42, 44, 45, 48, 51, 52, 69, 70, 77, 80, 89–93, 95–6, 144–5, 148, 152, 201, 208, 254, 256–8, 261–2
L. stoechas, 17, *18*, 19, 35, 37, 38, 40, 42, 44, 48, 51–2, 67, 69, 70, 77, 80, 82–3, 89, 91–3, 95–6, 152, 208, 254, 256, 258, 261
L. stoechas subspecies, *see L. stoechas*
L. stoechas Ssp. *Luisieri*, 18, 89
L. stoechas x *viridis*, 34, 69, *71*, 72
L. subnuda, 5, 7, 8, 9, 10, 27, 28
L. vera, 214–26, 35, 36, 37, 38, 41, 42, 48, 51, 52, 206, 227
L. viridis, 5, 7, 9, 10, 17, 18, 19, 51, 52, 71, 89, 91, 93, 144, 148, 255, 256–8
L. x allardii, 10, 32, 33, 227–31
L. x heterophylla, 10, 33, 227–30, 256–7, 259, 227–31
L. x intermedia, 2, 13, 53, 54, 55, 58, 68, 77, 81–5, 140, 234

mad dogs, 50
mail order, 62, 63
man: *see* human
marketing, 61, 77
massage, 1, 43, 45, 126–7, 141, 168, 185–90, 243
mathematical, 124–5, 165, 185, 189
membrane receptor, 141
memory, 125, 164–5, 189
mental depression, 43
mentally-challenged, 190
mice, 125, 128, 133–4, 141–2, 144, 162–3, 245
microsomal enzyme induction, 132
microwave extraction, 120
migraine, 37, 140, 201
miscellaneous uses, 200–5
mites, 210
mixtures of essential oils, 243, 248–9
monoterpenes, 87, 128, 131, 214, 245, 246–9, 257, 260
mood, 125, 164–6, 189
mosquito repellent, 204
moth repellent, 157–9, 163
motility in mice, 125, 142
munstead, 12, 54, 68, 81, 202, 236–40
myogenic, 141, 144–5, 150

National Plant Collection, 64
nervous system, 124
neurogenic, 141, 144–5
neurological, 187
neurophysiological measurements, 129
NMDA, 133
Norfolk lavender, 57
nursery management, 62

occult, 44
odour, 36, 42, 45, 46, 47, 125, 128, 133–4, 164–8, 185, 190, 208, 261
oil glands, 100
Old English, 14, 55, 69, 81
oleum spicae: *see* spike oil
oral, 130, 181, 185–6
organically-grown lavender, 79, 232–42
over-wintering, 75

pains (head), 38, 39, 40, 43, 45, 48, 155, 201, 243
pains (other), 131, 181, 190, 245
palpitations, 37, 38, 48
palsy, 35, 36, 38, 43, 47, 48
panic attacks, 37, 38, 48, 201
parasympathetic, 186, 244
parenteral, 140
Parkinson's, 37, 38, 47, 48
patch testing, 212
patents, 128, 133, 204

patients, 186–90, 246, 248–9
percentage of components, 246–8
percutaneous, 126–7
perfume, 1, 166, 194, 196–8, 200, 203, 243, 248–9
perfumery uses, 1, 194–9, 119–20, 124, 133, 140, 214
perillyl alcohol, 131, 208–9
perineal discomfort, 133, 168, 187
periodontal disease, 132
peristaltic activity, 143
pests and control, 55, 66–7, 79, 232–3
pharmaceutical products, 171
pharmacological action, 1, 144, 148, 151, 162, 163, 165, 169, 247, 249
pharmacology, 123, 140–54, 207, 243–50
phenolics, 86, 98, 216–17
phenols, 172
pheromones, 211
phosphodiesterase inhibitors, 146
photosensitization, 130, 190
phrenic nerve diaphragm, 129, 131, 141, 150, 163, 208, 245
physical work, 133
physically-challenged, 190
physiological, 125, 129, 140, 187, 243
phytochemistry, 86–99
pigmentation, 130, 190
pigs, 127, 167
pillows, 1, 126–7, 140, 200, 203, 212
placebo, 130, 155, 168, 187–8
plant growth, 133
pleasant, 155, 158, 165–7, 208–9
poisonous creatures, 50
post-synaptic, 129, 163
potassium channels, 147–8, 244
pot-pourri, 74, 202–3
potting, 66, 74
pregnancy, 190
propagation, 58, 64
pruning, 75–6
psychological effects, 46, 129, 155–70, 187, 209, 243, 249
psychopharmacological effect, 128
psycho-physiological response, 125, 209
PUFA's, 132
pulmonary congestion, 36

rabbit, 141
radiocarbon and stable isotope analysis, 236, 241
rats, 128, 141, 162, 208, 245
receivers/separators, 111–13
red lavender, 42, 43, 156–7, 200–1

relaxant and relaxation: *see also* spasmolytic effect, 1, 76, 125, 127–8, 133, 140, 144, 157, 164, 184, 190, 201, 202, 243, 245–8
respiratory effect, 129, 168, 190
retail industry, 57–9, 60–75, 76–9
retention time, 236, 245–8
root rot, 55, 118
rosmarinic acid, 96, 207, 214–26
rotation crop, 206

safety of essential oils, 141, 190
section *stoechas*, *see L. stoechas*
sedative, 125, 127–9, 130, 133–4, 142, 162–3, 167–9
seed propagation, 65, 117
senile dementia, 168
sensitization, 130, 188
sesquiterpenes, 120, 131, 194, 246–8, 257–60
sexual attraction, 243
skeletal muscle, 123, 129, 141, 150–2, 243
skin, 129, 180–1, 185–8, 209, 243
skin penetration, 133
skin treatment, 204
sleep, 126, 128–9, 140, 163, 167–8, 190, 211
smell: *see* odour
smooth muscle, 1, 123, 141, 143, 145–8, 150–2, 243–4, 247–9, 261
snake bite, 50
soil conditions, 66, 74, 77
solvent extraction, 119–21
sores, 43
spasmogenic effect, 129, 141, 144–5, 148, 153, 244–5
spasmolytic effect, 1, 128–9, 141–52, 244–5
spike lavender, 1, 35, 36, 40, 42, 117, 119, 120, 133, 140, 152, 175–6, 194, 203
spike oil, 37, 39, 44, 88, 133, 140, 201–2
spirits of lavender, 156–7
standardization, 117–23
steam distillation and theory, 103–6, 115, 117, 119, 120, 121, 174, 235, 258–61
still rooms, 45, 195
stills, 37, 45, 102, 103, 105–7, 108, 110, 206
stimulate/stimulation: *see also* contraction, 127, 243
storage life, 176
stress, 127, 167, 190, 201, 243, 245
structure of flowers, 6
subcutaneous route, 245
super, 81–2, 85, 236–40, 251–2
supercritical CO_2 extraction: *see* carbon dioxide extraction
suspension culture, 216–23

sympathetic nervous system, 186, 243–4
synthetic, 1, 119, 187, 243

taxonomy, 2–34
teas: *see* tissanes
tender lavenders, 73
terpeneless oil, 119
terpenoids: *see also* monoterpenes, 171–3
theta wave, 129–30, 164
tissanes, 200, 202
tomography, 142
toothache, 39, 40, 43, 49, 201
topographical brain maps, 129
toxicity, 1, 131, 133, 173, 187, 189
tracheal muscle, 149
transdermal absorption, 133

travel sickness, 127, 167
triterpenoids, 90

unilateral supratentorial brain tumours, 134
ureteric stones, 185
uterus, 141, 149

vacuum distillation, 119
very hardy lavenders, 67–9
voice loss, 39, 40, 43, 49, 156
volatile oils: essential oils

water-soluble extracts, 148, 208, 261–2
weed control, 232–3
women's periods, 40, 140
wound, 156, 181, 204